THE INTERNATIONAL POLITICS OF NUCLEAR WASTE

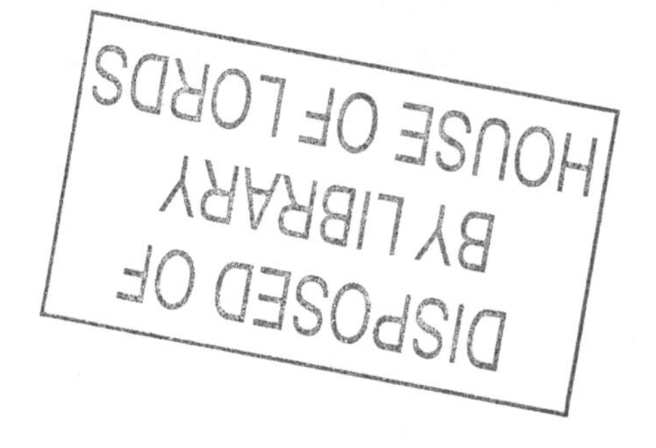

Also by Andrew Blowers

THE LIMITS OF POWER
SOMETHING IN THE AIR
THE FUTURE OF CITIES (*editor with C. Hamnett and P. Sarre*)
INEQUALITIES, CONFLICT AND CHANGE (*editor with G. Thompson*)
URBAN CHANGE AND CONFLICT (*editor with C. Brook, L. McDowell, P. Dunleavy*)
NUCLEAR POWER IN CRISIS (*editor with D. Pepper*)

Also by David Lowry

ISSUES IN THE SIZEWELL 'B' INQUIRY (*editor with Ian Gamble, in six volumes*)

Also by Barry D. Solomon

GEOGRAPHICAL DIMENSIONS OF ENERGY (*editor with Frank J. Calzonetti*)

The International Politics of Nuclear Waste

Andrew Blowers

Professor of Social Sciences
The Open University

David Lowry

Visiting Research Fellow, Energy and Environment Research Unit
The Open University

Barry D. Solomon

Energy Policy Analyst, Global Climate Change Program
US Environmental Protection Agency

MACMILLAN

First edition 1991

Published by
MACMILLAN ACADEMIC AND PROFESSIONAL LTD
Houndmills, Basingstoke, Hampshire RG21 2XS
and London
Companies and representatives
throughout the world

Typeset by Footnote Graphics,
Warminster, Wiltshire

Printed in Hong Kong

British Library Cataloguing in Publication Data
Blowers, Andrew, *1940–*
The International Politics of Nuclear Waste.
1. Nuclear waste materials. Disposal. Political aspects
I. Title II. Lowry, David, *1956–* III. Solomon, Barry D. (Barry David), *1955–*
363.7′28
ISBN 0–333–49363–X (hardcover)
ISBN 0–333–49364–8 (paperback)

For Gill, Rosalind and Diana
and to the future generations
who inherit our legacy

Contents

Acknowledgements

The authors are grateful to the following for permission to reproduce the original artwork of cartoons and photographs:

Ch 2: *Daily Mirror*, *West Highland Free Press*, Scottish Campaign to Resist the Atomic Menace (SCRAM), 'Colin' and the UK Atomic Energy Authority, and Amigos de la Tierra (Friends of the Earth) Spain.

Ch 3: Bedfordshire County Council.

Ch 4: *Bedfordshire Times*, Bedfordshire Against Nuclear Dumping (BAND), 'Griffin' and *Daily Mirror*, *The Times* and Syndication International Ltd.

Ch 5: Bob Taylor and Faith Kenrick.

Ch 6: WISE, Paris, Folkkampanjen mot Karnkraft och Karnvapen (FMKK).

Ch 7: Faith Kenrick, Chris Madden and UK Green Party.

We also gratefuly acknowledge the following sources from which figures, maps and cartoons have been derived:

Ch 1: NIREX, International Atomic Energy Agency, National Radiological Protection Board and UK Atomic Energy Authority.

Ch 2: *The Observer*, *Nature*, *The Sunday Times*.

Ch 3: NIREX, UK Atomic Energy Authority, Bedfordshire County Council.

Ch 4: NIREX.

Ch 5: US Department of Energy, Radioactive Waste Campaign, US Nuclear Regulatory Commission, Faith Kenrick.

Ch 6: Svensk kärnbränslehantering AB (SKB).

Ch 7: Faith Kenrick.

List of Acronyms and Abbreviations

AEC	Atomic Energy Commission (US)
AERE	Atomic Energy Research Establishment (UK)
AFR	Away from Reactor (storage facility)
AGR	Advanced Gas-Cooled Reactor
AKA	Swedish Nuclear Waste Review Commission
ALARA	As Low as Reasonably Achievable
ALATA	As Low as Technically Achievable
ANDRA	Agence National pour la Gestion des Déchets Radioactifs (France)
BAG	Barrow Action Group (UK)
BAND	Bedfordshire Against Nuclear Dumping (UK)
BAND	Billingham Against Nuclear Dumping (UK)
BAS	Bundesamt fuer Strahlenschutz – West German Office for Radiation Protection
BMN	West German Ministry for the Environment, Nature Conservation and Nuclear Safety
BNFL	British Nuclear Fuels Limited (now plc) (UK)
BOND	Britain Opposed to Nuclear Dumping
BPEO	Best Practical Environmental Option
BRC	Below Regulatory Concern
BIs	Burgerinitiativen (citizen action groups, West Germany)
CANDU	Canadian Heavy Water Reactor Design
CDU	Christian Democratic Union (West Germany)
CEA	Commissariat à l'Energie Atomique (France)
CEGB	Central Electricity Generating Board (UK)
CFCs	Chlorofluorocarbons
CFR	Code of Federal Regulations (US)
CLAB	Centrala Lagret för Anvant Karnbrasle – Swedish Interim Spent Fuel Storage Facility
COGEMA	Compagnie Générale des Matières Nucléaires – French nuclear fuel cycle company
CORE	Cumbrians Opposed to a Radioactive Environment

DAD	Decide, Announce, Defend
DoE	Department of the Environment (UK)
DOE	Department of Energy (US)
DWK	Deutsche Gesellschaft fur Wiederaufarbeitung von Kernbrennstoffen – West German spent fuel storage and reprocessing company
EAND	Essex Against Nuclear Dumping (UK)
EC	European Commission
EdF	Electricité de France
EEC	European Economic Community
EEG	Environmental Evaluation Group (State of New Mexico)
ENEA (NEA)	European Nuclear Energy Agency
EP	European Parliament
EPA	Environmental Protection Agency (US)
ERDA	Energy Research and Development Administration (US)
ERL	Environmental Resources Ltd (UK)
EURATOM	European Atomic Energy Community
FBR	Fast Breeder Reactor
FMKK	Folkkampanjen mot Kärnkraft och Kärnvapen – Swedish National anti nuclear power and weapons movement
FoE	Friends of the Earth
FRG	Federal Republic of Germany (West Germany)
GAO	General Accounting Office (US)
GDR	German Democratic Republic (East Germany)
GETT	Grants Equal to Taxes
GIR	Gorleben International Review
GWe	Gigawatts (electrical)
HAND	Humberside Against Nuclear Dumping (UK)
HLW	High-Level Waste
HSE	Health and Safety Executive (UK)
IAEA	International Atomic Energy Agency
ICI	Imperial Chemical Industries

ICRP	International Commission on Radiological Protection
IGS	Institute of Geological Sciences
ILW	Intermediate-Level Waste
INEL	Idaho National Engineering Laboratory (US)
KASAM	Samrådsnamnden för Karnavfallsfrågor – Swedish Consultative Committee on Radioactive Waste Management
KBS	Kärnbränslesäkerhet – Swedish Nuclear Waste Management Project
KFDK	Karlsruhe Federal Nuclear Research Centre (West Germany)
LAND	Lincolnshire Against Nuclear Dumping (UK)
LDC	London Dumping Convention
LLRWPA (1980)	Low Level Radioactive Waste Policy Act of 1980 (US)
LLW	Low-Level Waste
LULU	Locally Unwanted Land Use
MAFF	Ministry of Agriculture, Fisheries and Food (UK)
MAGNOX	Magnesium Oxide Fuelled Nuclear Reactor (UK)
MOX	Mixed Oxide (fuel)
MRS	Monitored Retrievable Storage
MWe	Megawatts (electrical)
MEP	Member of the European Parliament
NAS	National Academy of Sciences (US)
NEA	Nuclear Energy Agency (of OECD)
NEPA (1970)	National Environmental Policy Act of 1970 (US)
NERC	Natural Environmental Research Council
NIA	Nuclear Installations Act of 1965 (UK)
NII	Nuclear Installations Inspectorate (UK)
NIMBY	Not in my Backyard
NIREX (UK NIREX Ltd)	Nuclear Industry Radioactive Waste Executive
NRC	Nuclear Regulatory Commission (US)
NRPB	National Radiological Protection Board (UK)
NUS	National Union of Seafarers (UK)
NWPA (1982)	Nuclear Waste Policy Act of 1982 (US)

NWPAA (1987)	Nuclear Waste Policy Amendments Act of 1987 (US)
OECD	Organisation for Economic Cooperation and Development
PAMELA	Western German designed vitrification plant at Mol, Belgium
PCBs	Polychlorinated Biphenyls
PEON	Commission Consultative pour la Production d'Electricité d'Origine Nucléaire (France)
PERG	Political Ecology Research Group (UK)
PRAV	Programrådet för Radioaktivt Avfall – Swedish nuclear waste research council
PTB	Physikalisch Technische Bundesanstalt – West German National Institute for Science and Technology
PWR	Pressurised Water Reactor
RAF	Red Army Faction (FRG)
RBMK	Soviet Military Reactor Design, used at Chernobyl
RCEP	Royal Commission on Environmental Protection (UK)
RSA	Radioactive Substances Act of 1960 (UK)
RWMAC RWMC	Radioactive Waste Management Advisory Committee (UK)
SDO	Special Development Order
SFL	Slutförvar för Langlivat Avfall – Final Waste Repository (Sweden)
SFR	Slutförvar för Reaktoravfall – Final central repository for LLW and ILW (Sweden)
SKB	Svensk Kärnbränslehantering AB – Swedish Nuclear Fuel and Waste Management Company
SKI	Statens Kärnkraftinspektion – Swedish Nuclear Inspectorate
SKN	Statens Kärnbränslenamnd – Swedish National Board for Spent Nuclear Fuel
SPD	Social Democratic Party (West Germany)
SRP	Savannah River Plant (South Carolina)

SSI	Statens Stralskyddsinstitut – Swedish National Institute of Radiation Protection
SSK	Strahlenschutz Kommission (West Germany)
STAND	Serious Texans Against Nuclear Dumping (US)
THORP	Thermal Oxide Reprocessing Plant at Sellafield (UK)
THTR	Thorium High Temperature Reactor (West Germany)
TMI	Three Mile Island (Pennsylvania)
TN	Transnuklear – nuclear materials transport company
UKAEA	United Kingdom Atomic Energy Authority
UNSCEAR	United Nations Scientific Committee on the Effects of Atomic Radiation
USAEC	US Atomic Energy Commission
USDOE	US Department of Energy
WIPP	Waste Isolation Pilot Plant (New Mexico)
WAK	Wiederaufarbeitungsanlage – pilot reprocessing plant at Karlsruhe (West Germany)
WAW	Wackersdorf Reprocessing Plant (cancelled, West Germany)
WISE	World Information Service on Energy

List of Figures

List of Tables

'. . . two things only remain for us to attempt: to send it over the Sea, or to destroy it.'

'But Gandalf has revealed to us that we cannot destroy it by any craft that we possess,' said Elrond. 'And they who dwell beyond the Sea would not receive it: for good or ill it belongs to Middle-Earth; it is for us who still dwell here to deal with it.'

'Then,' said Glorfindel, 'let us cast it into the deeps . . . in the Sea it would be safe.'

'Not safe for ever,' said Gandalf. 'There are many things in the deep waters; and seas and lands may change. And it is not our part here to take thought only for a season, or for a few lives of Men, or for a passing age of the world. We should seek a final end of this menace, even if we do not hope to make one.'

. . . 'Then,' said Erestor, 'there are but two courses, as Glorfindel already has declared: to hide the Ring for ever; or to unmake it. But both are beyond our power. Who will read this riddle for us?'

'None here can do so,' said Elrond gravely. 'At least none can foretell what will come to pass, if we take this road or that. But it seems to me now clear which is the road that we must take . . . . We must send the Ring to the Fire.'

The Council of Elrond from *The Fellowship of the Ring*, Part One of *The Lord of the Rings* by J. R. R. Tolkien, Ballantine Paperback edn., pp. 349–50. Originally used by Amory Lovins in an unpublished 1978 review of proposed US policy on nuclear waste management.

Preface

In Tolkien's *The Lord of the Rings* Erestor declares that only two courses are possible for dealing with the menace of the Ring, 'to hide the Ring for ever; or to unmake it. But both are beyond our power'. Precisely this dilemma applies to nuclear waste. The technical solution generally espoused is to bury nuclear waste (or at least the most dangerous and long-lived part of it) for ever in deep geological repositories where it will remain isolated from the 'accessible environment'. Enormous scientific effort has been expended in Europe and North America researching and demonstrating the proposition that such repositories will be safe, for all practical purposes, for ever. Yet clearly the assertion is preposterous. The safety of an untried method cannot be proven until repositories have been constructed and monitored over many generations and the radionuclides have decayed to safe levels. Running such an empirical experiment is literally inconceivable. Sophisticated geological analysis, risk assessment or modelling of repository behaviour must rest upon heroic assumptions and are no substitute for empirical knowledge. Scientific predictions for periods of 10 000 years or more lie in the realm of fantasy, not rationality. In conditions of such uncertainty it must be concluded that there is no technical solution to the problem of radioactive waste. Without an agreed and acceptable technical solution, nuclear waste must remain an issue over which there is political conflict.

Nuclear waste management may well be a political problem without a political solution. Yet, as was the case in Tolkien's Ring, we cannot unmake nuclear wastes. They exist and are steadily accumulating. They emit radiation that can threaten life and which can remain harmful over thousands of years. They are being managed somehow and methods must be found to ensure their continued management even if a political solution based on the minimal technical uncertainty that is acceptable to all interests extending down the generations is unattainable. This book examines the political conflicts that have been fought and the compromises that have been forged in the search for acceptable solutions to an intractable problem.

Nuclear waste management involves communities, the places where the waste is stored or buried. In our research for this book we travelled widely visiting nuclear sites and the people who manage them or live near them. Our study is based on first-hand experience of the attitudes that influence the political conflict over nuclear wastes.

Among the sites we visited were those we have described as 'nuclear oases', those places with a heavy dependence on the nuclear industry. In the UK, Sellafield, the civil and military reprocessing centre with its 11 000 employees in an area with few alternative opportunities, stores nearly all the country's 'high level' wastes and has the main surface burial facility for 'low level' wastes. Sellafield has also been identified as a possible site for a deep nuclear waste repository. It has become a controversial site with revelations of accidents, cover ups and fears of leukaemia clusters. It is a site that has tenacious support especially from its dependent workforce, but also attracts widespread opposition from local and national anti-nuclear groups.

In the USA the Hanford nuclear reservation, the heart of the country's military nuclear industry, covers 570 square miles in the semi-desert of south-central Washington in the Pacific North West and dominates the economy of that region of the state. The banks of the Columbia River are littered with closed plutonium reactors and reprocessing plants and high-level liquid wastes in stainless steel tanks are leaking into the surrounding environment. Local enthusiasm for the nuclear industry manifested in support for a possible high level waste repository at Hanford is overwhelmed by the state-wide hostility particularly concentrated in the big cities of Seattle and Spokane over 200 miles away.

Elsewhere there is the large nuclear oasis at Cap de la Hague near Cherbourg in northern France, the major reprocessing and waste management centre in the country. In some of the smaller nuclear oases the threat of closure of nuclear facilities can produce acute economic problems such as are now being faced at Dounreay in northern Scotland and at Barnwell in South Carolina. Nuclear oases can also occur in places with no previous nuclear history. Such a case is Carlsbad in the remote desert of southeastern New Mexico where the Waste Isolation Pilot Plant (WIPP) has been constructed (though, at the time of writing, not yet opened) in a salt formation for the disposal of long-lived military wastes. The economic distress caused by the closure of the local potash industry, combined with New Mexico's uranium mining and nuclear research commitment, provided propitious political circumstances for the project.

As the opposition to the nuclear industry has focused on nuclear waste, governments and industry have been in retreat to these nuclear oases where they can be assured of a welcome by at least part of the community. Apart from isolated examples such as the WIPP, nuclear waste repositories have been rejected whenever they have been proposed in greenfield locations. The background and historical context for the contemporary political geography of nuclear power is set out in chapters 1 and 2 which use original sources to trace the emergence of radioactive waste as a political problem and the gradual narrowing of options for its management as opposition developed. The central story of the book (chapters 3 and 4) chronicles the process of retreat in the UK. Written by one of the participants in the conflict and using unpublished sources, this case study reveals how political power was discovered, developed and deployed by the opponents of the nuclear waste proposals during the 1980s. The impact of growing opposition leading to the search for politically acceptable solutions to the problem is the theme carried into the comparative studies of the United States (chapter 5) and Western Germany, Sweden and France (chapter 6). These chapters, too, are based on visits, discussions and observations by all three authors in order to gain insights which complement the analysis from our published sources.

Many questions are prompted by the case studies. Among these are: Why is nuclear waste acceptable to some communities and not to others? How have some communities succeeded in defeating proposals for nuclear waste sites in their areas? How important is familiarity with the nuclear industry in securing its acceptability? Is geographical remoteness important in repository site selection? How far are policies and outcomes a reflection of the political institutions of individual countries? What is the significance of geographical or technical factors in the choice of sites? What role does the distribution of costs and benefits play in the selection of strategies and sites? How far is and should account be taken of the risks to future generations?

Our studies spanned six years spread over two continents. One of the authors, Andrew Blowers, became involved as a local politician when Elstow was identified as a potential site for shallow burial of nuclear wastes in 1983. Throughout 1983–87 he was a participant in the 'battle of the dumps' described in chapters 3 and 4. He gained a comparative perspective through visiting the major nuclear waste sites in the USA, France and Sweden. David Lowry's perspective originally came from the anti-nuclear protest movement. He gathered material from many countries with a special focus on the USA, Germany and Sweden. Barry Solomon was working as an economic geographer at the US Department of Energy when this book was written though his views are his and not those of the US Government. His broad perspective and knowledge of the United States provides a counter to any UK bias. The collaboration between three different personalities with three distinctive perspectives was never comfortable but usually complementary and always challenging. In the end, as in the political world which we describe, compromises were struck in our search for agreement and truth. We cannot claim the book to be disinterested but it is borne out of a deep and often vigorous disagreement. The result, we hope, is not a partisan polemic but a reasoned and reasonably balanced account of the state of nuclear waste politics.

During our studies we were helped by many people, too numerous to mention all by name. Among those to whom we are especially grateful are our academic colleagues and those many people in government, anti-nuclear and citizens groups, scientists, journalists and consultants who gave help, hospitality and different viewpoints. They include David Albright, Maj-Britt Andersen, Roger Anderson, Don Arnott, Diane D'Arrigo, Maureen Barker, Luther Carter, Sten Cedarquivst, Lokesh Chaturvedi, Frank Cook, Susan Cutter, Karl Davies, Scott Denman, Rob Edwards, Tove Fihl, Peter Floyd, Paul Flynn, Gary Fowler, Marianne Fritzen, Miles Goldstick, Laura Haight, Don Hancock, Rebecca Harms, Maggie Harrison, Julie Hazemann, Paul Helliwell, Barbara Hetherington, Colin Hines, Meinir Huws, Jerry Jacob, the late Richard Johnson, Eia Joss-Palmaer, Mitzi Karacostas, Roger Kasperson, Hannes Kempmann, Remi Langum, Paul Leventhal, the late Helène Liedegren, Leon Lowery, Patrick Malloy, Carol Mongerson, Tom Murauskas, Cordula Nowak, Neil Numark, Lee Oprea, Mike Pasqualetti, Caroline Petti, Edie Pierpont, David Pijawka, Nikolaus Piontek, Marvin Resnikoff, Tim Roberts, Brian Rome, David Ross, Mycle Schneider, Jeane Smith, Llewellyn Smith, Jonathan Spink, John Stewart, Dr Dafydd Elis Thomas, Bonnie Titcomb, Ray Vaughan, Ann Wieser, Jane Wildblood, Tim Williams, Lilo Wollny and Don Zeigler.

Throughout the writing of this book Alan Wheeler and Tony Griffin of the Bedfordshire County Planning Department provided a constant stream of material, knowledge and ideas. We are indebted, too, to officials in the nuclear industry who provided tours of facilities in the UK, US, Canada, Sweden, Western Germany and France and documentation and guidance on the issues as they perceived them. We also wish to express our gratitude to those with whom we worked during this study – to the NIREX officials with whom we were in constant debate and to the anti-nuclear groups and local politicians whose actions are recorded in the book. The research was generously funded by a research grant from the Open University and the Nuffield Foundation provided funds for travel to ensure a comparative perspective.

In completing this work we are greatly indebted to our secretaries Lyn Brinkley, Patricia Roe, Jenny Brown and especially to Maureen McManus who endured the problems of meeting changing deadlines in a transatlantic collaboration. Our cartographer, John Hunt's skills are clearly visible in the text. The authors' partners, Gill, Rosalind and Diana have given constant help and encouragement and to them, as always, we owe the greatest debt.

By the close of the 1980s the problem of nuclear waste was being perceived in a broader, more global context. Moreover the public in North America and Europe had also become sensitised to and worried over non-nuclear hazardous and toxic wastes, following a stream of media revelations over mismanagement and corrupt practice. As the hole in the stratospheric ozone layer and potential global warming from the burning of fossil fuels and forests gripped the public imagination, anxiety about nuclear energy and wastes began to recede. Indeed, nuclear energy was being proffered as the solution to the 'greenhouse effect'. Yet, there would be enormous, probably insuperable, difficulties in attempting the siting, building and financing of the large number of nuclear plants needed worldwide to substitute for fossil fuel combustion. The costs and compensation would have to be achieved on a world scale. Quite aside from such unprecedented international cooperation, the risks of nuclear accident and nuclear weapons proliferation would be intensified and the burden of nuclear wastes would be immensely increased, spreading risks to new areas and increasing the hazards faced by future generations.

This book attempts an understanding of the contemporary politics of radioactive waste. Understanding is but the prelude to appropriate action. In the final chapter we relate the findings of our various studies to a practical question, namely, what are the political conditions necessary for the development of publicly acceptable policies for the management of radioactive wastes? In a situation where there is no ultimate political solution public acceptability becomes the test and the key to achieving the political legitimacy of nuclear waste policies. But, public acceptability raises some tantalising philosophical questions. In particular it does not follow that public acceptability guarantees the interests of nuclear communities who may unwillingly have to bear the costs conferred upon them by the majority. Nor does public acceptability in this generation secure the interests of future generations who will have to maintain surveillance over facilities that represent the technology of a bygone age. Nuclear waste poses profound moral questions which urge that we undertake those measures available, whatever the cost, to bequeath a safe environment to the future. For, as Tolkien's Gandalf observes, 'And it is not our part here to take thought only for a season, or for a few lives of Men, or for a passing age of the world. We should seek a final end of this menace, even if we do not hope to make one'.

Andrew Blowers,
David Lowry,
Barry Solomon,
Bedford, UK, November 1989

1 An Achilles Heel

A GROWING PROBLEM, FEW SOLUTIONS

Nuclear waste disposal is a growing international problem. In 1990 there were 416 operating commercial nuclear reactor units in the world, 26 per cent (110) of them in the USA and 37 per cent of the total (154) in Western Europe, all producing irradiated (spent) reactor fuel and other toxic wastes. Radioactive wastes are also generated from uranium mining and milling, spent fuel reprocessing, nuclear weapons production, and from industry, research and medical facilities. All nuclear wastes can be dangerous and even deadly, hence they must be isolated from the accessible environment. Yet the rapid proliferation in the production of civil nuclear power, weapons, and other nuclear products has not been matched by the development of safe methods of waste management. Until the 1980s the commitment by governments to the exploitation of the 'peaceful' atom was in stark contrast to their wilful neglect of the consequences.

Radioactive waste has now become an Achilles heel of the nuclear industry. Proponents of nuclear energy development recognise that unless solutions are found the future expansion of the industry will be threatened, as has already happened in the USA and a few other nations. A scientific and technical consensus has developed within the industry favouring a permanent solution to the problem by disposing of radioactive wastes in geological repositories. Some opponents of the industry have seized on the problem of nuclear wastes as a means to the end of eventually shutting down the industry altogether. Hence they resist plans for deep disposal and advocate long-term waste storage, thus ensuring the continuing physical and political visibility of the problem. As the gulf between opponents and advocates of the nuclear industry has widened the prospects for consensus over what to do with radioactive wastes have diminished. Yet solutions to the waste problem must be found since large volumes of nuclear wastes already exist and more is accumulating every day.

WHAT IS RADIOACTIVITY?

While this is not a technical book on nuclear waste disposal it is nonetheless essential that the reader understands the concept of radioactivity. Radiation is, after all, the reason for all the fuss over nuclear wastes. Nuclear power plants, whether civil or military, are by far the largest producers of radioactive wastes. This results from the fissioning (or splitting apart) of the uranium or plutonium atoms,[1] which results in new radioactive elements, 'radionuclides'. Some of these new radionuclides are higher in atomic number than uranium, in which case they are called 'transuranic elements'. Others are lighter than uranium and are called 'fission products'.

A typical fission reactor splits apart uranium-235; breeder reactors (which have been most developed in France) operate by converting non-fissionable uranium-238 into plutonium-239, following irradiation by neutrons in the reactor.[2] Nuclear fission splits the uranium-235 atom into two products or fragments of lower atomic weight. The fragments retain part of the nucleus (143 neutrons and 92 protons) and part of the electrons of the original atom. Transuranic elements are also formed when a uranium atom absorbs one or more neutrons and thus becomes a new element, such as plutonium or americium. Each of the more than 80 fission products so formed are capable of releasing 'ionizing radiation', which produces 'ionization' (the process of adding or removing electrons from atoms or molecules) by removal of orbital electrons. Two ions are created by this process, the negatively charged electron and the rest of the atom that now has a net positive electrical charge.[3] Radiation is the energy transferred by nuclear fission or similar physical processes as particles or waves through space or from one 'body' to another. The chain reaction that occurs inside a nuclear reactor also irradiates non-radioactive materials, yielding 'activation products' in the surrounding air, water, pipes and containment building. As these things absorb energy and change structure slightly in the process, they become radioactive. In turn, the activation products return to their normal stable state over time by releasing their own ionising radiation. The rate of radioactive decay of the unstable atomic nucleus in each radionuclide is measured by its half-life, which is the time it takes to decay to one-half of its original mass. The most common measure of the intensity of radioactivity is the curie (Ci), named after Marie Curie who co-discovered radium with her husband Pierre in 1898 (though the curie has been superseded by the bequerel in some countries). A Ci equals 37 billion disintegrations per second, roughly the rate of decay of one gram of pure radium. A single large nuclear reactor may contain over 10 000 megacuries of radioactivity.[4]

There are three types of ionising radiation (Figure 1.1). *Alpha*

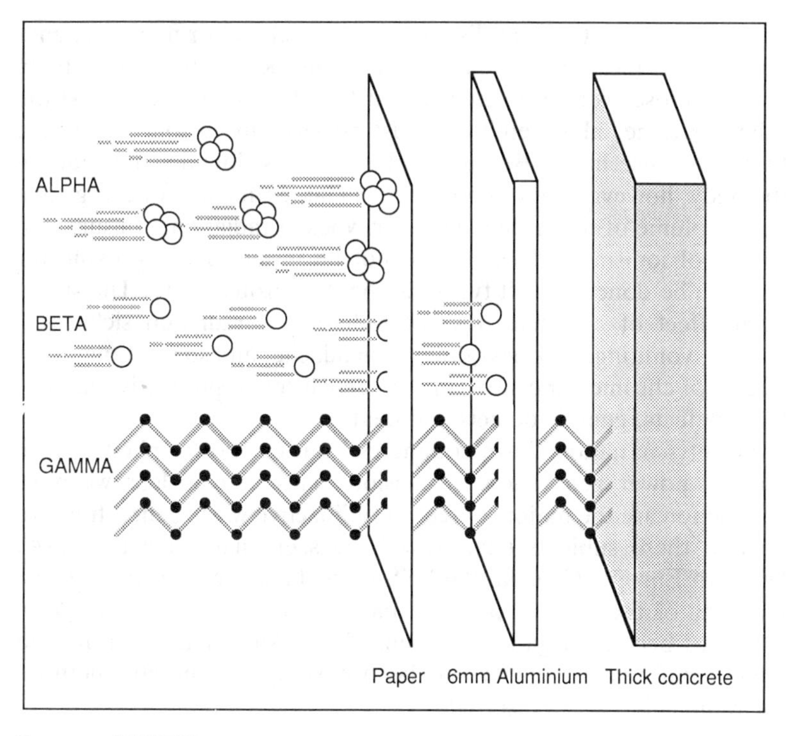

SOURCE NIREX
FIGURE 1.1 *The three types of ionising radiation*

particles are helium atoms positively charged from the loss of their two electrons, and which contain two protons and two neutrons. Alpha radiation is the least penetrating, but is the most densely ionising and thus potentially the most dangerous. Significant cellular damage can occur if an alpha-emitter such as plutonium enters the body, possibly leading to death. *Beta particles* are negatively charged and have the mass of one electron. Beta radiation, such as strontium-90 and caesium-137, can cause skin burns. *Gamma rays* are high-energy, short wave-length electromagnetic radiation, which usually accompany alpha and beta emissions. Though similar to X rays, gamma rays are so highly penetrating that several feet of concrete in the walls of a containment building are required for effective shielding. An example of a gamma-emitter is iodine-131.

All forms of ionising radiation are hazardous if inhaled, ingested, or otherwise incorporated into the body. Gamma rays can cause damage to living tissue at distances of several feet or more, depending on the intensity of the source. This occurs because gamma radiation loses very little energy as it travels through the air. At the other extreme, alpha emitters must be virtually in touch with live tissue to cause harm. Because alpha particles slow down rapidly in the body, however, they deposit a much larger quantity of energy in a given volume of tissue than do other types of radiation.[5] In sum, the source of ionising radiation need not be inside of living tissue for harm to be done, and all types can be dangerous to life. The short-term effect of an acute dose of radiation is radiation sickness – nausea, vomiting, dizziness, intense headache, and so on. Long-term effects of chronic exposure can include cancer, reproductive failure, birth defects, genetic defects and death.

Twenty-six nations have commercial nuclear power plants. Since nuclear power generation, expansion plans and nuclear weapons production are the major sources of radioactive waste generation, we turn to these subjects next. It will be seen that while the USA, Western Europe, Canada, the USSR and Japan account for the vast majority of the nuclear power capacity, radioactive waste disposal truly has become a global problem. Even Australia, which has no commercial nuclear plants, has been developing synthetic methods for radioactive waste disposal.

WORLD STATUS OF NUCLEAR POWER PROGRAMMES AND WASTE GENERATION

The era of commercial nuclear power began when the UK Atomic Energy Authority turned on their Calder Hall nuclear station in October 1956.[6] Calder Hall was designed primarily for military purposes, although it also fed electricity into the national grid. The Duquesne Light Company followed suit in the USA when they commissioned the first civil reactor at the Shippingport station in Pennsylvania the next year. This, too, was a modified military reactor. For the next two decades, nuclear power developers enjoyed almost uninterrupted expansion in Britain and the USA, unimpeded by the Achilles heel of nuclear waste. Before 1970, civil nuclear reactors had also opened in the USSR, Italy, Japan, France, Canada, Netherlands, Spain, India and Switzerland. By the late 1970s, however, the pace of expansion slowed in the US as concern over safety and economics clouded nuclear power. Even before the accident that occurred at Three Mile Island unit 2 near Harrisburg, Pennsylvania in March 1979, the nuclear energy option was effectively closed by market forces in the US. Although no new commercial orders have been placed since 1978, more than 40 nuclear reactors were to open in the USA in the 1980s, many extensively delayed because of financial and safety problems. Thus momentum had been so great that by 1990 the USA had 110 operating civil nuclear reactors and over 15 000 tons of spent nuclear fuel had accumulated (Table 1.1).

The UK was initially the world leader in nuclear power production. Indeed, its four gigawatts (GWe) of installed electric generating capacity by 1970 exceeded the amount developed in the US at that time. France's big push for the atom did not occur until after the Arab oil embargo of 1973. And despite the opening of the Latina nuclear station near Rome, Italy in May 1963, the Italians have been content to generate most of their electricity from foreign oil. Yet the largest development of nuclear power outside the USA has come in the industrialised democracies of the West. France, West Germany (FRG), UK, Canada, Sweden, Spain, Belgium and Switzerland are all major producers of nuclear power and thus spent fuel (Table 1.1).

The Soviet Union opened its first nuclear power plant, Troitsk unit A, 75 miles south of Chelyabinsk in September 1958. By 1990, the USSR had over 30 GWe of nuclear power capacity, ranking third in the world behind the USA and France. The catastrophic accident at Chernobyl unit 4 in the Ukraine in April 1986[7] has slowed but not stopped the Soviet Union's ambitious nuclear construction programme. Though at least eight reactors have been cancelled since 1987 in response to safety concerns and unprecedented public protest, many others are still being built. The turn away from the RBMK[8] reactor design used at Chernobyl may signal a greater concern for safety. Eastern European countries are generally proceeding with nuclear power as well. Existing capacity is being expanded in Czechoslovakia, Bulgaria and East Germany, while Romania and Poland (and also Cuba) are expected to join the nuclear club in the 1990s. Only in Hungary and Yugoslavia are the nuclear power programmes of this region in clear jeopardy.

Japan began nuclear power production in 1966 at Tokai-mura, and is the fourth leading producer of nuclear generated electricity in the world. The Chernobyl accident has only had a minimal effect on the industry there, which was protected by location from the main fallout in Europe. South Korea and Taiwan are also highly dependent on nuclear power, though in the latter there are currently no new plants

TABLE 1.1 *Civil nuclear power and spent fuel streams by country*

Continent Nation	*Installed Capacity (GWe)*	*Nuclear Reactors*	*Cumulative Spent Fuel Arisings (Metric Tons of Uranium)*[a]	
EUROPE	1990 (Jan.)	No.	1986	2000
France	51.4	52	2600[a]	20 000
USSR	30.8	45	NA	NA
FRG	22.3	21	2700	11 000
UK	13.7	41	700[a]	3000
Sweden	9.8	12	1500	5100
Spain	7.6	10	1000	2540
Belgium	5.5	7	1000	3000
Czechoslovakia	3.3	8	NA	NA
Switzerland	2.9	5	700	2000
Bulgaria	2.6	5	NA	NA
Finland	2.4	4	411	1520
East Germany	1.7	5	NA	NA
Hungary	1.7	4	NA	NA
Italy	1.1	2	330	NA
Yugoslavia	0.6	1	NA	NA
Netherlands	0.5	2	190	420
NORTH AND SOUTH AMERICA				
United States	97.9	110	15 700[b]	41 000
Canada	11.9	18	12 300[a,b]	33 900
Argentina	0.9	2	945[b]	5800
Brazil	0.6	1	48	1000
ASIA AND AFRICA				
Japan	28.2	38	4500	18 000
South Korea	7.2	9	60	4400
Taiwan	4.9	6	430	2600
South Africa	1.8	2	22	714
India	1.4	7	780[a]	5000
Pakistan	0.1	1	110	400

[a] Spent fuel arisings are for total production since the inception of civil nuclear power, and before the spent fuel is reprocessed. These data do not include spent fuel quantities in France (16 500 tons) and the UK (25 000 tons) that were generated by the older gas cooled reactors which have been used for *both* military and civil purposes. Spent fuel arisings are relatively higher in Canada and India because of their use of the unenriched 'CANDU' fuel, which has a lower burnup rate.

[b] By 1987. NA = Not available.

SOURCES: Compiled from *Nuclear News* (1990),[9] Numark (1987)[10] and Leigh (1988).[11]

being built. Argentina and Brazil have just three operating reactors between them, while there are seven in India; future developments will depend on the economic and political climates. The same holds true for the Philippines, which mothballed its first completed reactor at Morong, before it ever operated. The only operating commercial nuclear power plant in Africa is South Africa's two-unit Koeberg facility, though Egypt and Libya would like to build nuclear reactors.

The Chernobyl disaster has had a major effect on nuclear power in Central Europe. The most immediate impacts were the 31 initial deaths and the extensive contamination of water and land in the area surrounding the plant. Radioactive caesium-137 from the accident (that was detected around the world) has given humanity the equivalent of over 60 per cent of the dose from *all* the atmospheric nuclear weapons tests. The political fall-out from the accident has also been heavy.[12] Chernobyl has dampened the prospects for new nuclear power plants in the following European countries: Sweden, Finland, Denmark, FRG, Austria, Netherlands, Italy, Switzerland, Greece, Belgium, Hungary and Yugoslavia. The Swedes had passed a referendum in March 1980 to complete their existing plants but to phase out nuclear power completely by 2010 (see chapter 6). Talk in the mid-1980s of overturning this policy quickly disappeared when Chernobyl's fall-out hit central and northern Scandinavia. Nuclear issues are also highly politicised in the FRG. The government's efforts to add new reactors, waste facilities, and a commercial reprocessing plant have been continually challenged by the German Green Party (chapter 6). Since Chernobyl, many prominent leaders of the Social Democrats (SPD) advocate a nuclear phase-out too. Only in the USSR, France and Japan are major nuclear power plans going forward. But France has excess electricity and a moribund breeder reactor programme, while the Soviet and Japanese governments will have to deal with growing anti-nuclear movements.

The other major source of radioactive wastes is the manufacture of nuclear armaments.[13] Over 60 000 nuclear warheads have been produced in the USA alone since 1945, some of which have been dismantled.[14] Comparable production has occurred in the USSR. The UK, France, China, and India also have nuclear weapons capability, and many other nations are suspected of having the technical basis for producing nuclear weapons.[15] The future of nuclear weapons manufacture is highly uncertain in this age of 'glasnost'. Nonetheless, the military waste stream, whatever its size, requires disposal. Unfortunately, only in the USA are there accessible data on the military nuclear wastes (chapter 5). As we will argue below, the civil-military nuclear connection may well have much to do with a nation's political strategy on the nuclear waste problem.[16]

WHAT ARE RADIOACTIVE WASTES?

> Unlike the disposal of any other type of waste, the hazard related to radioactive wastes is so great that no element of doubt should be allowed to exist regarding safety.[17]

Radioactive wastes are a major source of radioactivity and therefore a key problem. Wastes are created at each stage of the nuclear fuel cycle (Figure 1.2). The first stage, uranium mining, involves the extraction of uranium ore from sandstone, shale or other rocks, with no more than five pounds of uranium extracted from each ton that is milled. The result is huge, bulky tailing piles that create liquid and solid wastes at the mill site, with long-lived radionuclides such as radium and thorium. These mill tailings can be dispersed off-site by wind and rain, and can cause contamination as during the July 1979 Church Rock, New Mexico (US) accident, when a dam broke at United Nuclear Corporation's uranium mill tailings pond, sending 1100 tons of tailings and 100 million gallons of radioactive water down the Rio Puerco River. Though the company claimed that the dam was engineered soundly, 'the dam was not constructed in accordance with plans and specifications. We made no regular inspections; their engineers were responsible for that', according to New Mexico State engineer Steven Reynolds.[18]

The next two stages in the fuel cycle, uranium hexafluoride conversion/fuel enrichment and fuel manufacture, create relatively minor waste quantities. The following stage (four), electricity production at nuclear reactors, creates the majority of radioactive wastes in the open (without reprocessing) fuel cycle.[19] Examples of these wastes include spent fuel, ion exchange resins (which purify the cooling water), sludges and less intensely contaminated trash. Chemical reprocessing of spent fuel (stage five) for plutonium and breeder fuel, chiefly in Britain and France, gives rise to hot liquid radioactive wastes and large volumes of solid nuclear wastes. Finally, when nuclear power plants are permanently closed, there is the unresolved problem of decommissioning wastes.[20]

Radioactive wastes are classified according to sources, disposal routes and levels of radioactivity. There is no internationally-agreed system of classification, though European nations and Canada recognise the following three broad categories:[21]

– *High-level wastes* (HLW). Heat-generating wastes arising from spent fuel reprocessing or in the form of spent fuel. These wastes

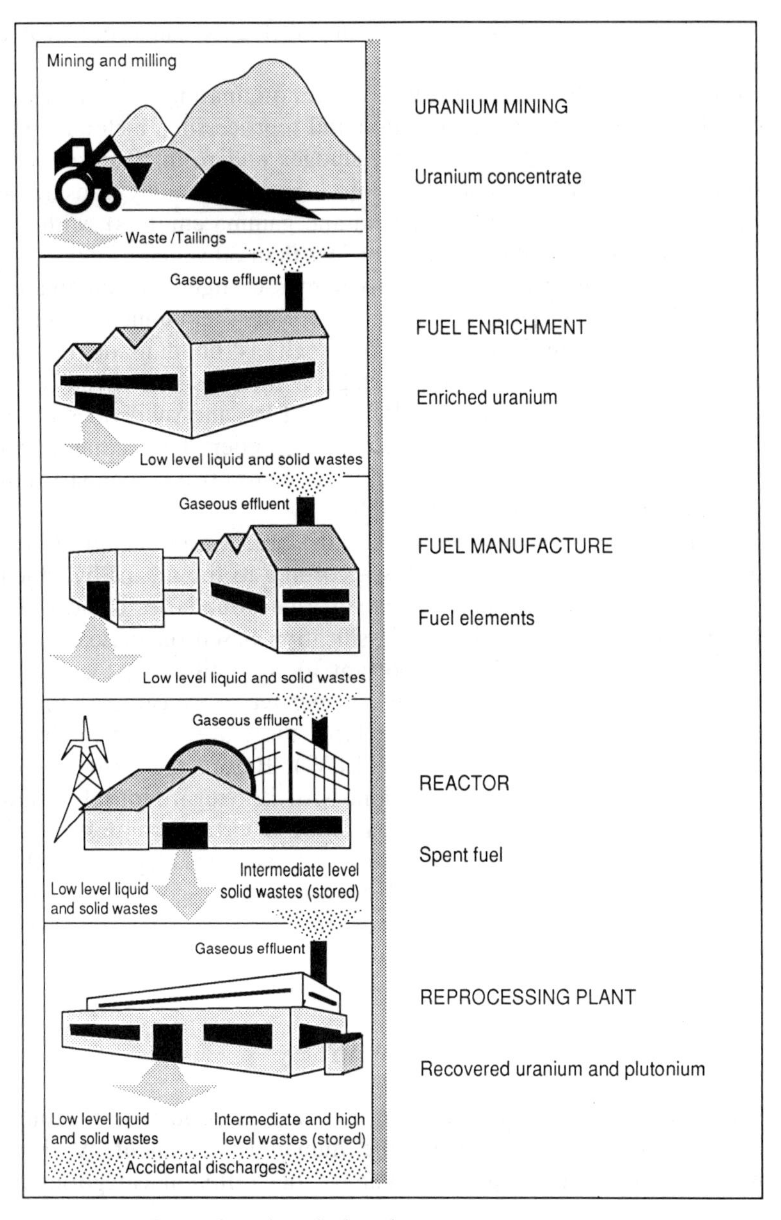

FIGURE 1.2 *The civil nuclear fuel cycle*

contain the most radioactivity and are all highly dangerous (Table 1.2).

– *Intermediate-level wastes* (ILW). Wastes originating from processes closely related to energy production and reprocessing, including fuel cladding and control rods, filters, sludges and resins from cooling systems. The ILW are subdivided into *long-lived* (mainly transuranic wastes) and *short-lived* (mainly beta and gamma emitters), with the division based on half-lives more or less than 30 years.

– *Low-level wastes* (LLW). High-volumes of lightly contaminated materials (clothes, papers, plastics, laboratory equipment, and so on). Most decommissioning wastes such as building debris are considered LLW. In some countries LLW includes short-lived ILW. The USA classifies virtually all ILW as LLW, and subdivides LLW into classes A, B and C. In countries undertaking commercial reprocessing, the preponderance of LLW comes from secondary and tertiary contamination in such plants.

There are, of course, problems with these waste classifications. Britain originally classified wastes according to the means by which they were disposed of, and this meant that boundaries could shift. There is a need for consistent classification based on radioactivity, half-life, and toxicity. It is important to recognise that the radioactivity from a nuclide is the same irrespective of the category that it is classified in (for example, in the USA there are long-lived alpha-emitters found in both HLW and LLW). Before the 1970s, LLW disposal in the USA and UK was done with little regard for safety and composition of the waste stream, making it virtually impossible to determine the environmental hazard of old sites. To avoid this problem in the future, it would be prudent for government and industry planners to isolate all radionuclides into a few internally consistent categories before their disposal in new waste sites.

MANAGING RADIOACTIVE WASTES

Radioactive wastes require care in handling, transport, storage and disposal to isolate them from the human environment. Various methods are currently used to manage the different categories of wastes:

HLW. Where reprocessing is used the HLW is in liquid form or vitrified. The HLW is stored until it is cool enough for geological emplacement, that is, in 20 to 50 years. Spent fuel reprocessing

TABLE 1.2 *Major nuclides in HLW, half-lives and specific activity levels*

Nuclide	*Half-life (years)*	*Grams/Curie*	*Curies/Gram*
Fission products			
Hydrogen-3	12.3	0.0001	3600
Cobalt-60[a]	5.26	0.0009	1140
Strontium-90	28.1	0.0071	140
Zirconium-93	900 000	390	0.0026
Technetium-99	210 000	58	0.017
Ruthenium-106	1.0	0.0004	2500
Iodine-129	17 000 000	6 100	0.00016
Caesium-134	2.05	0.00074	1350
Caesium-135	2 000 000	1300	0.00077
Caesium-137	30.0	0.011	91
Cerium-144	0.78	0.00032	3120
Promethium-147	2.62	0.0011	910
Samarium-151	87	0.036	28
Europium-154	16	0.071	141
Europium-155	1.8	0.00078	1280
Actinides			
Uranium-234	247 000	160	0.0062
Uranium-235	710 000 000	460 000	0.000002
Uranium-236	24 000 000	20 000	0.00005
Uranium-238	4 510 000 000	3 000 000	0.0000003
Neptunium-237	2 100 000	1400	0.0007
Plutonium-238	86	0.057	17.5
Plutonium-239	24 400	16	0.062
Plutonium-240	6600	4.4	0.23
Plutonium-241	13.2	0.0088	114
Plutonium-242	380 000	260	0.0038
Americium-241	458	0.031	32.2
Americium-242m	150	0.1	10
Americium-243	7950	5.4	0.18
Curium-243	32	0.022	45.4
Curium-244	17.6	0.012	83.3

[a] Cobalt-60 is not a fission product but is formed by neutron activation in the fuel cladding.

NOTE Fission products are the fragments of radionuclides resulting from nuclear fission of a larger atomic nucleus, and nuclides formed by the radioactive decay of these fragments. Actinides are the series of elements occupying atomic numbers from 89 to 105 in the periodic table, which include uranium and the transuranics.

SOURCE Adapted from Lipschutz (1980, p. 180). Reprinted with permission of R. D. Lipschutz and Union of Concerned Scientists.

reduces the volume (though not, importantly, the radioactivity) of HLW but creates greater volumes of ILW and LLW.[22] The US, Canada and Sweden are the only major nuclear states that do not have their commercial spent fuel reprocessed. The other western nuclear states and Japan contract for their reprocessing services with Britain and France, though a European reprocessing centre used to operate in Belgium. Japan is now building its own reprocessing plant, while the USSR manages spent fuel from Eastern Europe.

Unreprocessed spent fuel is stored in pools at power plants or in a centralised storage site (as in Sweden at the 'CLAB', discussed in chapter 6). Final disposal in a deep geological repository is the method proposed by most governments and the nuclear industry for spent fuel and vitrified HLW, though some analysts prefer sub-seabed disposal. Sites are currently under investigation in the US, France, FRG, Sweden, Finland, Switzerland, Argentina, and South Africa. Permanent surface storage is the option favoured by many protest groups. As yet no repository has been finally chosen though 'preferred' sites exist in the US, FRG, Argentina and South Africa.

ILW. Long-lived ILW are currently stored, except for a very small volume that is buried at shallow disposal sites in the USA and UK (which is no longer practised in the latter). Deep burial is the preferred disposal method in most countries. So far only the 'WIPP' site in New Mexico has been built for possible permanent disposal of military transuranic wastes (see chapter 5), though there are other sites under consideration in several countries.

Short lived ILW is managed in a few ways. Some countries have opted for co-disposal with other wastes: either in deep geological repositories (proposed in Britain, FRG and Switzerland, though an 'experimental' facility has operated in FRG); or in shallow land burial trenches (practised in France and the USA). Sweden disposes of these wastes in its new offshore sub-seabed repository, the SFR (chapter 6).

LLW. As with ILW various strategies are followed including co-disposal in either deep, sub-seabed (for example, Sweden's SFR) or shallow land facilities. In some cases such as the Drigg site in Britain shallow burial is currently practised for LLW alone. Very low-level wastes have such low radioactivity that they are 'below regulatory concern' in American parlance, and are disposed of in the same way as other toxic wastes or are incinerated. In addition, for many years LLW and some ILW were disposed of in the oceans (USA in the Atlantic and Pacific until 1970, and the UK on behalf of itself,

Belgium, Switzerland and the Netherlands in the Atlantic until 1983). Some wastes are emitted into the air, rivers or seas, as from the reprocessing plants in Britain and France.

Waste management strategies reflect, in part, the geological and geographical characteristics of individual nations. Most countries have a variety of geological regimes but increasingly tend to avoid political conflict by opting for geographically remote disposal sites.[23]

Radioactive wastes remain predominantly in storage awaiting final management solutions. There have been problems of leakage at several storage sites. In the Soviet Union, Medvedev[24] reported a major accident that involved radioactive wastes sloppily stored near Kyshtym in 1957 (chapter 2). In the USA three shallow land burial sites have been closed down. Major leakages of liquid HLW have occurred in the USA at Hanford (Washington), Idaho Falls (Idaho), the Savannah River Plant (South Carolina), and West Valley (New York). In Europe, Sellafield has been a major source of concern and there was a major scandal over the transport and labelling of wastes from the FRG into Belgium in the late 1980s. These problems have aroused growing concern and increased the salience of nuclear wastes as a technical, political and geographical problem, as later chapters will reveal.

WHY ARE RADIOACTIVE WASTES A PROBLEM? PERCEPTUAL ISSUES AND RISK ASSESSMENT

Nuclear power generates relatively little energy (roughly 3 per cent of the world's end use energy) but great anxiety. Opinion polls rate it as one of the public's greatest fears.[25] The fear is consistent over time and is perhaps growing, enhanced by media attention, by dramatic incidents (especially the accidents at Three Mile Island [TMI] and Chernobyl), and by reports of health effects (leukaemia, cancers, genetic effects, and so on). Unlike some phobias or environmental concerns, anxiety about radioactivity is apparently not subject to the whims of fashion. While it varies among different groups in the population (especially between the general public and workers in the nuclear industry) and between countries, the evidence of widespread anxiety is an inescapable problem for governments and industries committed to nuclear energy.

The typical reaction of the nuclear industry has been to discount, dismiss or ignore these fears as somehow irrational, born of ignorance

or perhaps motivated by simple hostility toward the industry. They point to the excellent safety record of the industry and the very low statistical probabilities of accidents or cancers. They urge greater public awareness and education to overcome morbid fears. And, indeed, there does *appear* to be evidence of irrationality and ignorance in the public's attitude toward nuclear energy as the following points demonstrate:
– people with the least fear of radiation are among those most familiar with the nuclear industry.[26]
– people do not discriminate – their fear of radiation is generalised. A LLW dump often provokes as much concern as a nuclear power station.
– people are not reassured by the low probability of danger, though there is a very low risk of health hazards and the risk diminishes with time following plant closure since radioactivity decays;
– public anxiety has increased over time while safety measures have also increased;
– public fear of radioactivity is greater than anxiety generated by other, statistically far more dangerous activities, such as driving a car, smoking, coal mining and so on.

Confronted by such apparent irrationality and ignorance the nuclear industry adopts various postures. Among these are:
– to advocate *greater public awareness* through education, demonstration, assertion of safety, and the like. This often becomes counterproductive. 'Often the only accomplishment of public information campaigns (and, one surmises, perhaps also their principal aim) is to leave the public more confused than ever.'[27] There can be an inverse relationship between efforts to encourage understanding and public resistance to it.
– to emphasise the *benefits of nuclear energy*. It is suggested that it is cheap, clean and safe compared to other energy sources. The benefits outweigh the costs;
– to stress the *lack of economic large-scale alternatives to nuclear energy*; the problem of the lights going out in the next generation;
– to stress that *nuclear energy is less destructive than fossil fuels*, which cause acid rain and deposition, urban smog, and global warming (the 'greenhouse effect').

The *pro-nuclear argument* is that nuclear power is necessary for the survival of electricity generation; the *anti-nuclear argument* is that a nuclear moratorium is necessary for the survival of future generations. Public fear of radioactivity has an emotional strength based on

rationality. 'The actual socioeconomic impacts of a nuclear waste repository will depend on "the facts" that actually get through to the public, not merely those that are believed in agencies and in the technical community,'[28] Public fear has the following components:
– it is a *low probability/high consequence phenomenon*. People may recognise that the statistical chances of death, cancer, and so on from radiation are low but the consequences are potentially catastrophic. It constitutes a 'megarisk'.
– compared to most other risks the *risk from nuclear energy is involuntary*, that is, individuals can exercise no control over their vulnerability.
– *it is an avoidable risk*. It is also a socially-induced risk. Although background radiation is unavoidable, the concentrated dangers that are inherent in nuclear reactors and radioactive wastes can be avoided. At least any increase in risk can be avoided by the use of energy conservation technologies and renewable energy sources.
– *it is an irreversible risk*. Once a nuclear reactor is commissioned, the production of plutonium and other transuranics begins and, though careful waste management can reduce risks, risk from radioactivity cannot be halted until decay reaches harmless levels. Fossil fuel use, of course, creates environmental, health and safety risks too, but some of these can be eliminated by stopping consumption.
– radioactivity is *dangerous by proximity*. It is invisible to the eyes and can be widely dispersed if a major accident occurs. (If it were coloured purple radioactivity would be even more resisted). Radioactivity creates a rational fear of the unknown.
– radioactivity is a *chronic risk* extending over long time-scales. The half-lives of some radionuclides extend over millions of years and they remain hazardous for even longer periods (for example, uranium-238, iodine-129, neptunium-237 – see Table 1.2). Such time-scales are unimaginable and the phenomenon of radioactive decay is overshadowed by problems of long-term security and safety (the problem extends into the future far further than human civilization has existed in the past).
– radioactivity is *a risk whatever its source*. Nuclear reactors, nuclear waste facilities and nuclear bombs can all be deadly. The concentrations, volumes and dispersal of radiation may vary. But if an accident occurs, the impact of a given amount of radiation is the same and therefore it is irrational for people to discriminate among the different radiation sources. According to this view, 'the peaceful atom is a bomb'!

– *absolute safety and security cannot be guaranteed.* Statistical probabilities reflect uncertainties and already many accidents and some major 'incidents' have occurred. There can be no ultimate assurance that the problems of melt-down and radioactive leakage, let alone sabotage, can be avoided. As nuclear proliferation increases so the probability of catastrophe increases.
– *health risks of low-level radiation have been continually revised upward.* Recent reassessments of the radiation doses received by the atomic bomb survivors at Hiroshima and Nagasaki, Japan, will only fuel the debate further.[29]

Some of these characteristics radioactivity shares with other environmental problems. Many toxic wastes are recalcitrant and present long-term hazards (for example, PCBs, dioxins, mercury, cadmium), and must be isolated from the human environment. Many other risks are potentially catastrophic. Fossil fuel combustion may cause global disaster through climate change within a few generations and CFCs are apparently breaking up the ozone layer, also creating potential global disaster. The existence of other large risks, of course, is no justification for pursuing nuclear technology with its evident dangers, though the nuclear industry often claims otherwise. The relative risks of energy alternatives need to be examined in a commensurate manner for a defensible energy policy.

In certain respects a radioactive material may be regarded as unique:
– *it is chemically different from other environmental contaminants.* It is 'immune to outside influence. Each radionuclide decays and emits radiation at its own particular rate regardless of temperature, pressure or chemical environment and continues to do so no matter what is done to it'.[30] A radioactive material can only be rendered harmless by decay, which may take a very long time.
– *it is highly toxic in very small amounts*, similar to some heavy metals with which it shares the quality of indestructability.
– *its effects are cumulative* and may not be seen for an exceptionally long time; indeed, genetic effects may not occur for multiple generations.
– *some radiation is indeed dangerous by proximity* as well as through inhalation or ingestion. Moreover, an increase in radioactivity may well be matched by an increase in deaths. Thus, health effects may be unavoidable.
– *it is inextricably connected to the production of plutonium for military purposes.*[31] The link between 'peaceful' and destructive

purposes is a singular source of fear and public opposition to nuclear energy.
– *it engenders greater and more widespread fear than any other form of socially-induced risk* (since radiation can cause leukaemia, cancers, genetic mutation, and the like). In that respect nuclear power is certainly a unique energy source.[32]

The risks from radioactivity emanate essentially from five sources (Figure 1.3):

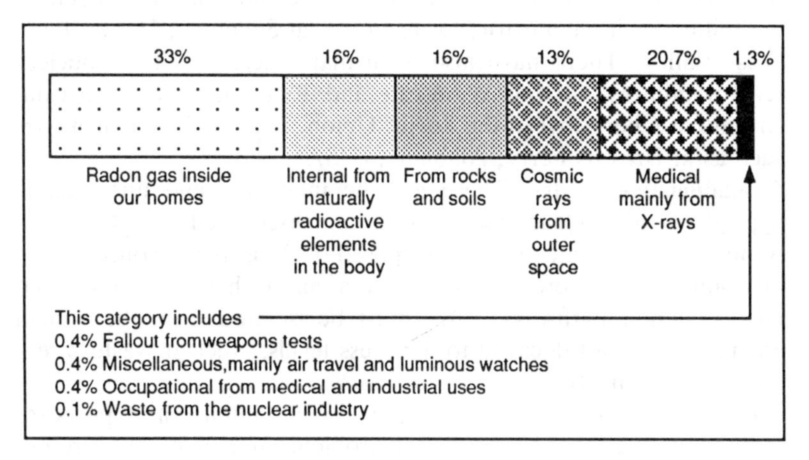

SOURCE Adapted from IAEA and NRPB original illustrations.

FIGURE 1.3 *Contributions of radiation in the environment by source – a nuclear industry view.*

1. **Background radiation**, such as from radon gas, cosmic rays from space, and so on, is greater in certain areas and rock types than others (for example, at high altitude, and near sandstone, shale, granite and phosphate). There is nothing that we can do about this except to avoid such areas. It is a natural environmental hazard but we do not need to add to it, since any radioactivity is cumulative. By comparing its own radioactivity with the quantities of background levels the nuclear industry imples that radiation is an unavoidable risk, and at the same time minimises the amount it produces. The nuclear industry makes much of the fact that background levels account for a very high proportion of the total but *it is the concentration of radioactivity that matters*.

2. **Medical applications**. Primarily radiation from diagnostic x-rays

from doctors and dentists, and occupational exposures, where the dosage has been greatly reduced.

3. **Nuclear weapons**. The fallout from early atmospheric testing of atomic weapons and the bombing of Hiroshima and Nagasaki (Japan) at the end of World War II.

4. **Deliberate or accidental discharges from nuclear plants**. These include routine emissions within allowable standards into the atmosphere, rivers, seas and oceans. In addition there have been accidental discharges of radioactive effluents above the limits through engineering failure or human error, for example, at Sellafield, Hanford and West Valley. There have also been major accidents from nuclear reactor failures, such as Windscale and TMI, one full-blown catastrophe at Chernobyl in the Soviet Union,[33], as well as near misses such as at Browns Ferry, Alabama, (USA).

5. **Radioactive wastes**. These are the inevitable and unavoidable byproducts of nuclear fission. Some are discharged (see 4 above), some are stored and some are disposed of. Wastes are concentrated in volume and therefore represent a major hazard if they find environmental pathways. They must be isolated from the human environment until decayed to harmless levels. Nuclear wastes are a unique problem, that is:

– *a problem of indefinite duration*. To all intents and purposes radioactive waste is a permanent problem created by the present generation but inflicted on all future generations.[34]

– *an indefinable problem*. The relationship between radioactive wastes and pathways to the accessible environment are difficult to fathom. The research being undertaken thus far in several countries indicates the scientific uncertainty inherent in deep geological disposal. The consequences of radioactive wastes entering the human environment are difficult to predict. The costs of correcting the problem are impossible to forecast with any accuracy. It is a problem of great uncertainty.

– *an insoluble problem*. Despite scientific consensus on the merits of deep geological disposal with engineered barriers there can be no certainty that the problem can be solved.[35] There is little empirical evidence that the radioactivity can be adequately contained in perpetuity and great reliance must be placed on predictions of waste containment whose uncertainties increase over time.

In this book we attempt to explain the politics of nuclear waste and to identify the obstacles and opportunities that must be considered in the search for solutions. At one level, radioactive waste is a **technical** problem. Any solution must ensure safety and security over the lengthy time-scales required for radioactive decay. Achieving agreement on technical and scientific criteria will prove difficult enough since the uncertainty increases the further we look into the future. Yet disagreement over technical matters reflects profound differences in moral attitudes and so it is an **ethical** problem too. Technical and ethical issues inform political choices and create conflicts – ultimately, and inevitably, nuclear waste is a **political** problem. We have considered some of the technical issues and will look at the ethical problems before outlining the key political issues on which our analysis is based. Nuclear waste disposal involves choices between different locations, trade-offs between employment and health, the short-term and the long-term, the present and the future. It is the interrelationship of these three dimensions – technical, ethical and political – that explains the conflicts explored in the chapters that follow.

THE ETHICAL DIMENSIONS – EQUITY OVER SPACE AND TIME

Interregional Inequity

A nuclear waste facility is a LULU (locally unwanted land use) except in what may be called 'nuclear oases', those places where waste sites or other nuclear facilities already exist and are generally welcomed. Radioactive waste disposal is a negative externality imposing costs on local communities that far outweigh the local benefits. In some respects nuclear waste disposal is little different from other LULUs in that there is inevitably some discordance between the local population that bears most of the costs and burdens (economic and psychological) of the facility, and those who benefit from it. For instance, the local benefits include both the employment associated with the waste facility and its tax revenues, which can be substantial for some rural areas. These benefits contrast with the local costs due to the blighting impact of the waste site, and the environmental and health concerns about radiation leakage. Young children are at a higher risk near a plant since they are the most vulnerable to leukaemia. The wider 'community' or perhaps the whole nation receives the benefit of 'clean' electricity production from nuclear power without bearing localised costs, though they may contribute toward disposal costs.

As with other unwanted facilities a trade-off must be made between regionally distributed benefits and locally concentrated costs. This is a classic example of local versus regional or national interests. The extent of national benefit and the magnitude of local costs varies and judgements in nuclear waste management must be made by decision-makers, implicitly or explicitly, using principles of equity and need.[36] It should be noted that the benefits and costs are not entirely incongruent since electricity consumers may live close to waste sites, and workers in the industry must also face the increased hazards from radiation.

The uneven (inequitable) distribution of costs and benefits must be justified on the basis of the needs of the wider community or society. In the case of radioactive wastes this assumes a necessity for the facility to be put somewhere. The need for a radioactive waste disposal or storage site must take into account: existing waste volumes and forms, projected waste quantities (and thus future energy policy), and trade-offs between geological disposal and surface storage.

There are contrasting views on the question of need. The nuclear industry and its supporters argue that nuclear power generation is necessary to assure present and future (growing?) electricity needs. They generally assume that permanent waste disposal offers the appropriate form of management by isolating and containing wastes; costs thereby fall on relatively few communities. This 'solves' the waste problem and 'ensures' the future of nuclear power by mending the Achilles heel. Opponents of nuclear energy claim that energy conservation and alternative power sources are safer, cleaner and cheaper, and therefore the nuclear option should be foreclosed, leaving a definable volume of nuclear wastes to be managed (see the chapter 6 discussion on Sweden). Need and equity are thus ethical questions, based on values.

Intergenerational Inequity

'The most obvious class of persons who will probably bear higher health and financial risks, because of the creation of nuclear waste, are members of future generations.'[37] Radioactive waste involves intergenerational inequities[38] – most benefits of nuclear fission are relatively short-lived (except for technology transfer and fossil fuel conservation) and largely accrue to the present consumers. There is no certainty that nuclear power will be 'needed' in the future. In contrast, the costs of nuclear electricity in the form of radioactive wastes will fall primarily on future generations. These costs include monitoring, surveillance, and potential hazards. Radiation risks may increase over time if knowledge fades and leakages occur. The short-term benefits of nuclear power are far easier to estimate than the long-term costs. Moreover, the political benefits are short-term while problems in the distant future have little or no effect on contemporary political perspectives. Thus, the tendency by current decision makers is to heavily discount the needs of disenfranchised future generations.

Nuclear waste thus represents a gamble with the future, 'systematically short changing our children'.[39] Radioactive waste management assumes that the problems posed for future generations will be no greater than those experienced by the present generation. But we are confronted with profound uncertainties compounded by long time-scales. The problem is rationalised by the arguments that either the waste management technology will work or that future generations will be able to take care of themselves. The statistical probability of catastrophe may be remote but the possibility exists. The uncertainty does not provide an excuse for persisting with nuclear technology. Rather, the certainty that if a problem occurs it could be catastrophic and irreversible provides a convincing ethical case for limiting such impacts on future generations.

Ethics and Politics

Compensation. The inequitable distribution of the costs of nuclear wastes over space and time can be partly addressed by financial compensation or 'bribe', depending on one's viewpoint. Monetary payments might be provided as an incentive for communities to accept a nuclear waste facility. Communities already hosting nuclear facilities or those anxious to attract jobs might be prepared to accept lower compensation than other places. This could raise the dilemma of poor areas bidding for minimum compensation. Such differential rates of compensation might be regarded as inequitable by society. Payments might be made as recognition of the costs imposed on localities. The rate of payment should be the same regardless of whether communities are willing or unwilling to host a waste site. In this case the inequity arises because compensation is acceptable to some and not to others; non-monetary incentives may thus also be required (information sharing, independent monitoring, public control over facility closure, and so on).

The form and amount of compensation is difficult to calculate. It might be a tax rebate, a grant, an investment in community facilities, environmental improvements, participation in decision-making, or some combination thereof. Compensation might be offered before site selection or negotiated with specific communities. Some forms of compensation, of course, will be regarded by some people as bribes, but by others as essential and necessary to ensure development. Compensation to future generations is even more problematic. While it is possible to provide information and warnings it is unrealistic to offer monetary payments indefinitely. It is impossible to guarantee any political arrangements (such as public participation or control over a waste facility) for more than a relatively short period.

Compensation is widely regarded as a key to achieving public acceptance of radioactive waste facility proposals.[40] The justification is usually, however, based on grounds of political expediency rather than ethical principles.

Accountability. Nuclear facilities involve restrictions on civil liberties. They are safeguarded to prevent access and to protect people from the radiation. Surveillance and monitoring is a routine requirement. Radioactive waste facilities may well require a permanent para-military force to keep out intruders and a cadre of people who are 'highly skilled and can be motivated sufficiently to perform continuously at extraordinarily high levels of reliability, even though it is likely that the job will generally be boring and routine on a day-to-day basis'.[41] These requirements imply permanent removal of nuclear reservations from effective democratic accountability (that is to say, more Sellafields and Hanfords). Restrictions on access and information, and failure to disclose some problems is also regarded as necessary by the state. Ironically, full disclosure of health and safety data may be required for the public to accept a new waste disposal facility. Last but not least, it will be necessary to maintain vigilance for several hundreds or thousands of years at nuclear waste facilities, until the radiation hazard subsides.

There are major problems of accountability here. While ultimate authority rests with government, day to day operation is controlled by the industry to whom responsibility for safety and security is effectively delegated. Accountability to the affected public is remote and to future generations is non-existent. Kasperson, *et al.*[42] recommend a 'public defender for the future' to participate in current decisions on waste management, though choosing the 'correct' defender is obviously problematic. We may conclude that 'if technological decisions cannot be made democratically, then there is a strong argument suggesting that they should not be made at all'.[43]

The ethical issues of equity, compensation and civil liberties are inherent in political decisions about radioactive wastes. These issues are rarely confronted explicitly, but moral judgements are made by decision-makers nonetheless. The moral issues suggest that radioactive waste management policies and proposals should be subjected to two tests – is it right? is it fair? These are underlying questions in our political analysis, which are most often raised in the context of siting proposals for nuclear waste disposal facilities.

THE POLITICAL PROBLEM – NUCLEAR WASTE DISPOSAL AND SITING CONFLICTS

> The containment and storage of radioactive wastes is the greatest single responsibility ever consciously undertaken by man.[44]

While many people may disagree with Senator Baker, it is clear that the technical and ethical problems that underlie conflicts over what to do with radioactive wastes must be resolved by the political process. The conflicts require conscious choices to be made, choices that reflect the interests of different groups. The outcome of conflict will reflect the relative power of these competing interests. Conflict over siting of radioactive waste facilities is part of a broader conflict over the full range of issues surrounding nuclear power and nuclear weapons. Ultimately the two sides (those for and against a siting proposal) are irreconcilable because of a clash in values, and the government must favour one or the other; this maxim clearly holds true for nuclear waste disposal. On specific aspects of radioactive waste management government may support different sides, or try to effect a compromise between them.

The power of the conflicting interests in nuclear waste disposal varies among the world's nuclear states, and specific conflicts will be explored in chapters 3 to 6. It is possible, however, to offer some broad ideas at the outset, which will be analysed against the empirical evidence of different countries. In the last chapter we will return to these to see what generalisations can be made about the future of this major Achilles heel of nuclear power.

Pro-Nuclear Interests

These interests comprise the components of the industry, its dependents and supporters: companies engaged in stages of the nuclear fuel cycle; trade associations; nuclear defence contractors; workers and communities with a stake in the jobs and wealth created by the industry; and supportive politicians and government agencies.

While management of military nuclear wastes is the responsibility of the central government, the costs and implementation of policy for civil radioactive wastes are frequently devolved to the industry. Military wastes are a matter of national security. In some countries such as the USA (chapter 5) and West Germany (chapter 6) HLW disposal is the direct responsibility of government agencies or laboratories. In most European nations executive waste management authorities covering all or part of the waste stream have been established by the nuclear waste generators, and are accountable to the central government (see chapters 3 and 6). In the USA, LLW disposal and facility siting is the responsibility of the individual states and is managed by the private sector.

Nuclear waste management policy in most countries has historically tended to be incremental rather than comprehensive. It exemplified the characteristics of hidden decision making, and privileged access for technocrats so typical of the nuclear state.[45] The nuclear industry or relevant government agencies have typically unveiled waste management proposals before consulting the public, adopting a strategy that might be described as 'Decide, Announce, Defend' (DAD). Consequently several disposal sites have been developed, with little fanfare. Eventually problems with existing sites, greater public awareness, and defeat of new proposals have led to much greater government involvement with radioactive wastes, along with a slow shift toward more open, rational strategies of policy making. In many cases, laws or referendums have been passed that force this 'Achilles heel' onto centre stage. A major reason for this change has been the success of the anti-nuclear opposition in opening up the issues.

Anti-Nuclear Interests

These interests are usually dispersed, uncoordinated and often disunited. They range from national and international groups such as *Greenpeace* and *Friends of the Earth*, who oppose all things nuclear, to local protest groups who may oppose a specific nuclear facility in their own backyard while accepting nuclear power and nuclear weapons production as long as it is done elsewhere. Thus broad opposition to the nuclear industry has historically been relatively weak, whereas opposition to specific proposals at the local level is vigorous and frequently successful. Except for *Greenpeace*'s protests that contributed to Britain's cessation of ocean dumping of LLW in the North-East Atlantic in 1983 (chapter 2), and US environmentalists' campaign to prevent HLW repository siting in Utah, most major opposition to nuclear waste disposal has been organised at the local level (as our detailed case study in chapters 3 and 4 describes).

Anti-nuclear groups tend to use populist, legal and parliamentary methods to achieve their objectives. Some nuclear opponents in the FRG (and in the late 1970s in France), however, have occasionally resorted to violence, usually after provocation by armed police. More commonly the opposition has high visibility through a media with an appetite for good copy; uses tactics to pressure the political system; and relies on public opinion to support its cause. It has shown itself capable of organising broad coalitions of protest over certain issues that cut across conventional divides of class, community and politics. This is, perhaps, its greatest potential strength. Local rather than national anti-nuclear groups have garnered sufficient political power to change the waste management policy of the state. Indeed, recent anti-nuclear protest in Britain has shown a subtlety and flexibility in using tactical opportunities to defeat LLW facility proposals (chapter 4). There is also evidence of a more positive stance to seeking solutions for problems such as radioactive wastes while maintaining the aim of eventually shutting down the industry. We will return to the topic of 'solutions' in the last chapter of this book.

Each side in the conflict over nuclear wastes represents a variety of interests – professional advancement, electricity sales, energy security, public health and safety, environmental protection, economic development, and so on. Each side is intent on securing political clout and public acceptance of its arguments. In this book we shall explore the conflict over radioactive wastes in a variety of political contexts and try to explain why certain outcomes have been achieved, as well as attempt to indicate future prospects.

Nuclear waste disposal and siting conflicts have emerged as major political problems in the mid to late 1980s. Political conflict over radioactive wastes takes many forms. It may be a general conflict over national policy, the criteria for facility site selection, or over assumptions about safety. Or the conflict may be related to specific sites, either concerned with problems at existing sites or over a proposal for a new site. The outcome achieved by different interest groups (government, industry, environmentalists, anti-nuclear groups) will depend on the political context in which *political power*, the key dimension of conflict, is exercised. This context is defined by three main factors. First is *political institutions*, in particular the degree to which the system is centralised, or decentralised and fragmented. Second is *political participation*, the extent to which the political system is accessible or 'open' to the various interests involved, or conversely closed to participation in decision making. And third is *waste management policy*, the degree to which policy has the characteristics of "rationality" and synoptic vision, or is not clearly defined and is incremental. A fourth dimension is *political geography*, the spatial distribution of social and environmental constraints and opportunities for waste facility siting. This is relatively less important since, in most cases, there is rampant opposition to nuclear waste facilities in all but nuclear oases. In geographically large countries such as the USSR, Canada and the USA, however, the sheer extent of territory offers opportunities for sites not encountered elsewhere, and so geographical and geological factors can be a crucial consideration in siting.

The political context varies within and between countries. Our method for explaining the nuclear waste problem is a two-pronged approach. In a detailed case study designed to comprehend the political processes at work in Britain, we will show how these factors explain the development of nuclear waste conflict in the UK. We shall also provide comparative, though more general, studies of the USA and other western industrialised democracies in an effort to reach generalisations (contrasts or parallels) about the relationship between political context and outcomes of the nuclear waste problem.

Nuclear waste is not the only factor affecting decisions on nuclear power, as Chernobyl and the arms race demonstrate. Yet nuclear waste management and disposal have already become an Achilles heel of the nuclear industry in the USA, Sweden, FRG, Switzerland and Austria (where the only reactor, Zwentendorf, was mothballed before it ever opened), and may become so in Britain soon. What are some of the key differences in political culture and processes in the major nuclear states that explain their approaches to the waste problem and its likely resolution? These are the questions that we turn to next, and try to answer in the book.

POLITICAL CENTRALISATION AND NUCLEAR WASTE MANAGEMENT

The major factor of the political context of a country that appears to influence outcomes on radioactive waste disposal is political institutions. Nations with a strong central government (that is, centralised political system) tend to link civil nuclear power with military nuclear power (where it exists) and as such rely heavily on spent fuel reprocessing. Examples include France, UK and the USSR, as well as Belgium and Japan (which do not have nuclear weapons). Political centralisation as well as pro-nuclear consensus results in a strong, robust nuclear industry, which can generally withstand public opposition to nuclear power stations. Dissolution of this consensus leads to cancellation of power stations, as has recently occurred in the USSR and may yet happen in Britain. In these countries, reprocessing centres provide an oasis for potentially all levels of radioactive wastes, *greatly lessening the urgency for finding new waste sites.* Reprocessing centres exist in France, UK, Belgium (though closed) and the USSR that can store all levels of wastes; reprocessing facilities in the USA, in contrast, are restricted to military wastes. Japan is building a reprocessing plant (and LLW facility) at Rokkashomura on northern Honshu, while its current HLW stream is stored at the centres in the UK and France until the early 1990s.

Sweden also has a centralised political system, though in contrast to the others it has a high degree of openness and participation by its citizenry, and a strong anti-nuclear sentiment. For example, lengthy public hearings and the high voter turnout for the 1980 referendum (that phases out the nuclear industry) has given a legitimacy to government plans that is not found in any other major nuclear state. This legitimacy, combined with the lack of domestic reprocessing facilities, has paved the way for a spent fuel storage site and long-term geological repositories in Sweden (chapter 6).

Canada is a country with strong provincial government, though it is an anomaly since its decision-making on nuclear issues is controlled by two entities, the central government's Atomic Energy Canada Ltd, and Ontario Hydro. Moreover, 16 of its 18 operating reactor units are located in two sites in southern Ontario, Pickering and Tiverton. While citizens are usually afforded ample opportunity to participate in public hearings and meetings on nuclear matters, the major factor affecting nuclear waste decisions is apparently geography. With a huge land mass, much of which has few or no people, Canada (and also the Soviet Union) has more potential disposal areas and thus a greater opportunity to search slowly for waste sites away from populated areas than do the other nuclear states.

Political systems that are decentralised include the USA, the FRG and Switzerland.[46] Federalist systems allow extensive autonomy of decision-making (in general) in the states, lander or cantons. The nuclear industry in these countries is relatively weak, as the central governments are unwilling or unable to stifle anti-nuclear protests, and opposition in the courts. The USA, FRG and Switzerland (and also Sweden) have various laws that hinge nuclear power developments on plans for radioactive waste disposal. The political context in these nations is thus one of fragmentation of decision-making, no pro-nuclear consensus, a relatively weak industry, strong and active opposition, and extensive opportunity for public participation through administrative and judicial channels that can thwart proposals for nuclear facilities. This has been the case in the USA especially on spent fuel storage and HLW repository siting (chapter 5), and in the FRG on spent fuel storage, reprocessing plants, and facilities for all levels of radioactive wastes (chapter 6).

Political deadlock on nuclear power, however, has been a major reason for the central government in federalist countries to accelerate the search for waste sites; for what better way to legitimise a nuclear programme in trouble than to eliminate an Achilles heel? These countries also lack reprocessing plants to store civil HLW; FRG's plan to build a reprocessing centre at Wackersdorf, and previously at Gorleben, ran into strong opposition (and was finally shelved in mid 1989), as has its plan for dry spent fuel stores at Wackersdorf, Gorleben and Ahaus (chapter 6). Nonetheless, the nuclear agencies in the federalist polities hope that development of both HLW and LLW sites will help to revitalise a struggling industry, and siting proposals have thus come relatively early and in large number. Indeed, the pressure to 'solve' the HLW problem is so great in FRG and Switzerland that a proposal to send spent fuel for storage or disposal to China's Gobi Desert has been revived.

The evidence to support our political theory of radioactive waste management, that is, that both centralised and decentralised government systems alike must engender real public participation and trust for nuclear waste disposal to ultimately succeed, is presented in chapters 3 to 6. The second chapter provides a history of civil nuclear waste management policy, beginning with its military origins in the USA and focusing on the UK. Chapters 3 and 4 are an extensive case study of nuclear waste policy and political conflict in Britain, especially over the 1983 to 1987 period. Chapter 5 gives an overall review of radioactive waste management policy and political experience in the USA, while chapter 6 does the same for the FRG, Sweden and France, the leading producers (with the UK) of nuclear power in Western Europe. The final chapter of the book tallies the lessons learned, and charts out "the way forward" for what many believe to be an insoluble problem.

NOTES

1. An atom is the smallest part of an element having all of the properties of that element, consisting of protons, neutrons and electrons (Figure 1.4). A proton is an elementary particle that is a fundamental constituent of all atomic nuclei, having a positive charge equal in magnitude to that of the electron; the number of protons in the atom determines which element it is (for example, uranium contains 92 protons per atom and is thus atomic number 92). A neutron is an elementary particle with no charge, which is found in all atomic nuclei except that of hydrogen, and with a mass slightly greater than that of a proton. An electron is negatively charged, and is found outside the nucleus of an atom. The number of electrons in an atom usually equals its number of protons, making that atom electrically neutral. Atoms with an unstable number of neutrons are radioactive isotopes of an element, which seek stability by giving off energy through radioactive decay.

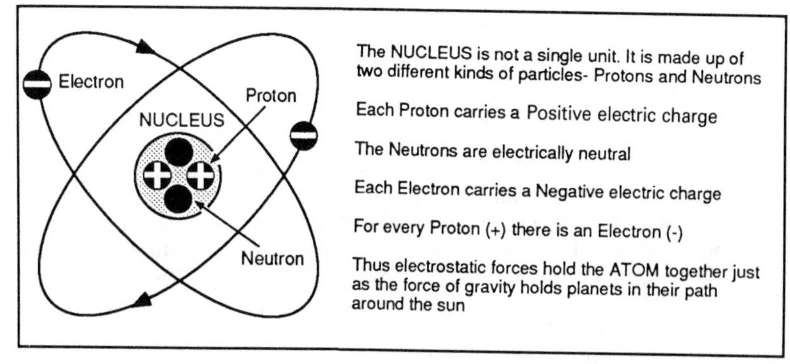

SOURCE Adapted from UKAEA original.
FIGURE 1.4 *A schematic representation of the structure of the atom.*

2. France, FRG, Belgium, the UK, Switzerland and Japan are developing 'mixed oxide' or MOX fuel, which is nuclear reactor fuel that combines plutonium and uranium oxide (see n. 10). The sole reason for this is that a 'surplus' of plutonium exists from the reprocessing operations, which was originally to be used in the breeder reactors that are now well behind schedule.
3. Bertell, R., *No Immediate Danger: Prognosis for a Radioactive Earth* (Summertown, Tennessee, The Book Publishing Co., 1985), p. 19.
4. Lipschutz, R. D., *Radioactive Waste* (Cambridge, Mass., Ballinger, 1980) p. 11.
5. Ibid., p. 8.
6. Patterson, W. C., *Nuclear Power*, 2nd edn (London, Penguin, 1983).
7. Bojcun, M. and Haynes, V., *The Chernobyl Disaster* (London, Hogarth Press, 1988).
8. The RBMK 1000 is a graphite moderated, plutonium-production reactor that the USSR also uses for power production. The reactor's overly complex cooling system is archaic and is considered a plumbing nightmare (see n. 7).
9. *Nuclear News*, 'World list of nuclear power plants', Vol. 33, No. 2, February 1990, pp. 63–82.
10. Numark, N. N., 'Analysis of factors influencing national spent fuel management strategies, and overview of worldwide activities', paper presented at the International Atomic Energy Agency's International Conference on Nuclear Power Performance and Safety, Vienna, Austria, October 1987.
11. Leigh, I. W., *International Nuclear Fuel Cycle Fact Book*, PNL–3594, Rev. 8. Prepared by the Battelle Pacific Northwest Laboratories for the US Department of Energy (Richland, Washington, 1988).
12. Flavin, C., *Reassessing Nuclear Power: The Fallout from Chernobyl* (Washington, DC, Worldwatch Institute Paper Number 75, 1987).
13. In addition to nuclear reactors and other nuclear fuel cycle facilities, and nuclear weapons plants, radioactive wastes are generated by various industrial, research and medical activities. Though the latter categories are relatively minor contributors to the waste stream, incomplete and inconsistent international data on nuclear waste production makes quantification extremely difficult.
14. Hansen, C., *U.S. Nuclear Weapons: The Secret History* (New York, Orion Books, 1988).
15. Lovins, A. B. and Lovins, L. H., *Energy/War: Breaking the Nuclear Link* (New York, Harper & Row, 1980).
16. Cf. Solomon, B. D., 'The politics of nuclear power and radioactive waste disposal: from state coercion to procedural justice?', *Political Geography Quarterly*, Vol. 7, No. 3, July 1988, pp. 291–8.
17. US National Academy of Sciences committee, quoted in Lash, T., 'Radioactive waste: nuclear energy's dilemma', *Amicus*, Vol. 1, No. 2, Fall 1979, pp. 24–34.
18. Fuerst, I., 'New Mexico's radiation river: aftermath of the Church Rock spill', *Not Man Apart*, Vol. 9, December 1979, p. 9.

19. Resnikoff, M., *Living Without Landfills* (New York, Radioactive Waste Campaign, 1987).
20. Pollock, C., *Decommissioning: Nuclear Power's Missing Link* (Washington, DC, Worldwatch Institute Paper Number 69, 1986).
21. Heath, M., 'Deep digging for nuclear waste disposal', *New Scientist*, Vol. 108, No. 1480, 31 October 1985, pp. 30–2.
22. Albright, D. and Feiveson, H., 'Why recycle plutonium?', *Science*, Vol. 235, 27 March 1987, pp. 1555–56.
23. Solomon, B. D. and Shelley, F. M., 'Siting patterns of nuclear waste repositories', *Journal of Geography*, Vol. 87, March/April 1988, pp. 59–71.
24. Medvedev, Z., *Disaster in the Urals* (London, Angus & Robertson, 1979).
25. Slovic, P., Fischoff, B. and Lichtenstein, S., 'Rating the risks', *Environment*, Vol. 21, April 1979, pp. 14–20, 36–9.
26. Van der Pligt, J., Eiser, J. R. and Spears, R., 'Nuclear waste: facts, fears and attitudes', *Journal of Applied Social Psychology*, Vol. 17, No. 5, 1987, pp. 453–70.
27. Goodin, R., 'No more nukes', *Ethics*, Vol. 90, April 1980, pp. 417–49.
28. Freudenburg, W. R., 'Rationality and irrationality in estimating the risks of nuclear waste disposal', in Post, R. (ed.), *Waste Management 87*, Vol. 2 (Tucson, AZ, University of Arizona, College of Engineering and Mines, 1987) pp. 109–15.
29. Roberts, L., 'Atom bomb doses reassessed', *Science*, Vol. 238, 18 December 1987, pp. 1649–51.
30. Lash, op. cit., p. 27.
31. Lovins and Lovins, op. cit.
32. Slovic, P., Fischoff, B. and Lichtenstein, S., 'Images of disaster: perception and acceptance of risks from nuclear power', in Goodman, G. T. and Row, W. D. (eds), *Energy Risk Management* (London, Academic Press, 1979) pp. 223–45.
33. Bojcun and Haynes, op. cit.
34. Cameron, D. M. and Solomon, B. D., 'Nuclear waste landscapes: how permanent?, in Cullingworth, J. B. (ed.), *Energy, Land and Public Policy* (New Brunswick, NJ, Transaction Books, 1990), pp. 137–86.
35. Lovins, A. B., 'Comments on the 10/78 draft of the IRG report to the President (TID–28817 Draft)', unpublished memorandum (1978) to the Interagency Review Group on Nuclear Waste Management, available from the Rocky Mountain Institute, Old Snowmass, Colorado.
36. Kasperson, R. E., Derr, P. and Kates, R. W., 'Confronting equity in radioactive waste management: modest proposals for a socially just and acceptable program', in Kasperson, R. E. (ed.), *Equity Issues in Radioactive Waste Management* (Cambridge, Mass., Oelgeschlager, Gunn & Hain, 1983) pp. 331–68.
37. Shrader-Frechette, K., *Nuclear Power and Public Policy* (Dordrecht, Netherlands, D. Reidel, 1980).
38. Kasperson, *et al.*, op. cit.
39. Goodin, R., 'Uncertainty as an excuse for cheating our children: the case of nuclear waste', *Policy Sciences*, Vol. 10, 1978, pp. 25–43.

40. Miller, C., 'Efficiency, equity and pollution: the case of radioactive waste', *Environment and Planning A*, Vol. 19, 1987, pp. 913–24.
41. LaPorte, T. R., 'Nuclear wastes: increasing scale and sociopolitical impacts', *Science*, Vol. 201, 7 July 1978, pp. 22–8.
42. Kasperson *et al.*, op. cit., p. 363.
43. Goodin, 1980, op. cit., p. 433.
44. Former US Senator Howard Baker, quoted in Shrader- Frechette, op. cit., p. 49.
45. Blowers, A. and Pepper, D. (eds), *Nuclear Power in Crisis: Politics and Planning for the Nuclear State* (London, Croom Helm, 1987), and see chapter 2.
46. Deese, D. A., 'A cross-national perspective on the politics of nuclear waste', in Colglazier, E. W. (ed.), *The Politics of Nuclear Waste* (New York, Pergamon, 1982) pp. 63–97.

2 From Scientific Serendipity to Scientific Problem

NUCLEAR WASTE – THE ORIGINS AND GROWTH OF THE PROBLEM

We have already seen in Chapter 1 that radioactive wastes may be divided up into various categories by source – each a legacy from society's use of radioactive materials. These wastes include the mining and milling residues from uranium, spent nuclear fuel, either in its original processed solid form, or in a combination of liquids, solids and gases; 'operational' contaminated on-site solid wastes from nuclear plants, nuclear fuel cycle facilities or bomb tests; discharges in liquid and gaseous form from reactors or processing facilities and also the solid, liquid and gaseous wastes from industrial, medical and military nuclear operations; and lastly the contaminated reactor itself which will become the main source of 'decommissioning' wastes. It is perhaps with the original waste legacy from experimentation with uranium that we should start, for it is the use of uranium and its radioactive derivatives that has created the legacy of radioactive waste. Admittedly, there are naturally occurring pockets of concentrated radioactive materials that have the same or similar form to radioactive wastes, such as the ground soil contamination of the Oklo[1] natural reactor site in the Gabon or in the Kerala region of India that has concentration levels of 'background' radiation higher than some so-called low-level radioactive wastes created by human endeavour. However, these are exceptions: the vast bulk of the nuclear waste is produced by social rather than natural processes.

Radioactive materials were first used at the turn of the century in medical experiments, including the use of X-rays. The luminous property of radium, a decay product of uranium, stimulated the commercial exploitation of radioactive materials in luminous paints for clocks and watches. Production took place in unregulated factories where workers and the local community were exposed to high radiation levels causing severe health effects.[2] The factories were converted, left derelict or demolished without any precautions being taken. All of this material should have been classified as radioactive waste, carefully packaged and isolated from the human environment. Exploitation of radioactivity without much concern for the management of its waste products was a characteristic feature of the later developments resulting from the use of nuclear fission, the atomic bomb and nuclear energy.

This chapter attempts to set the political and to a lesser extent technical background to the key decisions to be taken on radioactive waste in the 1990s. It traces the origins of scientific appraisal of the nuclear waste issue, examining evidence drawn from a period of roughly 40 years – the mid 1950s to mid 1980s. Although the primary focus of the book as a whole is on political issues, in this chapter we discuss the various technological developments that sometimes precede, sometimes supersede political choice. The chapter concentrates particularly on the UK and provides the context for the detailed political analysis of the conflict over shallow burial which is the subject of chapters 3 and 4. The early history also covers the United States. The choice of the UK and US is justified both because of the accessibility of data and the fact that both countries also have extensive military nuclear programmes which have impinged substantially upon waste burdens, management strategies and technical problems. They provide a basis for discussion of other countries later in the book. The chapter is broadly chronological. As a result of this approach, the earlier part of story will at times switch from one side of the Atlantic to the other, to compare or contrast the differing ways radioactive waste has been handled in technical and political terms. Discussion of detailed policy developments in the United States from the mid 1970s is discussed in chapter five. This chapter therefore only traces US developments to that point.

Over the forty year period, it will be shown that radioactive waste has changed from being practically a non-issue to one of central political significance. The sociological constituents of this political change are examined, as the issue of nuclear waste emerges from being merely a technical problem to one which demands intervention at both the highest political level (cabinet or executive government) and the lowest (individual or community involvement).

THE WASTE LEGACY OF THE ATOMIC BOMB PROJECT

Although this chapter mainly concentrates upon the management problems and policies in regard to commercial radioactive wastes from power reactors, reprocessing, and to a lesser extent medical and industrial wastes, it is important to explore briefly the progenitor role played by the atomic bomb project. The style of 'hands-on' management developed under duress by scientists, the military commanders and some politicians during this period of intense effort left a political and organisational legacy that meant decision-making over atomic energy was for the next three decades (1945–75) dominated by three components: secrecy, centralisation and technocracy.[3] These led respectively to lack of political or popular participation; to institutional self protection; and to a lack of properly debated policy.

A unique collection of scientists and engineers were scattered in teams across the continental United States and Canada over the six and half years from January 1939 (the month after atomic fission had been demonstrated in Berlin) to July 1945, when the world's first experimental atomic bomb (code-named Trinity) was exploded in the New Mexico desert at Socorro, about 80 miles from Alamogordo. The Manhattan Engineering District or 'Manhattan' Project had been initiated in August 1942 to produce the atomic bomb. The project involved over 150 000 personnel. Three secret atomic cities were established in the United States: at Los Alamos, New Mexico the bomb design centre; at Hanford, Washington State, to produce plutonium and at Oak Ridge, Tennessee, to create enriched uranium. An additional plutonium production centre was established at Chalk River in Ontario, Canada. So secretive were these organisational arrangements that only a very small number of the Manhattan project staff knew they were working on the production of an atomic bomb at all.[4]

Even today it is hard to establish the number and location of sites in the Manhattan Project. It is nevertheless an important question, for this project bequeathed inestimable quantities of radioactive contamination at the production sites. It also produced massive radioactive contamination at the A-bomb test sites in New Mexico, and later in Nevada and in the Marshallese and Aleutian Islands in the West and Northern Pacific Ocean respectively. In chapter 5 we will look in further detail at the military nuclear waste problem outstanding in the USA today. A study by the US Department of Energy (USDOE) from September 1980 (see Figure 5.1) suggested that there were at least 76 separate sites, the highest single number being in New York State (14). Later evidence from the US General Accounting Office (GAO) identified nearly 1500 inactive sites holding nuclear waste.[5] The total US investment in nuclear warhead production from the start of the Manhattan Project has been around $89 bn ($230 bn in 1988 dollars).[6] This sum, for the production of around 25 000 warheads made of plutonium or enriched uranium, may be compared with the predicted clean-up costs of radiotoxic wastes at the present 17 USDOE military nuclear production sites. A report prepared by the USDOE for the Senate Armed Services Committee and the Senate Committee on Governmental Affairs,[7] made public on 1 July 1988, estimated that clean-up costs might reach US$110 bn by 2045 – that is nearly half the total investment cost in the whole programme, 1942–88. Even this figure is likely to prove a conservative estimate. Of all the facilities, Hanford is seen to pose the most acute problems with a forecast cost of $46 bn to clean up all the contamination, taking 30–50 years to complete. In comparison the total money spent on the 'Superfund' to restore thousands of non-military sites across the US contaminated by toxic and hazardous wastes, for 1980–88, was only around $5 bn.[8] Moreover, other government agency estimates for the full Hanford clean-up – involving restoration of the 570 square mile complex – reach as high as $100 bn.[9]

A recent report on the environmental contamination and containment record of the US Departments of Defense and Energy concluded that the military facilities were 'amongst the worst violators of the hazardous waste laws', and added that the Defense Department's attitude varied between 'reluctant compliance and active disregard for the law'. One clear example cited – though in this case of a USDOE managed plant – was the Fernald uranium fuel processing plant in Ohio, which was said to dump annually 109 million gallons of highly radioactive wastes into storm sewers.[10]

Weldon Spring, a similar case, is a uranium processing plant in St Charles County, Missouri.[11] Four miles to the southwest of the abandoned plant is a nine acre quarry now filled with nearly 100 000 cubic feet of hazardous wastes, including radioactive rubble from the earlier Mallinckrodt plant which in the early 1940s processed uranium for the first successful nuclear fission experiment at Chicago in 1942. US government reports admit that ground water has been contaminated from leakage and seepage from the quarry since 1960. By the mid-1980s the 'clean-up' of defence wastes was finally being given serious attention. Four decades on it is useful to compare this changed attitude with the lax policies and perspectives adopted by the atomic establishment in the 1940s towards radioactive wastes and contamination.

Whereas today a great expense is devoted towards the proper security and containment of radioactive materials, including wastes, during the Manhattan atomic bomb project the use of radioactive wastes as radiological weapons was being seriously considered. In May 1941, a special panel of American scientists proposed that radioactive waste weapons be developed as a top priority by the US. Atomic power for a nuclear propelled navy was credited the second priority, and the atomic bomb the third priority. A top secret report – dated 10 December 1941 – concluded that:

> the fission products produced in one day's run of a 100 000 kilowatt chain reacting pile might be sufficient to make a large area uninhabitable.[12]

The aim was not to separate out, by reprocessing, the plutonium for the manufacture of atomic bombs, but to make use of the high-level waste in the form of a 'cancer bomb'. In 1951, James Conant, President of Harvard University and the American Chemical Society, expressed unusually pessimistic views with the prophecy that the disposal of atomic waste products would present gigantic problems to be lived with for generations; that in his view the problem was insoluble; and that people would decide regarding continued atomic development that 'the game is not worth the candle and a self denying ordinance is but common sense'.[13] The other key scientist involved was Dr Robert Oppenheimer, who masterminded the atomic bomb project. Oppenheimer's consideration of the use of radioactive wastes as atomic poisons was noted by the British government scientific adviser Professor P. M. S. Blackett in his controversial book published in 1948.[14] In his chapter on 'future technical developments' Blackett sketches in the detail thus:

> 'Of more importance perhaps as a weapon [compared to underwater bursts of atomic bombs] is the possibility of spreading over enemy country the radioactive materials (fission products) which are a by-product of the manufacture of plutonium in a pile. For every kilogramme of plutonium produced, about a kilogramme of fissile products are also produced with lifetimes ranging from minutes to years. If these fission products could be spread by spraying or in some other way over cities or the countryside, all personnel remaining there for a considerable time would be killed. No statement seems to have been published indicating the area that could be made lethal in this way by a kilogramme of fissile

> material, but a comparison with the known intensity of radiation from the fission products with the maximum safe dosage of radium rays allowed in medicine, shows that the area must be extremely large.'

In the peculiar exigencies of wartime, it is perhaps not surprising that such plans as developing radiological weapons based on nuclear 'waste' dispersal over enemy forces and territory were seriously considered and researched. What is both surprising and disturbing is the discovery which came to light in February 1986 when the USDOE issued, in response to public demand, 19 000 pages of hitherto classified documents on the operations at Hanford in the 1940s and 1950s. In brief, the US military had conducted radiological experiments on local people, without their permission or knowledge.[15] In 1945 alone, the Hanford facility, charged with the secret wartime production of plutonium, released routinely through the exhaust stacks of its reactors 340 000 curies of radioactive iodine-^{131}. This was, to put the figure in context, 339 480 curies *more* than officials were to consider 'safe' a decade later, and present standards permit less than one curie of iodine-^{131} release each year at the plant.[16]

In the immediate post-war years, Hanford began to tighten its standards of airborne radiological release. Nevertheless, the contamination spread into the deep, confined aquifers below Hanford and into the Columbia River. The radioactive wastes stored on-site, in mostly single shell underground steel tanks steadily increased. In late 1949 the fourth plutonium reactor (out of eight that were to be built by the mid 1950s at Hanford), was opened. Political pressure was exerted on Hanford to boost the preparedness of the US military to confront the Soviet Union which had just tested its first atomic bomb. It was this that led to the 'Green Run' experiment three months later, on 2–3 December 1949.

'Green Run' was conducted by General Electric's Nucleonics Department for the U.S. Atomic Energy Commission (USAEC). It constituted the deliberate release of 5500 curies of iodine-^{131} (and an inventory of still classified fission products) to the atmosphere over a 24-hour period. The resultant 200-by-40 mile plume from Spokane, Washington to Dallas, Oregon was secretly measured. The resulting contamination exceeded the USAEC's own tolerance levels (which in the 1940s were considerably greater than today's permissible exposure standards) by 11 000 times on the Hanford reservation and hundreds of times in downwind communities. The aim of the experiment was according to an official of the US Department of Energy, present day successor the USAEC,

> related to development of a monitoring methodology for intelligence efforts regarding the emerging Soviet [Military] nuclear programme [in 1949].[17]

Prior to these revelations, the single most serious release of iodine from Hanford officially reported was from the PUREX plutonium purification plant in 1963, when 63 curies were accidentally ejected.

There is now substantially documented evidence of a continuous series of radiological accidents and experiments at the Hanford plant in the 1950s, including the releases of ruthenium-106 from the reduction oxidation (REDOX) plant from 1952, throughout its operational lifetime to 1967. Ruthenium particles contaminated local farmers' fields and soldiers of the 518th AAA (anti-aircraft artillery) division stationed at Camp Hanford from 1951–61.[18] Hanford's chief radiation protection officer, in a then classified report in September 1954, recorded that the REDOX plant had been routinely littering the region with visible particles contaminated with radioactive ruthenium. Parker noted: 'There is a definite probability that information or rather misinformation on the offsite condition will leak to the public in the near future', and predicted 'Not all the residents will be as relaxed as the one who was recently quoted as saying that living in Richland (the company town for Hanford) is ideal because we only breath tested air.'[19]

It is evident that the USAEC, under strong cold war military pressures, was prepared to release radioactive fall-out, and effectively countenance the use of radiological weapons creating a delayed 'cancer bomb' on its own citizens, many of whom had made considerable personal sacrifice to support the atomic developments at Hanford; but was not prepared to release information on its activities to the same people. As one local farmer, Tom Baillie, born in 1947 near Hanford, and who became a strong campaigner against Hanford, put it, 'I'm not anti-nuclear, but what they did to us was industrial recklessness and stupidity'.[20] These experiments were made possible by the secrecy which shrouded the nuclear industry in all countries in its early years.

ACCIDENTS AND NUCLEAR WASTE

The world's first major reactor accident, at the Canadian NRX reactor pile at Chalk River, in Ontario's northern wilderness, occurred on 12 December 1952. The accident created the first major quantities of radioactive wastes from unintended reactor operations.[21] The chief of the reactor division, Fred Gilbert, admitted that a 'calculated risk' had been taken. Coolant water was allowed to flow into the basement of the containment building housing the reactor. D. O. Woodbury described the situation as 'growing more serious by the minute, as contamination seeped into cracks and crannies, permeating even the concrete itself. The pile was out of danger but now the building had to be saved. Doggedly, the directors decided on an emergency pipeline to a disposal area over a mile away. . . . Chalk River faced a new year of cleaning up, a scrubbing job without precedent. Despite the relatively lax radiation exposure levels in force at the time, the fear was strong that the accumulated absorbed radiation dose received by clean-up workers would exceed the limits many fold. By the summer of 1953, the clean-up was completed. It produced atomic debris that was buried in disposal areas and heaped with sand after containment in canvas bags to confine loose dust. The accident indicated that 'the worst problem of the atomic age may turn out to be the drab matter of ash handling and disposal'.[22]

In the winter of 1957/58 there was an incident in the Soviet Union involving the probable explosion at a reprocessing plant, where on-site nuclear wastes were poorly handled, and which killed an unknown number of people and certainly contaminated a substantial area around Kyshtym, near Chelyabinsk in the Urals. Unlike the Chernobyl accident three decades later, the radioactive contamination remained relatively local; and hence the Soviet government and atomic authorities were able to keep all details secret. Wider knowledge of the extent of the accident existed only in rumour and scanty press accounts until November 1976 when Zhores Medvedev published the first English language scientific account of the accident in the magazine *New Scientist*.[23] The Soviet authorities finally confirmed the accident in June 1989, interestingly, as part of a new nuclear glasnost when trying to get local public support for further atomic development in the area.[24]

There is evidence that there had been a deliberate cover-up of the accident by the western nuclear authorities. This was because, having learned about the accident from CIA sources soon after it took place, it was felt that if details were to be made widely known about such an accident at a waste site so soon after the major fire in the Windscale Piles, in October 1957 (the first serious accident to set back the British atomic project,[25]) then knowledge of the Soviet accident could badly dent public confidence in the atomic energy programmes in western societies that then enjoyed widespread public support. But the matter of a cover-up, which if true would undoubtedly be a disreputable political collusion, is by no means clear cut. According to one analysis of the cover-up thesis,[26] Soviet visitors to the Brussels Science Fair in the summer of 1958 talked openly of a 'great explosion' near Chelyabinsk with scores of dead. It is certainly true that less than a year later stories appeared in the Western press concerning an accident.[27]

Two decades later, a few months after Medvedev had published a second article on the Kyshtym accident,[28] a US environmental group used the US Freedom of Information Act to release internal CIA papers on the event. Although considerably censored, the declassified papers[29] did confirm CIA knowledge of some sort of 'unspecified accident' at the Kasli atomic plant in the winter of 1957 and indicated that precautions against radiation exposure had been instituted. The original source was deleted. More recently the USDOE has published two reports in 1979 and 1982[30] reviewing the evidence on the accident. Just two months before the first of these, Medvedev himself published a full length study[31] explaining from Soviet primary sources and a detailed literature search, what he thought had really happened. What is politically interesting is that even two decades or so after the accident, the western nuclear establishment were still keen to underplay the significance of the event, as evidenced by dismissive letters and articles in the press and scientific journals in subsequent months.[32] These were sent to offset the deleterious effect on public confidence in the plans to develop nuclear energy and consolidate nuclear waste disposal strategies.

THE FOUNDATION OF INTERNATIONAL NUCLEAR INSTITUTIONS

Nuclear waste was a relatively minor issue on the agendas of the growing number of international nuclear institutions established during the late 1950s. These followed the original United Nations conference on the peaceful uses of atomic energy (the first fully international gathering of atomic scientists) in Geneva in 1955, and the establishment in the same year of the UN Scientific Committee on the Effects of Atomic Radiation (UNSCEAR). Only one of these organisations was truly global in its scope: the International Atomic Energy Agency (IAEA), established as an autonomous inter-governmental agency within the UN system. Its prime purpose was made clear in article II of its statute (objectives) approved in 1956.

> The Agency shall seek to accelerate and enlarge the contribution of atomic energy to peace, health and prosperity throughout the world. It shall ensure, so far as it is able, that assistance provided by it or at its request or under supervision or control is not used in such a way as to further any military purpose.[33]

In a detailed recent history of the IAEA, Scheinmann[34] makes virtually no reference to the consideration of radioactive waste in the setting up or development of the IAEA. The main contribution of the IAEA has been on the control of fissionable (explosive) nuclear materials, ie safeguards; and technical assistance on nuclear facilities and research (including 'nuclear medicine'). An early US draft for the IAEA statutes suggested that the Agency would specify disposition of by-product fissionable materials and waste. Despite this the 1986 IAEA annual budget, for instance, devoted over ten times the financial resource to safeguards development and implementation than to radioactive waste management.[35]

The oversight of nuclear waste was not eventually included in the statute. The agency has since developed its radioactive waste programme, which has been primarily one of information exchange through a series of publications and the hosting of international conferences and national programme reviews, rather than original or primary research. It has concentrated on regulatory aspects of the whole waste management process (handling, treatment, conditioning, underground disposal and environmental consequences of discharges), and has developed internationally accepted guidelines, standards and codes of practice for use by national authorities. Especially important have been those developed for transfrontier transport of radioactive materials.[36]

In addition the IAEA has developed several guideline principles to govern the management of nuclear waste.[37] These are:

(1) 'Dilute and disperse' waste to the environment in effluents containing radionuclides in amount below authorised radiological protection limits (based on the recommendations of the International Commission for Radiological Protection, the ICRP).
(2) 'Delay and decay' those wastes which contain only short lived radionuclides.
(3) 'Concentrate and confine' those wastes which contain significant amounts of long-lived radionuclides.

However, it was only in late 1988 that the IAEA established its first formal radioactive waste policy body – the International Radioactive Waste Management Advisory Committee.[38]

In the same year (1957) that the IAEA formally began operations, two other inter-governmental bodies were founded that have played a role in nuclear waste oversight. The Organisation for European Economic Co-operation and Development, which became the OECD in 1960, set up a European Nuclear Energy Agency (ENEA) – which became simply NEA in 1972 when Japan fully acceded to the OECD. Although the ENEA set up Committees on Radiation Protection and Public Health, and on the Safety of Nuclear Installations at the outset, it was only in 1975 that the 24 member state organisation established a Radioactive Waste Management Committee (RWMC) which, as with the IAEA, facilitated the sharing of member state information as well as sponsoring research on land and sea disposal.[39] Earlier, in 1965, the ENEA did begin a study 'aimed at developing safe and economic methods' of oceanic disposal of radioactive wastes, which resulted in the development of 'framework guidelines' for these operations, in 1967.[40] The ENEA/NEA role since has been to provide legal and technical assistance and 'in the absence of any specific legal framework, international surveillance on a purely voluntary basis'.[41]

The third body set up in 1957 was the European Economic Community's (EEC) nuclear agency, Euratom. Its full title was the European Atomic Energy Community and it had similar aims to the IAEA and ENEA. A main difference between the agencies was that member states of Euratom were bound by legal treaty obligations; and that technically, at least, Euratom was to be the owner of commercial nuclear materials in the EEC, as well as the designer/operator of certain nuclear facilities. In addition Euratom was given a watchdog role over nuclear safety, covered by chapter III of the

Euratom Treaty. Article 37 of the treaty addressed nuclear waste as follows:

> Each member state shall provide the Commission with such general data relating to any plan for the disposal of radioactive waste in whatever form as will make it possible to determine whether the implementation of such a plan is liable to result in the radioactive contamination of the water, soil or airspace of another Member State.[42]

However, for nearly the first two decades of its existence Euratom concentrated almost entirely on research, development and promotion of nuclear energy. Only in 1975 did the EEC Council of Ministers adopt its first (five-year) programme for Euratom on the management and storage of radioactive waste. Even then this remained relatively technical. The waste issue began to be politicised within the European Community, through the European Parliament, in the early 1980s. By the late 1980s this became an issue of major political consequence as environmental directives were issued and scandals and cover-ups of malpractice emerged (see chapter 6).

There was one further international agreement in the late 1950s that set the framework for national governments to control sea disposal/discharges of radioactive wastes: this was the UN Convention on the High Seas, agreed in Geneva on 9 September 1958. Article 25(i) reads in part

> Every state shall take measures to prevent pollution of the seas from the dumping of nuclear waste, taking into account any standards and regulations which may be formulated by the competent international organisation.

But although the IAEA and other international institutions drew up guidelines and facilitated information exchange through conferences, it was national governments that were responsible for policy and execution. It was really only after the Chernobyl accident in 1986 that a truly international acceptance of transfrontier regulation became more accepted.

EARLY PERCEPTIONS OF NUCLEAR WASTE AS A PROBLEM IN THE UK

In the UK nuclear waste was a matter of small political significance during the period 1945–75. As in the US, during wartime Britain planning for the disposal of radioactive wastes was a low priority.[43] Covered by secrecy, two land disposal sites were developed, one at Amersham (Buckinghamshire), site of the radiochemicals centre for the dumping of radium and barium sulphate residues; and another in Barton-le-Clay (Befordshire), in a cave where the Medical Research Council had its medical radon plant from 1939 to 1948.[44] The low activity liquid and gaseous wastes were routinely discharged to the sea, local rivers and the atmosphere from atomic facilities. For the low activity solid wastes, a disposal site was considered using abandoned mine shafts in the Forest of Dean, about 80 miles from Harwell. The UKAEA's plan was opposed, Bertin records,[45] by a local group known as the 'Free Miners' who had ancient rights in the Forest. Bertin believed there was 'no sound scientific basis for their attitude'. But it nevertheless prevailed, and the atomic authorities were forced to continue with sea disposal, which had begun in 1949, in the deep waters of the north east Atlantic ocean.[46] Additionally, from 1950 to 1963, the British government authorised the dumping of radioactive waste packages into the shallower waters of the south west English Channel, in an area known as the Hurd Deep, about ten miles north of Alderney (one of the Channel Islands), at a site also used to dump surplus and degraded high explosives from World War II.[47]

The destiny of the solid radioactive wastes 'that could not be dealt with by normal methods of storage and eventual disposal' was, according to Bertin, 'one of the most vexed questions of atomic energy'. The interim 'solution' was the trip taken several times a year by one of the Royal Fleet Auxiliaries several hundred miles out to sea 'to dump a number of concrete-covered metal canisters into the deep water beyond'. Bertin proclaimed that these canisters were 'undoubtedly safe for as long a time as their contents will remain active', although the deep sea dump procedure was very expensive.

For the first 15 years or so of atomic energy development there was little apparent political interest in the subject of radioactive waste. The official historian of the British Atomic Energy project, Professor Margaret Gowing, makes no reference whatever to radioactive wastes in the first volume of her history covering 1939–45. In her discussion of the second period, 1945–52, Professor Gowing, having had privileged access to all original internal documentation, stated that 'Atomic energy risks were treated as a technical not a political or social matter, and were firmly in the hands of experts, with politicians showing little interest except in the Radioactive Substances Bill of 1948'.[48] Some concern was shown by Sir Harold Hartley in his presidential address in 1950 to the British Association on 'Man's Use of Energy',[49] where he commented:

> It is clear from the discussion at the Fourth World Power Conference that the difficulties to be overcome before the use of nuclear energy becomes economical are most formidable, and their solution will require intense effort over a long period . . . the removal and disposal of the radioactive products from the reactors will be a costly and puzzling task.

One of the first indications of political concern over nuclear waste in the UK came in 1950 when the Minister of Works, Richard Stokes, although not departmentally responsible, showed an interest and wrote to the Minister of Supply about the effluent problems at atomic sites. His particular concern was the highly active liquids that might be stored 'for a thousand years' in tanks at Windscale. He enquired what might happen if the system of containment failed. Sir Christopher (later Lord) Hinton, the chief engineer and manager at Windscale felt the tanks to be 'a grave responsibility' giving grounds for considerable anxiety. An anodyne reply was sent to the Minister of Works describing the 'elaborate precautions' for the Windscale tanks, but conveyed no suggestions of Hinton's anxieties. The letter accepted that some risk existed in spite of the safeguards but the only way of ensuring absolute safety 'would be to abandon the entire atomic energy project'.[50]

EARLY WASTE MANAGEMENT OPTIONS IN THE UK

According to an early study,[51] investigations were 'being vigorously pushed' in the mid 1950s at the Atomic Energy Research Establishment (AERE), Harwell, on methods of manipulation, separation, storage and disposal of wastes that were predicted to increase enormously in volume. The authors concluded that 'it is clear from progress made already that there will be no insurmountable long term problem'. This optimistic viewpoint echoes the single paragraph[52] devoted to radioactive waste in the White Paper '*A Programme of Nuclear Power*' (Cmd 9389) which launched the British civil nuclear programme in February 1955 using the so-called 'Magnox' design. The government view was that 'the disposal of radioactive waste products should not present a major difficulty'. It conceded that the problem primarily would be with the chemical processing plants, of which there would only be a few, and not the power stations. Waste volumes would be small and 'great efforts' were being devoted to determining 'economic methods of storing or disposing of them'. It also reinforced the dilute and disperse principle, with the assurance that 'any material that is discharged will be tested to ensure that it is of extremely low radioactivity, so that it will be *harmless* and comparable in effect to the natural background radioactivity which is always present' (emphasis added).

In 1956 AERE scientist, R. Spence, stressed that in Harwell's research into reprocessing 'the first thing that had to be done, however, before any radioactive work could begin, was the development of a radioactive waste system, and especially in its analytical control, which was acceptable to the public bodies concerned'. In order to reduce to the minimum liquid effluent discharge from (the Windscale) reprocessing plant, a chemical separation process had been adopted that 'permitted the main fission product waste solution to be concentrated by evaporation to small bulk'. It would be stored on-site for many years 'in tanks of moderate capacity', though Spence conceded that 'the permanent storage of millions of curies of fission products in solution is not very desirable'.[53] The UKAEA sponsored a long term research programme to try to resolve this problem. The research underway to achieve safe and secure methods of storage and disposal included the preparation of concrete from liquid wastes, especially for small quantities of highly active materials; fusion of wastes in glass and ceramics (the process that would become known as vitrification); and fixation of the wastes in montmorillonite clay, which was known to be readily available, inexpensive, and highly stable.[54] It is clear that volume reduction for ease of safe management was the priority of British nuclear waste management.

By 1958, the British government recognised that it would have to formalise a plan for the overall management and disposal of nuclear wastes, as required under the Nuclear Installations (Licensing and Insurance) Bill.[55] Harwell was consulted to advise the Ministry of Housing and Local Government. This led to the 1959 government policy paper 'The Control of Radioactive Wastes' (Cmnd 884), which set the policy framework for the next 20 years in Britain. As the atomic energy programme began to develop, decisions were made without any real reference to public participation in the process.

The main interest of the nuclear authorities and government in the 1960s still remained in the promotion of nuclear expansion through the development of successful power reactors. Waste management was considered 'a trivial issue'.[56] In the UK practically no political interest was shown in the management of radioactive wastes in Parliament in the early 1960s, if entries in the official record (Hansard – whose Parliamentary answers are referred to in this book by date and column number) may be judged a reliable barometer of interest. In 1960 there was only one parliamentary question (pq) asked on nuclear wastes in the whole year and that was on the relevance of the planned national disposal service for contaminated wastes from nuclear submarines (6 December, col. 13). In 1961 two pqs were posed. One enquired about research progress on the vitrification (glassification) project. The minister for science who was then responsible stated that the project was 'progressing satisfactorily'.

A year later, in reply to one of only two pqs asked in 1962, the same science minister told Parliament that the UKAEA's vitrification plant had opened (16 July, col. 98). Although it operated for a year, the project was wound down because so small a volume of high-level waste had been accumulated that it was not felt further expenditure was justified.[57] At the end of 1961, the science minister told Parliament confidently (19 December, cols. 149–151) that the government were 'satisfied that the controlled disposal of (such) wastes after appropriate treatment' as currently carried out was 'the safest practicable way of dealing with them'. No detailed estimates as to the cost of managing low, intermediate or high-level wastes had been made, he said. In July 1962 the minister said that 'it was not feasible to give totals' of radioactive wastes being stored or disposed of, but that they were being managed 'in a variety of ways and at a number of places' under the stipulations of the 1959 white paper on controlling radioactive wastes. The British government policy in 1959 – which set the framework for the next two decades – was to find a disposal route.

Despite the lack of active political interest, nuclear waste was subject to increasing regulation. Until 1959 the disposal of radioactive waste in Britain had been subject only to the laws governing other forms of waste disposal, with the exception of radioactive wastes from operations at UKAEA sites which were subject to special control under the 1954 Atomic Energy Authority Act. Another law, the Radioactive Substances Act of 1948 had given appropriate ministers power to make regulations to control hazards from nuclear materials, including wastes. But it was the successor Radioactive Substances Act (RSA) of 1960, which implemented the recommendations of the 1959 White Paper, that provided the first comprehensive and specific (governmental) powers in regard to radioactive waste. Exemptions to the requirements of the RSA were crown establishments, UKAEA premises and certain licensed nuclear sites. The RSA prohibited the disposal of radioactive waste on or from all premises (exceptions apart) unless authorised by the appropriate department.

In 1965 the Nuclear Installations Act (NIA) was passed. It consolidated earlier legislation providing similar controls, ensuring that the Health and Safety Executive (HSE) had powers to prevent the installation or operations of nuclear facilities and the processing, storage or disposal of waste or associated radioactive matter, through its licensing control. In the same year as the NIA came into force, the Central Electricity Generating Board (CEGB) produced what turned out to be a highly contentious assessment of the benefits of the Advanced Gas Cooled Reactor (AGR), the reactor type it had decided was the most propitious for the second programme of nuclear power to succeed the Magnox plants. The CEGB/UKAEA report[58] provides an insight into what, in retrospect, is a clear lack of appreciation of the long term costs which the electricity supply industry faced. Yet it made no mention of radioactive waste disposal or eventual decommissioning costs,[59] which became critically important in the late 1980s, as a true evaluation of these costs was made for the first time as part of the privatisation process, and proved extremely problematic.[60]

By contrast, at the public inquiries held in the 1980s on the CEGB plans to build pressurised water reactors (PWRs), the reactor type that lost out to the AGR in 1965, the project appraisals included substantial information from the CEGB on radioactive waste and decommissioning in its voluminous proofs of evidence. But by this time the nuclear waste issue was firmly on the political agenda and public and pressure groups clamoured to participate.

THE EMERGENCE OF PUBLIC OPPOSITION: THE US EXPERIENCE

In the United States, the interim solution was as in Britain: sea-dumping of awkward solid wastes and discharges of liquids by pipeline. From 1946 to June 1970, when the practice was abandoned in favour of containment because of international environmental objection, the United States dumped around 95 000 curies of solid packaged LLW into the ocean. Of these 90 000 or so containers about

1 per cent were dumped into the Gulf of Mexico; 19 per cent (about 15 000 curies) off the Farrallon Islands in a site 25 to 60 miles west of San Francisco covering 500 or so square miles, and 80 per cent dumped off the eastern seaboard, at two Atlantic Ocean sites 120 miles and 220 miles respectively south east of Sandy Hook, New Jersey. Two major studies, by the USAEC, in 1954 and the US National Academy of Sciences Committee on Oceanography, in 1971, have attempted to collect together comprehensive data on this dumping programme. Apparently it has been impossible for the US Environmental Protection Agency to compile full records on the specific materials dumped, or the accurate radioactivity levels.[61] One study by the Marine Policy Program of the Woods Hole Oceanographic Institution in Massachussetts[62] concluded that in total 295 000 curies of radioactivity were dumped into the oceans by the US by 1971. The sea dumping programmes went ahead with virtually no public opposition for the first twenty years.[63]

Public concern about nuclear waste first emerged in the United States in the late 1960s. In the spring of 1968, with the western world confronted by social and political upheavals,[64] the USAEC undertook for the first time to develop a comprehensive policy for the disposal of commercial high level wastes. Within months the main components of the policy had been agreed within the Commission. Reprocessing companies were to pay the federal government a 'one-time' fee in return for the government assuming full responsibility for ultimate disposal. The High Level Wastes (HLW) at reprocessing plants were to be converted to a solid form, as approved by the USAEC, within five years of their creation. This solidified waste would then be delivered to a federal repository not later than ten years after the irradiated fuel bearing the HLW was reprocessed.[64] In 1969 notification of the halt to sea dumping by the US was given and the policy was formally promulgated, in November 1970. On 1 January of that year the National Environmental Protection Act (NEPA) had been signed into law by President Nixon. The Environmental Protection Agency (EPA), which was established with the NEPA soon clashed with the USAEC, which continued to have the task of developer and promoter of atomic energy, as well as being its federal regulator. According to Carter the USAEC was confronted with a new political demand which could not be met because nuclear power, then in full surge of development was 'a technology ahead of itself in regard to reactor safety, plutonium safeguards and waste management'.[66]

Another analysis of nuclear developments identifies the early 1970s as a key moment of transition.

> Until the early 1970s the nuclear industry basked in the sunlight of public approval and political patronage. Because it was a technically sophisticated industry with military linkages, it was allowed to be secretive and largely unaccountable either to economists or politicians. This induced a sense of patronising certainty about the competence of management, and allowed the industry to hide its mistakes and near misses.[67]

Up to this time, H. G. Slater of the American Atomic Industrial Forum (AIF) argued that public opposition to nuclear projects had had 'no serious effect at all'. He noted that in 1968 utilities had ordered 17 nuclear plants 'with virtually no adverse public reaction', and that which was manifested was concern on legal and economic grounds, not health or environmental issues.

Slater predicted that there were nonetheless a number of other issues, beyond operational safety of plants, which were at the time not prominent but might well become more so in the future as more and more large nuclear plants were started up. 'The reprocessing, transportation, and ultimate storage of radioactive wastes from these plants will most likely undergo increasing press, public and political scrutiny as the activities in those parts of the fuel cycle expand.'[68]

In the United States throughout the early to mid 1970s, the primary focus of opposition to nuclear energy was upon nuclear power plant safety and risk.[69] A subsidiary, although highly politicised concern was the emergence of the debate over the so-called 'plutonium economy'. This peaked in 1976 (an election year) with both the Democratic challenger for President, Jimmy Carter, and incumbent Republican President Gerald Ford making it clear they were opposed to reprocessing spent fuel.[70] This whole debate brought attention to problems with the 'back-end' of the fuel cycle, which was taken up too in Britain by the Royal Commission on Environmental Pollution (1976) and the hearings at the 100-day-long Windscale Public Inquiry into plans to build a new reprocessing plant (1977). Those political issues that did emerge specifically on radioactive waste were mainly local concerns, such as the State of Michigan's three year battle (1975–1977) with the Federal Government over plans to test drill for a radioactive waste repository.[71]

By the end of the 1970s – but before the Three Mile Island accident in March 1979 – opinion polls showed that the unresolved radioactive waste problem was the nuclear power issue of most concern to citizens.[72] Abbotts commented that:

> As exemplified by the Michigan experience, industry and federal government blunders, in incident after incident, have created problems long before the technical feasibility for long term waste management could have been demonstrated. Difficulties with institutional problems; political insensitivity and other non technical issues have served to cancel waste projects before they could get off the ground, and these difficulties contribute significantly to public skepticism that any viable waste management schemes can be developed.[73]

The 1970s ended with US radioactive waste policy in disarray. The US Department of Energy (USDOE) was 'in disagreement with most other interested departments and agencies on several radioactive waste policy issues. But the policies of the DOE and the other agencies could be said to have been almost in harmony compared to the extreme divergence of viewpoints in Congress'.[74] It was this chronic political impasse that led to the congressional moves to sort matters out through so-called interstate compacts for low level waste (Low Level Radioactive Waste Policy Act, LLRWPA, 1980); and to select a final federal repository site for high level wastes (Nuclear Waste Policy Act, NWPA, 1982). The politics and process of these new developments are taken up in chapter 5.

THE EMERGENCE OF RADIOACTIVE WASTE POLICY AS A POLITICAL ISSUE IN THE UK IN THE 1970s

During the 1970s radioactive waste began to emerge as an issue of political concern in the UK. This concern focused in particular on the import for commercial processing at Sellafield of spent nuclear fuel containing HLW, as well as plutonium and uranium. The concern was not initially reflected in the work of the various nuclear agencies dealing with nuclear waste. Discussion of nuclear waste had been largely confined to the technical literature.[75] Perhaps the first indication in the UK that it was a political issue was a letter to the *Guardian* on 7 January 1974 signed by 43 senior scientists, environmentalists and former governmental policy advisors – and printed under the evocative headline 'Nuclear dustbins for the centuries'. It brought attention to the decision in April 1973, by the Royal Commission on

Environmental Pollution, to investigate the safety and disposal of radioactive waste from nuclear power stations. The letter also commented that the vigorous debate during 1973 over the choice of a nuclear reactor system in the development of a nuclear power programme in Britain 'obscured the problems posed by the production of radioactive wastes which have to be kept away from the living environment for hundreds and some for many thousands of years'. These problems could be seen as representing 'a Faustian bargain, in which we are jeopardizing the safety of future generations and their environment for our short term energy benefits and the comforts that go with them'. The authors considered it immoral and unwise to pursue a technology which would 'leave such a dangerous legacy to posterity'. They called for a halt to the building of further power stations 'while there was no safe and proven method of disposal of the long lived radioactive wastes'. The energy shortage problem could and should be tackled by much more efficient energy use; the recirculating of 'waste' heat in buildings and power stations; and 'priority should be given to the development of non-polluting energy sources such as wind, waves and organic wastes'.

By the end of 1974 the significance of the radioactive waste problem was recognised publicly by the UKAEA's Chairman, Sir John Hill, in a statement at the Annual Press Conference in 1974:

> in addition to our responsibilities for the development of nuclear power, we have public responsibilities to do everything we can to make nuclear power safe and to make sure the environment is not damaged.
>
> We all know what the issues are. Nuclear plants have radioactive effluents and although we can, by spending enough money, control these to whatever level seems necessary, then (there is) waste disposal. The storage of fission products in high integrity tanks at Windscale is perfectly satisfactory at the present time, but we recognise that storage in a solid form would give increased security against possible dispersal ... All of us in the Authority recognise that these questions and environmental protection are difficult problems, not just technically but also philosophically.[76]

As part of a generally heightened awareness of environmental pollution problems, in 1974 the Control of Pollution Act had become law, extending earlier controls over waste disposal and water and air pollution. The Act, though not specifically directed at radioactive waste, did encompass some provision relating to discharges of radioactivity. An examination of the annual indexes of the UKAEA monthly magazine, *ATOM*, over the period 1969–74 confirms that the turning-point in political interest in nuclear waste came in the mid-1970s. In 1960 there was no mention of radioactive waste in the journal; in 1970 only three items on radioactive waste appeared, the same number in 1971; in 1972 only two items; but in 1973 the coverage leapt to ten items, and in 1974 14 items were published. The increase reflects the greater political interest shown, as most of the entries refer to Parliamentary replies and statements. If these figures are further compared to the mid 1980s when each year literally hundreds of Parliamentary questions on nuclear waste were being asked and some debates were also held in Parliament, then it is clear that the nuclear waste issue, at an institutional political level, was increasingly becoming a major political matter, having constituency and national policy implications. The growing concern was reflected in the early reports of the NRPB established under the Radiological Protection Act of 1970. It was an advisory body set up to ensure that the radiological effects of the nuclear power programme, including those from radioactive wastes, were minimised. Among its early reports was one on the implications of disposal in the Atlantic Ocean which concluded that the level of radioactivity in the dumped waste was 'far below that which could prove harmful'.[77] The NRPB's reports were initially regarded as disinterested and authoritative, a perception which changed in later years.

A shift in attitudes to nuclear waste occurred during the mid 1970s but it was not immediately reflected in any change in policies. At the beginning of the 1970s, radioactive waste management policy was summarised as follows:

> It is practicable to store the small volume of high level wastes in such a way that no leakage to the environment can occur. It is impracticable to remove completely the activity in low level wastes, which constitute the bulk of the arisings, and the policy in the United Kingdom is to reduce their activity to the point at which they can be safely discharged to the environment.[78]

By 1973, with the 'energy crisis' developing into a politically complex problem for government, Parliamentary probing on nuclear waste became more politically pertinent. One Labour MP, Dr 'Jack' Cunningham, whose constituency of West Cumberland took in the Windscale Reprocessing Works at Sellafield, asked about the pressures on land use of the growing storage of nuclear waste in Britain.

On the question of how much land area would be sterilised by use as dumping areas for radioactive waste he was told by the minister (26 March 1973):

> the areas currently reserved for this puprose total about 250 acres in West Cumberland – only a small proportion of which is actually in use – and 10 acres elsewhere. There are no plans at present to extend these areas.

Another Labour MP, Alex Eadie, on the same day requested information regarding volumes of radioactive waste arisings. The Minister replied,

> reprocessing is giving rise to concentrated liquid wastes at the rate of about 60 cubic metres a year and significantly contaminated solids at the rate of about 500 cubic metres a year. By 1980 the former figure is likely to increase to an average of about 100 cubic metres a year: the total volume in store at 1980 is not expected to exceed 1,200 cubic metres. The rate of accumulation of contaminated solids is dependent on the type of fuel to be processed, as well as on its volume, but might double over the same period.

To put these volumes into context, in 1972 the Royal Commission on Environmental Pollution (RCEP) cited figures of 3 000 000 cubic metres of industrial waste and 5 000 000 cubic metres of domestic sewage discharged to the sea along England's South and East coast alone over one year (figures which prompted a £1300M government clean-up programme for these discharges over five years). Even allowing for a tenfold increase the volume for radioactive waste remains comparatively minuscule when set against the volumes of toxic and hazardous industrial wastes and sewage processed and disposed of in the UK.

In 1973, there was cross-party support for the continuation of nuclear energy in Britain. Thus the Parliamentary questioning, though pursued in an adversarial forum, was inquisitive rather than challenging in nature. The Trades Unions also supported a substantial expansion of nuclear energy. With a political consensus on nuclear energy, nuclear waste remained a dominant issue. It was the debate over the Windscale reprocessing plant during the late 1970s that politicised the nuclear energy issue and paved the way for later political conflict over nuclear waste in the UK.

THE IMPORTANCE OF THE NUCLEAR REPROCESSING PLANT AT SELLAFIELD (WINDSCALE)

The Windscale reprocessing plant at British Nuclear Fuels Ltd's (BNFL's) Sellafield site had operated, until the autumn of 1973, with little public knowledge or interest beyond its immediate vicinity. (The 1957 Windscale fire occurred in one of the plutonium production piles, not the reprocessing works, bringing the name of Windscale to international attention for a short while.) But on 26 September 1973 an accident occurred at an old reprocessing plant, B204, which had been converted to a 'head-end' plant to cope with a different type ('oxide') of nuclear fuel. Thirty-five workers were contaminated, mostly by ruthenium–106. The plant was shut for repairs and modifications, but never reopened. An official inquiry was conducted by the Nuclear Installations Inspectorate (NII), which reported in 1974 with 17 recommendations aimed at improving the radiation protection measures, but generally exonerating BNFL from blame.[79] Although not specifically a waste disposal issue, the accident highlighted the role of Sellafield in processing nuclear materials. There was a brief flurry of Parliamentary questioning, over the on-site implications for spent fuel management and transport to Windscale of spent fuel elements, due to industrial problems accentuated by the September accident. But the main nuclear focus throughout 1973 was the debate over the future choice of nuclear reactor for Britain in the context of the escalating energy crisis. As Patterson[80] put it:

> Partly as a result of heightened public awareness of energy issues in general, the latest convulsion in British nuclear policy attracted public attention – the first time a civil nuclear issue had ever done so. Previous nuclear power controversies in Britain had, to be sure, been ferocious and bitter; but they had taken place essentially behind the scenes, between the immediately interested partners and their supporters. This time the controversy came into the open, and the public noticed. Another contributing factor was the rise of "the environment" as a matter for popular interest and concern.

It was in this context of growing concern that the Royal Commission on Environmental Pollution (RCEP) announced in March 1974 that it planned to conduct a major study into nuclear power and the environment. During the late 1970s Windscale became the main focus of British attention on nuclear waste, first through the publication of the RCEP Report on 6 September 1976, and subsequently through the inquiry into reprocessing which lasted a hundred days from June to November 1977.

As the Royal Commission carried out its work in private, its investigations provided no public interest until leaks of its conclusions came out in the early summer of 1976. The implications of developing reprocessing in the UK however had been spelled out by the NII's Chief Inspector who commented in 1974, 'the price for Britain of building a lucrative business worldwide in nuclear fuel services could be that it becomes the dumping place for the world's nuclear waste'.[81] The issue of the importation of spent nuclear fuel (often erroneously called 'nuclear waste') became the focus of objection in 1975 as public knowledge began to grow over the operation of Windscale and the possibility of Britain becoming a global nuclear dump.

The political significance of the development of Windscale was drawn to public attention in the popular press during 1975. The *New Scientist* observed that,

> Radioactive waste management and the possible diversion of nuclear materials were the key issues as far as the opposition to nuclear power is concerned.[82]

In April the *Observer* described the first delivery of spent nuclear fuel to Barrow-in-Furness for storage and reprocessing at Windscale.[83] In October The *Daily Mirror*, the tabloid popular paper supporting the Labour government, made the issue front-page news with the headline 'PLAN TO MAKE BRITAIN WORLD'S NUCLEAR DUSTBIN'. It followed up a month later with a half page headline 'Sign here for Japan's nuclear junk'. Media coverage generally was critical of the nuclear industry. *The Sunday Times* in an investigation of the problems of Windscale concluded that:

> the most intractable of these problems is what to do with the highly radioactive wastes produced in the plutonium extraction process. Unlike the effluent of other industries they are to all intents and purposes indestructible. It will be hundreds of thousands of years before some of their constituents can safely be released into the environment.[84]

The plan to expand reprocessing facilities at Windscale was reported initially to involve the processing of 4000 tonnes of Japanese spent

Daily Mirror

EUROPE'S BIGGEST DAILY SALE

5p Tuesday, October 21, 1975 No. 22,312

Shock report on our lethal imports

PLAN TO MAKE BRITAIN WORLD'S NUCLEAR DUSTBIN

By STANLEY BONNETT

A STORM of protest is growing over secret plans to import huge quantities of nuclear waste into Britain.

There are fears that Britain is fast becoming the nuclear dustbin of the world. The Japanese want to send us 4,000 tons of the lethal waste — because their own laws do not allow them to treat it.

But many top men in Britain's nuclear industry are totally opposed to the plan because the waste is so dangerous.

Last night Sir Kelvin Spencer, former Chief Scientist at the Ministry of Power, commented:

"This is horrifying news. We are poisoning the planet to an extent which is perhaps beyond return."

Britain is already a large-scale nuclear dustbin.

At the Windscale atomic centre in Cumberland we are storing waste from West Germany, Sweden, Spain, Italy, Switzerland, and Japan herself.

So Britain has already got a stockpile of 1,200 tons of near-indestructible waste.

Decision

Now, in secret talks, the Japanese are offering to advance £160 million so that Britain can build a much bigger plant to reprocess waste and make it safer. The deal could be worth £400 million over ten years.

The Japanese are pressing for a decision—and some sources close to the secret talks think the contract could be signed before Christmas.

Experts said last night that even if work started on the reprocessing plant tomorrow, it could not be operational before the mid-1980s.

But, if the Japanese have their way, from 1979 onwards a special fleet of British "nuclear dust-cart" ships will be bringing in this new stock-pile, every consignment labelled "irradiated fuel."

The fact that the big reprocessing plant was not operational would mean that virtually every ounce of

TURN TO PAGE 2

LETHAL STOCKPILE: Windscale atomic centre, where the nuclear waste is stored under water in steel drums.

WHAM!

PRINCE CHARLES seemed to be in a fighting mood when he attended a charity boxing show at London's Hilton Hotel last night.

But he was only hammering home a point while chatting with comedian Tommy Trinder—and proving, once again, that he doesn't believe in pulling his punches.

Picture by HARRY PROSSER.

BAM! CENTRE PAGES

MA'AM! PAGE 3

SOURCE *Daily Mirror*

nuclear fuel. The Secretary of State for Energy, Tony Benn, issued a statement saying that: 'the main concern is that the United Kingdom should not become a permanent respository for storing other countries' nuclear waste'.[85] He also admitted that although it was the government's intention to include return-to-sender clauses in their reprocessing contracts for Sellafield, ensuring that the waste products would not become a liability for Britain, contracts entered into up to that time for Japan and several European countries had included no such return-to-sender clauses. As if in mitigation of the criticism however, Mr Benn said such foreign contracts only constituted 'a small portion of reprocessing work done by BNFL'.[86] The local MP, Dr Cunningham, in a Parliamentary debate, stressed the stringent safety rules in operation, citing eight bodies with regulatory or advisory responsibilities.[87] The science correspondent of the *Observer* argued that whilst the waste management problem was comparatively clear cut, the plutonium proliferation problem arising from reprocessing was complex and deeply dangerous.[88] The proliferation issue along with radioactive waste, dominated discussion in the 1976 RCEP report and in the proceedings at the 1977 Windscale Inquiry (see Figure 2.1).

Commentators both broadly supportive of nuclear energy and broadly in opposition, agree that the issue of nuclear waste became an important political issue in British politics in the autumn of 1975. In Parliament it was by no means a left–right, Labour–Conservative split issue. The Conservative opposition spokesman on energy, Patrick Jenkin, said that Britain had 'for many years been safely reprocessing spent nuclear fuel at Windscale and storing the residues', but argued the government should 'come clean' if there were plans to step up such operations. The Labour Chairman of the Commons Select Committee on Science and Technology described the Japanese nuclear reprocessing deal as 'big business and good business'. It would be 'tragic if absurd statements based on misinformation that Britain was to become the nuclear dustbin of the world were to block or delay the deal'.[89] On the other hand another Labour MP, Eric Moonman, commented 'I am deeply alarmed by the fact that we are continuing to introduce radioactive waste and are stocking it at Windscale. There is no agreed long term plan for its disposal or destruction'.[90] The issue of the foreign reprocessing contracts was politically important in strengthening the emergent anti-nuclear movement. This opposition was beginning to build on localised protests against nuclear installations and to link it with action abroad to monitor the departure and arrival of ships carrying spent fuel. An international response was developing. But this inevitably took time. In January 1976 BNFL sponsored a public debate held at Church House, Westminster, on nuclear power which focused upon the issues of reprocessing and waste management. It

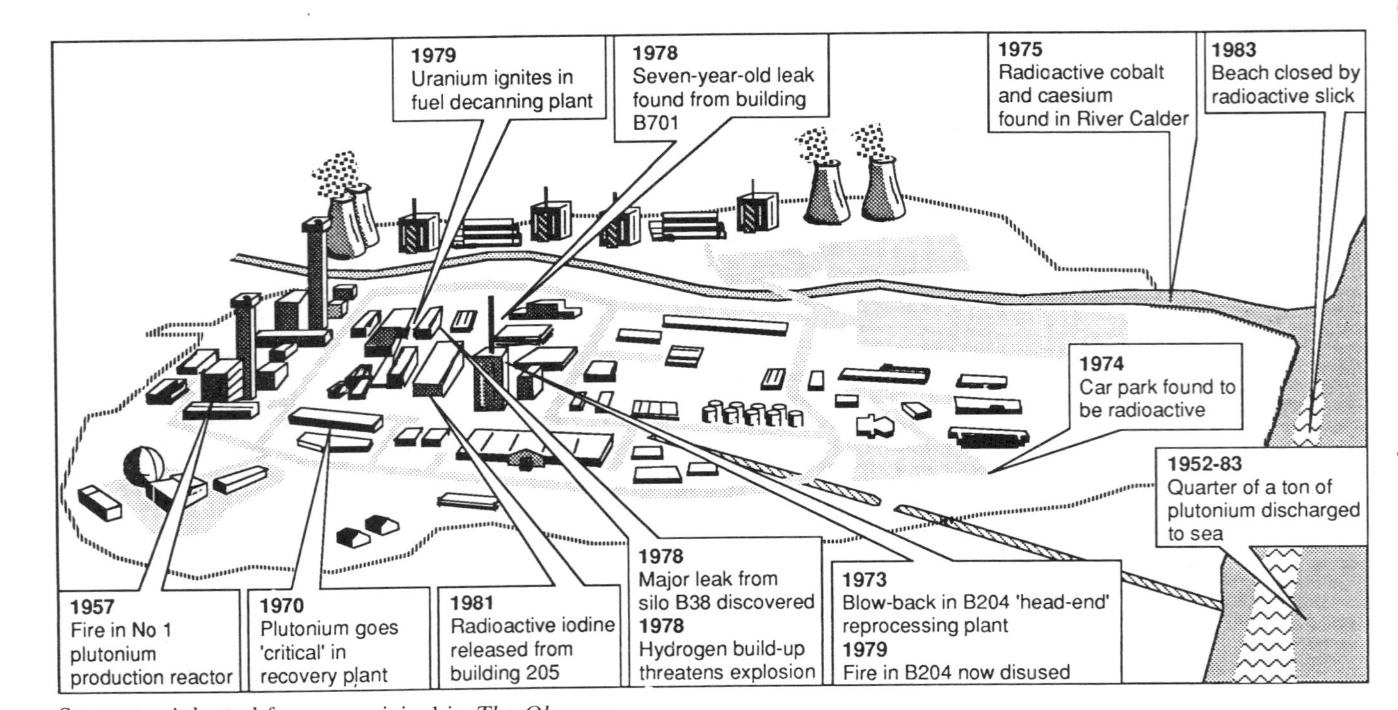

SOURCE Adapted from an original in *The Observer*

FIGURE 2.1 *Thirty years of trouble at Windscale*

was chaired by Sir George Porter, director of the Royal Institution, who was seen as independent. The meeting included BNFL, UKAEA, NRPB and conservationists and environmentalists such as Friends of the Earth, 'Half Life' and some local authorities and trades unionists. Tony Benn the Energy Secretary said that:

> it would have been preferable if this type of public debate had taken place in the 1950s, before the first nuclear power stations were authorised ... public debate halfway through an existing programme may seem fruitless. It is not, for fresh decisions have still to be made about the scale, timing and nature of future nuclear facilities.[91]

This public debate started a long drawn out conflict between environmentalists and those given official responsibility to manage nuclear wastes in Britain. In its early stages, the main opposition was advanced by national groups, such as Friends of the Earth, rather than by local activists at or near potential nuclear processing or waste disposal sites. For instance Dr Jack Cunningham told the debate he had received just one letter against reprocessing contracts. *New Scientist*[92] reported the Church House confrontation by saying that the anti-nuclear movement put on a 'poor showing' in a debate that raised no new issues.

The evolution of the debate leading up to the Windscale Inquiry, the minutiae of the exchange during the Inquiry itself and the subsequent political and Parliamentary debate are all important elements in the unfolding story of nuclear waste politics in Britain. This two-year period has fortunately been well covered elsewhere by various authors.[93] Here we only draw attention to some key highlights. There was a continuous stream of press stories on nuclear waste over this period, highlighting the unresolved existing problem of stored nuclear waste as well as predicting future problems. In March it was reported that all eight of Britain's nuclear power stations were having to store large amounts of radioactive waste because there was nowhere to process it. The issue was given public prominence on the BBC peak time TV programme *Nationwide*.[94] A week later Tony Benn in a Parliamentary written reply gave approval to BNFL to proceed with its £800M contract with Japan (which was to be shared with France). But he inserted the following important qualification,

> The company may subject to the negotiation of satisfactory terms, take on further work on the basis that the contract will include

> terms to ensure that the company will have the option to return residual radioactive waste and will not be obliged to retain it in this country for long term storage; and that suitable undertakings in support of the return option are realised between the UK government and the government of the Country concerned.[95]

In fact such undertakings were not enforced for LLW and ILW arisings. But politically it was important to make this assertion to counter the growing opposition.

It was evident that the government was shaken by the charge that by accepting overseas commercial contracts for reprocessing it was allowing the UK to be seen as the world's radioactive dustbin. Friends of the Earth claimed that even if contractual clauses precluded Britain from becoming the permanent dustbin, the deal inevitably meant that temporary status as 'world nuclear dustbin' would be achieved.[96] In April 1976 BNFL announced that serious corrosion had been discovered in some of the 10 000 steel drums in the ten concrete bunkers storing LLW at Drigg, the only nuclear disposal site then operating in Britain. There were also reports of an imminent shortage of storage space for Western Europe's spent fuel because reprocessing plants were not available to handle it.[97] The proposed dumping at sea of 6700 tonnes of LLW was described as 'increasingly controversial'[98] and press reports of leaks from drums dumped by the US in the Pacific and Atlantic Oceans increased the anxiety.

Meanwhile the nuclear industry was pressing ahead with its plans for expanding reprocessing facilities at Windscale. On the 1st June 1976 British Nuclear Fuels Ltd (BNFL) had submitted to Copeland District Council a planning application for the construction of a Thermal Oxide Reprocessing Plant (THORP), modification to the magnox reprocessing plant and construction of a vitrification research plant. These projects would lead to a manifold increase in the *volumes* of LLW and ILW, but a sevenfold decrease in the volume (though not the radioactivity) of HLW. Copeland District Council referred the matter to Cumbria County Council. The public inquiry was agreed to in December 1976 and was finally announced in March 1977. By the end of the year the Department of Energy had received an unprecedented 600 letters and 12 petitions on nuclear power matters.[99] Clearly nuclear energy was established as a matter for growing public interest. A decade later the number of objections from individuals and organisations at the inquiry into the Hinkley

Point 'C' PWR power plant was over 10 000. But, in the late 1970s, the issue of radioactive waste was still largely a matter of technical concern. Technical problems finally ensured that nuclear waste emerged as a prominent issue of public debate and political conflict.

TECHNICAL PROBLEMS EMERGE TO HINDER POLITICAL PROGRESS ON RADIOACTIVE WASTE MANAGEMENT

In mid-September 1976 the Royal Commission on Environmental Pollution (RCEP) published its (6th) report titled 'Nuclear Power and the Environment' (Cmnd 6618), usually known as the Flowers Report (after the Chairman Sir Brian, later Lord, Flowers). At the time, much media attention was devoted to Flowers' worries over the use of plutonium as a fuel when alternatives existed.[100] These concerns were an integral part of the parallel debates over the necessity of reprocessing and the future of fast breeder reactors. But Flowers also pronounced on waste disposal – Chapter VIII was devoted entirely to this issue. Out of the 50 recommendations in total, the report devoted 14 entirely to nuclear waste. Each on its own was important. Two new organisations were recommended. One, a national waste disposal company (NWDC), involving the constituent parts of nuclear industry, eventually became the Nuclear Industry Radioactive Waste Executive (NIREX) in July 1982, although not the body independent of the nuclear industry which Flowers had envisaged. The other, a nuclear waste management advisory committee, to advise on waste management policy, was established as the Radioactive Waste Management Advisory Committee (RWMAC) somewhat earlier, on 15 May 1978.

The most oft-cited recommendation of the Flowers Report was that,

> there should be no commitment to a large programme of nuclear fission power until it has been demonstrated beyond reasonable doubt that a method exists to ensure the safe containment of long lived, highly radioactive waste for the indefinite future (para 27).

The nuclear opposition took this to support their case that nuclear power should henceforth be halted. The nuclear industry concentrating on the words 'demonstrated beyond reasonable doubt', took it as an endorsement of their intention to continue exploration of practical geological disposal sites for the accumulating high level nuclear wastes.

The RCEP lent support to the latter interpretation through recommendations in paragraphs 31, 34, 35 and 36 which called firstly for a national land disposal facility for the more highly active wastes; and for 'a substantial UK effort in the field of disposal to geological formations on land' (for vitrified wastes). The required research should be carried out through the National Environmental Research Council (NERC) and the Institute of Geological Sciences (IGS). In paragraph 427 (p. 162) the Report commented:

> The picture that emerges from our review of radioactive waste management is in many ways a disquieting one, indicating insufficient appreciation of long term requirements either by government departments or by other organisations concerned. In view of the long lead times that will almost certainly be involved in the development of appropriate disposal facilities, we are convinced that a much more urgent approach is needed, and that responsibilities for devising policy and for executing it need to be more clearly assigned.

The Report's recommendations led immediately to the search for possible sites for deep geological disposal which became the first area of major political conflict over nuclear waste. The nuclear industry met opposition both from local communities concerned to prevent specific nuclear facilities from being developed in their area as well as from environmental groups engaged in a broader campaign against nuclear energy.

THE ORIGINS OF THE RESEARCH PROGRAMME FOR AN HLW DISPOSAL SITE

The search for potential disposal sites for HLW was a big political issue at the end of the 1970s and the early 1980s. The criteria to be observed in the search for a suitable geological disposal site for HLW were set out by Sir Peter Kent, chairman of NERC: avoidance of leakage within a time scale of hundreds of thousands of years; and obviating human interference. Acknowledging the RCEP guidelines for research, Kent concluded that what was needed was the will and funds, to add to the existing research expertise, in order to obtain the basic data.[101] The 1975 EEC nuclear waste research programme already committed Britain to investigate hard crystalline rock formations; and the UKAEA announced that it had commissioned the IGS to survey, using borehole drilling, granite rock sites in Scotland, in the Hebrides and in the Southern uplands, and in Cornwall in England, although the latter was soon declared unsuitable. The £600 000 project was supported with a grant of £200 000 coming from the EEC.[102] Sites in mid Wales and the Cheviot Hills in Northumberland were added subsequently. The announcement, predictably, caused widespread alarm in the districts concerned and protests were lodged.[103]

From the autumn of 1976 the HLW issue and reprocessing at Windscale became politically intertwined. As HLW in the UK is a product of reprocessing, so the HLW problems were a product of the increasingly unpopular politics surrounding Windscale. In November 1976 when the proposed test drilling programme emerged as an issue, a petition of 26 000 signatures collected in Cumbria opposing BNFL's expansion plans for Windscale, was presented to the Environment Secretary, Peter Shore.[104] He subsequently delayed his decision on the planning permission.[105] A further political complication arose within weeks when BNFL admitted that they had failed to inform the Department of Energy until 1 December about a leak discovered on 10 October in the soil surrounding silos containing HLW (officials in the Department of Environment apparently knew of the incident on 8 November).

Even the national safety watchdog, the Health and Safety Executive, was kept in the dark about the discovery until 22 October. Sensitive about their planning application lodged with the government and local planning authorities, BNFL feared negative repercussions of any early announcement of the leak.[106] Energy Secretary Benn reacted, making it clear that he was deeply concerned that there should be 'any disinclination to be quite open about these matters because there are big decisions to be made in the nuclear policy field'.[107] Subsequently Benn, who as the responsible minister played a key role in the nuclear decisions over the next few years, confessed that the leak of autumn 1976, followed by another at Windscale in December 1978, led him to become very sceptical about the merits of nuclear energy, particularly because of problems over nuclear waste.[108] Thus events in 1976 proved politically important in influencing decisions to come.

The issues of the borehole drilling and the Windscale Inquiry brought about two new developments in the environmental debate over nuclear energy. Firstly, they made radioactive waste a nationally recognised environmental problem. They provided the focus for direct confrontation between the nuclear establishment and environmentalists, who began for the first time, to make use of specialist counter-expertise. Secondly, a consequence of the borehole drilling programme was to mobilise local people to defend their communities against the perceived threat of unwelcome nuclear waste and the attendant disruption. For instance, the selection of the area of Mullwharchar Hill (which means 'hill of awesome grandeur') in the Carrick Forest to the South of Loch Doon in Ayrshire, Scotland (Figure 2.2), in a wilderness area of outstanding natural beauty, led to the formation of several community groups with three main aims: self-education (the need to develop counter-expertise), public education (conducted primarily through local meetings) and persuasion of the local council to veto the test bores (or at least to oppose the planning application at any public inquiry). At this time, it was HLW alone that was at issue. It presented the immediate 'threat'. LLW and ILW were regarded unproblematic by both the nuclear industry and its opponents.

Although the Not in my Backyard (NIMBY) response was the major driving force behind local campaigns, some communities, especially those with their own counter-expertise, did recognise the need for a research programme, without commitment to geological disposal.[109] Proposals for the long term management of existing HLW were also advanced by opponents of the IGS drilling strategy. They favoured vitrification and thereafter air-cooled storage for 600 years.[110]

The official planning application procedure for borehole drilling progressed very slowly after the initial announcement in late 1976. It was not until January 1978 that the UKAEA submitted their first application for planning to a local council, Kyle and Carrick, covering the Mullwharchar Hills in Ayrshire, Scotland. The Council's planning committee convened a special meeting for a public discussion. Opinion polls showed 80–90 per cent of the local population opposed test drilling. Subsequently the full council rejected the UKAEA's planning application. The UKAEA appealed in April 1979, just after the Three Mile Island nuclear accident had heightened global nuclear concerns and reinforced local worries over nuclear issues. The Scottish Secretary of State announced in reply, on 13 June, that a public inquiry would be held. Full details of the inquiry were announced on 5 November 1979. Despite appeals to the Minister that the inquiry remit was too narrow, the public inquiry began in February 1980, lasting 21 days in all.[111]

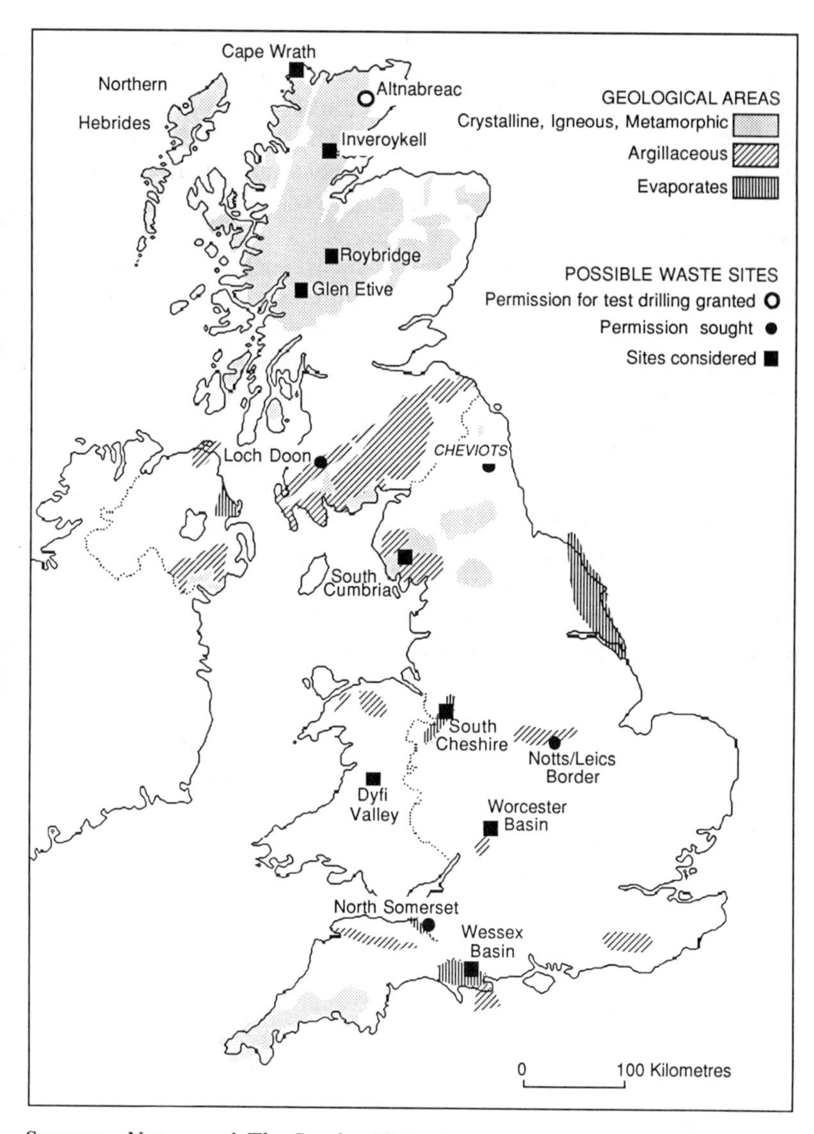

SOURCE *Nature* and *The Sunday Times*

FIGURE 2.2 *Sites considered for HLW disposal in the United Kingdom in the late 1970s*

Evidence from the UKAEA and government officials showed that out of 127 potentially suitable areas across Britain, it had been decided to seek consent for drilling in about 15–20 locations, covering seven different geological environments. The main types of geological strata were unfissured granites, argillaceous and hybrid clays, and salts. The outstanding geographical feature was the remoteness of the potential sites from population centres.

In the meantime, other local councils such as the Highland Regional Council, had been more accommodating to requests for permission to drill boreholes. Research began in Caithness in November 1978, and was completed in May 1979. This same arca was identified a decade later as suitable for LLW/ILW disposal as we shall see in chapter 4. But two other applications for drilling in the Chillingham and Usway Forests in the Cheviot Hills submitted in February 1978 were rejected by the Northumberland County Council. The local opposition campaign there was strengthened by local council support.[112]

The broad policy considerations behind the research programme into disposal options for HLW were set out in a statement on 24 July 1979, by the new Conservative Secretary of State for the Environment, Michael Heseltine, who had been in office barely two months. Heseltine stated, inter alia,

> Desk studies carried out by the Natural Environmental Research Council at its Institute of Geological Sciences indicate that potentially suitable rocks lie under about 16 per cent of the land area of the country. Specific areas ... have been identified for research purposes so that information can be collected about a wide range of rocks. The next step is a programme of geological research involving test drillings to examine fully the properties and characteristics of different geological formations in situ ... only when full information is available, and has been properly evaluated, will it be possible to judge whether or not disposal deep underground is an option to be pursued; and, if it is, which of the rocks would be most suitable ....
>
> ... when research has been conducted for about ten years the Government expect to have obtained sufficient information to enable decisions to be taken about the development of demonstration disposal sites underground or on, or under, the ocean bed. These would be fully engineered and in the case of land facilities would involve the construction of access shafts deep into the selected formation ... At most, two or three sites would be selected.[113]

With this statement the Department of Environment (DoE) published a provisional list of 11 areas identified as suitable for investigation: Cheshire, Cumbria, Grampian, Gwynnedd-Powys, Highlands, Leicestershire-Nottinghamshire, Northumberland, Somerset, Strathclyde, Western Isles and Herefordshire-Worcestershire.

By 1979, two years after the first Scottish protest march at Loch Doon in Ayrshire, some 80 local protest groups had been organised. A national Anti-Nuclear Campaign (ANC) was founded in the autumn of 1979 when the new Conservative government was correctly rumoured to be planning a large expansion of the nuclear power programme; ANC produced a regular newsletter for four years (January 1980–84) providing a convenient vehicle for the sharing of information between the various local groups opposing the Government's chosen disposal strategy.

The possibility of extending the area of search for sites was aired in a widely reported paper delivered at the British Association for the advancement of Science meeting in Edinburgh in September 1979 by Harwell scientist Keith Johnson, who indicated that many of Britain's small islands could provide ideal sites for HLW disposal. He reported that a preliminary study of 24 of the 'best prospects' had been

SOURCE *West Highland Free Press*

launched by the UKAEA.[114] The policy of listing a whole range of prospective sites proved disastrous for the government strategy in the aftermath of Heseltine's July 1979 announcement. Instead of the ten years of site research predicted by the Minister, the programme lasted barely over two years: and only one site was specifically evaluated geologically by borehole testing. A few years later NIREX adopted the opposite strategy in its public presentation of potential sites, restricting public release of the location of potential sites only to those specifically intended for examination.

The choice of so many potential sites in Scotland fuelled nationalist fervour. The Scottish National Party (SNP) had already given strong support to the five year long campaign against the new Torness nuclear plant, site clearance for which had begun, despite protests, in the autumn of 1978. They used the emotive slogan – 'No to Scotland becoming England's nuclear dump' – gaining support from all political parties. A spokesperson for the Scottish Conservation Society put it evocatively, saying that grass roots opposition to the drilling proposals was stronger than on any other single issue in Scotland. 'The English have not woken up to the nuclear issue yet. But the Scots certainly have.'[115] A similar Scottish response arose in opposition to NIREX's LLW and ILW option for disposal in Scotland a decade later (see chapter 4).

It seemed as if a dialogue of the deaf was developing over the nuclear waste management problem. In late September 1979, barely two months after Heseltine's research programme announcement, the DoE published a new report 'The Control of Radioactive Wastes', which reviewed the existing legislation, based primarily on the 1959 White Paper, and concluded that research and development of methods of vitrification and the disposal of HLW should be 'pursued vigorously'. It pronounced its expert group of authors (which did not necessarily reflect government policy) 'certain that safe methods of disposal can be established'.[116] Shortly after this report was issued a review article was published by the IGS on the strategy for burying Britain's HLW. It included a map (see Figure 2.2) pinpointing the areas under examination, with scientific justification for the programme, concluding

> The feasibility of the disposal of high level radioactive wastes within the geological framework of the United Kingdom remains to be demonstrated. The immediate need is for comprehensive site

specific data from a range of potentially appropriate geological environments both to test the criteria on which these environments were chosen and to provide a data base from which potential disposal sites may be identified. The feasibility studies suggested represent a major step towards these objectives and need to be initiated as a matter of urgency.[117]

A major block to the development of feasibility studies was the scepticism of local communities. The one exception was Caithness. The majority simply did not believe any IGS drilling programme, if it found their locality geologically suitable, would not lead to their area being politically and pragmatically seen as a *fait accompli* disposal site.[118] Such statements as 'drilling will carry no commitment to further works' by the DoE were received as unconvincing by the local people. There was some recognition that the technical management of nuclear waste required the political problem to be solved. But this problem was never adequately addressed. A review of nuclear waste disposal methods by the DoE's directorate of research, released at the end of 1979 stressed that public disquiet over nuclear issues must be taken into account, particularly over HLW.[119]

The transfer of responsibility for HLW research from the UKAEA to the Natural Environmental Research Council (NERC) early in 1980 made no difference in public perception of the drilling programme. NERC's applications, in October 1980, to drill at Puriton in Somerset, Wymeswold in Leicestershire, Brent Knoll south of Bristol in Avon and Ratcliffe-on-Soar in Leicestershire all met with rejections by local planning authorities. NERC, as with NIREX a few years later, chose government-owned sites where possible. The Puriton site was that of a Royal Ordnance Factory and Wymeswold a former Royal Air Force base owned by the Ministry of Defence.

At the Cheviot Inquiry at the end of 1980 Dr Lewis Roberts, director of the UKAEA's Harwell research establishment, complained that planning inquiries were delaying the research programme. 'We have lost a lot of time', he said, 'it was necessary and desirable for the nuclear industry to know what option of disposal and management would be decided by the Government'.[120]

The borehole drilling programme lasted five years (December 1976–December 1981). When the Secretary of State for the Environment announced its abandonment in Parliament on 16 December 1981 in a written reply (which in this form avoided any follow-up debate) he stressed he was taking the advice of RWMAC whose second annual report published in the spring had suggested that:

> serious consideration should be given to the possibility that containment in an engineered storage system either above ground or sub-surface, for which technology already exists, might be the best way to deal with solidified high-level wastes for at least 50 years and possibly much longer.

RWMAC had also expressed concern in the report that the IGS had experienced serious delays in its drilling programme because of difficulties in obtaining planning permission. The minister's announcement practically ignored this matter, stating it would seek the advice of RWMAC on the applicability of work carried out in other countries, whilst 'desk studies' continued in Britain. The statement said that the government's objective had been to establish *in principle* the feasibility of deep underground emplacement (emphasis added). This assertion was treated with scepticism from the experts advising the anti-borehole opposition.[121] 'Nothing has emerged to indicate that it would be unacceptable', the ministerial reply added. Further geological fieldwork might still be needed, and if the planned 50 year storage experienced unexpected difficulty, then this must require a change of plan at some future date. But the immediate effect of the decision was that the appeals for drilling in the Cheviots and the other pending appeals and planning applications were withdrawn.

Dr Roberts was concerned that the system of planning law that governed the applications for test drilling was working against the longer term interest of the country and the nuclear industry. One of the leaders of the Cheviot Hills protest, Anthony Murray, predicted that political rather than technical considerations would prevail, stating 'My feeling is that the number of people who are standing up and saying "no" to test drilling will be the deciding factor.'[122] His prediction was proved correct when the abandonment of the HLW testing programme was announced in 1981.

The abandonment of the test drilling programme was in a sense, a double edged political sword.

> By taking away one focus of the anti-nuclear movement, the government may have done something to damp down the controversy surrounding nuclear power. However, the way in which the latest announcement was made can only confuse the issue. In

> the nuclear industry the fear is that the anti-nuclear movement will now attack nuclear power because waste disposal is not proved 'beyond reasonable doubt'.[123]

One of the key outcomes of the HLW site-drilling campaigns was the recognition that it was possible to mobilise local communities in opposition to an external threat. Moreover, although first instincts were NIMBY instincts, politically it was recognised that co-operation and collaboration with other threatened communities were the keys to power in halting the drilling programme. At national level, the political parties, and indeed Parliament, were marginal actors. It was in the localities, and between the localities as participation grew and strengthened that victory was ensured. The nuclear waste industry began to recognise that 'greenfield' sites were practically impossible to secure for research leading to nuclear waste disposal. With this recognition we can see the beginnings of a shift in political power leading to further conflict and defeat for government initiatives during the 1980s.

SOURCE SCRAM

The abandonment of the HLW test-drilling programme was the first of a series of retreats on nuclear waste management policy made by the government during the period 1976–87 which are discussed here and in chapter 4. For the nuclear industry's opponents the decision was claimed as 'our greatest victory ever'.[124] The government's climb-down was described in the Press as one based on confusion for the Department of the Environment had told neither the NERC or the UKAEA of its decision beforehand. The government's position on the future management of HLW was uncertain. On the one hand the Secretary of State for the Environment maintained that despite the RWMAC's recommendation there was a 'need for research on all the ultimate disposal routes currently under investigation'.[125] On the other hand, prior to announcing the abandonment of the drilling the Department of the Environment claimed that the research programme 'is being kept continuously under review in the light of additional knowledge and advice'.[126]

Such uncertainty in the face of political conflict was characteristic of government policy over nuclear waste during the following decade. This uncertainty contrasted sharply with the developing confidence and organisation of the opposition to the nuclear industry. The controvery over high level test drilling illustrates the shifts in political attitudes and power that were to shape later conflicts over radioactive waste issues.

HOW SEA-DUMPING WAS ENDED

The opposition campaign against sea disposal plans for LLW and ILW necessarily took a different form, as there is no readily identifiable 'backyard' for NIMBY to flourish in. The campaign during the 1970s, and early 1980s overlapped in time the conflict over HLW borehole drilling. The problem of sea disposal of nuclear waste was described as early as 1958 by H. J. Dunster, a UKAEA research scientist, to a United Nations Conference on the Peaceful Uses of Atomic Energy,

> Radioactivity is a recent addition to the polluting agents in the sea ... large scale discharges of radioactivity can be made without affecting the total radioactivity of significant volumes of the oceans (which) contain substantial amounts of the natural radioactive minerals such as uranium and radium ... Radioactive materials

may concentrate in marine organisms such as fish and seaweeds, or on suspended particles of sand, mud and organic detritus. These processes tend to reconcentrate the activity in various stages of food chains leading to man and may also deposit radioactivity on the sea shore. It is these concentrating processes that make strict scientific control essential if large quantities of radioactivity are to be discharged to the sea.[127]

Such concerns were not evident when in March 1977, the NRPB published a report[128] prepared for BNFL, which optimistically concluded that

no overriding reason connected with radiological protection considerations has been identified which would preclude the disposal of suitably conditioned high-level waste on the ocean floor.

The London Dumping Convention (LDC) expressly forbids ocean bed 'emplacement' of HLW, a fact mentioned in the appendix to the NRPB report, which also said that further research was required on the properties of vitrified waste, the corrosion of its containment in sea water, and on physical oceanography and aspects of marine biology. Shortly after, the NRPB issued a second report evaluating the ocean dispersal of LLW, which included the description of a two part – short and long term – mathematical model for estimating maximum concentrations of radioactivity in water for continuous disposal.[129] Compared to the considerable fears being developed over the HLW test bore drilling, and the imminent Windscale Inquiry, these NRPB reports barely received any attention from pressure groups or the media. Only slightly more attention was paid to the announcement, in July 1977, that the OECD's Nuclear Energy Agency (NEA) had agreed a new code for surveillance and consultation[130] on sea dumping, which obliged participating countries to take advice on where and under what circumstances dumping could be carried out at sea. The NEA's environment committee was to police the code and regularly update standards.[131] These standards did not appear until 1976 ten years *after* the NEA oversaw their first sea dumping operation.

In early 1978, the Department of the Environment took over responsibility for nuclear waste management, and instigated a new Waste Management Division under the direction of Dr Frank Feates, previously manager of UKAEA's environmental safety group at Harwell. For much of the previous year Feates had talked at meetings in communities earmarked for bore hole drilling. He gained first hand knowledge of local hostility, and hence the political problems ahead, if the DoE was to achieve success in policy implementation. The location of waste management policy at the DoE instead of the Department of Energy followed a positive commitment made a year earlier in the government's White Paper response to recommendations in the RCEP (Flowers) report that responsibility for radioactive wastes be located in a 'department concerned to protect the environment, not one concerned to promote nuclear power'.[132]

The Role Of Greenpeace as Campaign Fulcrum

Opposition to sea dumping was led by the international environmentalist pressure group *Greenpeace* which was founded in North America in 1971, and established an office in Britain in July 1977.[133] When *Greenpeace* learned in Spring of 1978 that sea-dumping of waste was taking place from the Severnside port of Sharpness near Bristol, using the ship the M.V. *Gem*, they dispatched a camera team from London to shoot action and still film of the loading. Opposing

. . . and Colin adds his comment

SOURCE ATOM News, July 1983

descriptions of the cargo were given by the *Gem* captain ('spent rods from Royal Navy nuclear submarines') and the UKAEA ('Ion exchange resins').[134] *Greenpeace* held a press conference releasing film of their attempts to stop the MV *Gem* tipping the 5,500 barrels of radioactive waste into the sea, using inflatable dinghies off the *Rainbow Warrior*. The film showed a seven tonne barrel rolled from the tipping platform onto one of the two inflatable dinghies, throwing the crew into the water. *Greenpeace* claimed the dumping violated the rules of the London Dumping Convention, a charge denied by the UKAEA, saying the Royal Navy waste was 'no more active than a similar size load of luminous dial wrist watches'.[135] The argument was unresolved; what mattered politically was the salience that *Greenpeace* was bringing to the issue of sea dumping in the UK.

Greenpeace in the first year of its campaign to halt sea dumping had little chance to do its own monitoring or scientific research to assess the prospective dangers of the dump. But they argued that to dump broke the spirit of the LDC, which asked contracting members to view sea dumping as 'the least desirable option'.[136] Nonetheless, the Harwell director Dr Lewis Roberts, speaking to the British Nuclear Energy Society in November 1978, said that there were 'many technical attractions' to the possibility of putting (packaged) HLW on the ocean floor bed. Work was being undertaken to discover

> whether or not the ocean floor can be considered a safe repository for much higher levels of activity . . . whilst acknowledging that the LDC presently prohibits the disposal of highly active waste and major amounts of alpha bearing waste.[137]

In the summer of 1979, *Greenpeace* again used the *Rainbow Warrior* with its flotilla of dinghies to shadow the *Gem* in the Atlantic 600 miles off Lands End in opposition to the annual sea dumping. The attempt to repel the protesters using firehoses from the *Gem* was covered on national television news. It provoked a statement of concern, co-ordinated by *Greenpeace* and signed by 53 MPs, scientists and senior academics from many countries, suggesting this opposition had a base far wider than committed environmental activists.[138]

The new Conservative Government was pro-nuclear and intended to expand the nuclear power programme (and retain, and update, the military nuclear programme) and therefore needed to keep open the sea-dumping route. In September 1979, the publication of an interdepartmental review of the 1959 White Paper (Cmnd 884) setting out Britain's nuclear waste strategy gave revealing insights into the philosophy prevailing within government circles on sea dumping. The following selected paragraphs are illustrative:

> There is no evidence of irradiation of man following these disposals at sea . . . we strongly recommend that the UK should continue to use this disposal route (para 5.19)
>
> The UK also should begin gradually to increase the sea disposals in accordance with the London Convention and supported by appropriate environmental investigation (para 5.21)

The political pressure was clear. Dumping was to increase, even *before* research evaluation on its ecological impact was complete. The review report recognised that the international climate was such that more scientific evidence than was then available was needed to justify any substantial increase in sea disposal. It also reiterated the government's plans for continued research on a number of possible disposal routes (on or under the ocean bed, or in stable geological formations on land for HLW).

Early in January 1980, *Greenpeace* published a scientific report it had commissioned from the Oxford-based Political Ecology Research Group (PERG), evaluating the range of potential risks from the international maritime transport of spent nuclear fuel and radioactive wastes. The report was the focus for a public meeting, called by *Greenpeace* which was held in Barrow. BNFL used Barrow as the receiving port for foreign spent fuel shipped for storage and processing at Sellafield. Further meetings were held in the area and a local protest organisation, the Barrow Action Group (BAG) – which later became Cumbrians Opposed to a Radioactive Environment (CORE) – was founded. BAG, with some financial sponsorship from *Greenpeace*, became established in campaigning against radiation risks from the nuclear industry, especially sea and air discharges from Sellafield. In 1980 local activist groups became concerned about the rail transport of spent nuclear fuel and radioactive materials across Britain from nuclear power plants to Sellafield.[139] In response, the CEGB and British Rail made efforts to counter what they described as the 'lurid rumour' promoted by the anti nuclear groups.

An attempt by *Greenpeace* to blockade the BNFL ship the *Pacific Swan* from bringing its 49 tonne radioactive cargo from Japan to Barrow was prevented by the High Court. In June 1980 protestors from BAG chained themselves to the unloading crane at Barrow to stop the spent fuel flasks from the *Pacific Swan* being unloaded. On the same day in Belgium, five local protesters boarded the Belgian freighter *Andrea Smits*, in Zeebrugge, to try to halt it leaving to dump in the Atlantic. This followed similar protests in the Netherlands. Next month, on 9 July, seven anti-nuclear protestors in Gloucestershire halted a train taking nuclear waste to the Sharpness Docks for seadumping, by erecting a large scaffold platform across the railway line. The action again captured national media attention. Protesters pointed out that the radioactivity of Britain's waste dumped in the Atlantic had risen from 1600 curies in 1967 to 108 000 in 1980.[140] The UKAEA argued this was a 'very small fraction of the levels on which international controls are based'.[141]

Greenpeace combined international diplomacy with direct action. In 1981 *Greenpeace International* was allowed observer status at the London Dumping Convention (LDC), thus giving encouragement to some smaller nations, such as the Pacific Island Republic of Kiribati to challenge the prevailing wisdom of the larger established LDC contracting countries. It appeared that for the first time in its short existence the LDC was to have a fight on its hands.[142] In 1982, *Greenpeace* produced documents criticising the 'de minimis' criteria (that is, the level at which some materials with low radioactive content could be considered non-radioactive for the purposes of the LDC), investigating circumstantial evidence on deep sea currents, exposing the inadequacies of monitoring and surveillance programmes, and drawing on US empirical evidence that undermined the dispersal philosophy upon which the dumping operation is based. Further support was given by a resolution in the European Parliament in September 1982, calling for a halt to sea dumping.

Within Britain, *Greenpeace* courted the support of the trades unions, especially those involved in transport, as the key to ultimate success in halting dumping. This was difficult since the trades union body had initially adopted a pro-nuclear stance. At its annual conference in September 1979 the Trades Union Congress had endorsed a bigger programme of nuclear power than even the pro-nuclear Conservative government had wanted. But it soon became known from leaked Cabinet Minutes that the Government were keen on nuclear power because it removed 'a substantial portion of electricity production from the dangers of disruption by industrial action by coal miners or transport workers'.[143]

The government's posture was a factor that helped *Greenpeace* to secure the support of the leading transport unions to stop sea dumping. In 1983 the scientific and lobbying work within the LDC, and the close cultivation of the trades unions were consolidated. During the meeting of the LDC in February 1983 *Greenpeace* demonstrators in radiation suits brought three mock barrels of nuclear waste each daubed with the slogan 'Dump the AEA' along the River Thames in inflatable dinghies and addressed them to the British LDC delegation. A crowd of about 150 demonstrators waved protest banners outside the conference. On the same day in Madrid, protesters from North West Spain gathered outside the British Embassy and demanded an end to nuclear dumping in the Atlantic.[144]

Detailed technical and tactical briefings had been carried out with the Spanish and Republic of Ireland delegations to the LDC, whose primary concern was the effects of British nuclear waste being discharged or dumped. But it was the two tiny Pacific Island states of Kiribati and Nauru, aided by *Greenpeace*, who put forward a resolution to amend article II of the LDC to prohibit the dumping at sea of any nuclear materials. The Spanish delegation submitted a resolution calling for the suspension of sea-dumping for two years. This was supported by Iceland, Finland, Norway and Denmark, who submitted an amendment calling for the phasing out of sea disposal by 1990. To the surprise of many observers, the LDC voted 19 to six with six abstentions, in favour of the Spanish amendment, whose call for a moratorium also argued for the constitution of a specialist scientific committee which would report back having reviewed 'the scientific and technical considerations relevant to the proposed amendments'. The British delegation tried to avert a division by making an offer that Britain would stop dumping *if* it could be shown to be harmful to the environment.[145] But the LDC vote had decided that the burden of proof was now on the would-be dumping nations to demonstrate their actions were not harmful: it was thus up to the objecting six states, Britain, the US, Japan, Netherlands, South Africa and Switzerland to respond.

The UK did not, formally, have to observe the moratorium but it was placed in a difficult position. It was acknowledged privately by government officials that 'the international position was becoming increasingly difficult and the annual dump would certainly be watched very closely in future ... under these circumstances it was essential that the UK should comply scrupulously with international standards'.[146] On 22 February the junior Agriculture minister, Alec Buchanan-Smith, told Parliament in response to a question as to why the British representatives at the LDC had voted against the moratorium:

> We did so because the relevant resolution involved action that was against the policy of Her Majesty's Government and against the spirit of the convention. In particular, it called on member states to act in advance of receiving authoritative scientific advice: to do so would be contrary to the provisions of Article XV (2) of the convention.

The moratorium was especially disturbing for Britain, which already dumped 90 per cent of the total radioactivity disposed in the Atlantic Ocean, and had plans to increase its total by 20 per cent in the July dumping voyage. Nonetheless, the nuclear industry still maintained there was no problem and that the scale of dumping could be 'increased by a large factor before any significant addition could be made to the natural radioactivity of the sea'.[147]

More difficulties appeared for the government's dumping policy when in early April 1983, at a meeting organised by *Greenpeace*, senior officials of three transport unions – Transport and General Workers (TGWU), National Union of Railwaymen (NUR), and Associated Society of Locomotive Engineers and Firemen (ASLEF) gave their support to the National Union of Seamen's (NUS) decision at its conference in March, to ban nuclear waste dumping in the Atlantic Ocean. An NUS official said that they were 'greatly encouraged by the support given'. Peter Wilkinson, the then *Greenpeace* director, described the outcome as 'a major step in the fight against nuclear dumping'.[148]

In June 1983 *Greenpeace*'s trawler *Cedarlea* set sail for a further campaign against radioactive pollution of the seas, with the focus on the Irish Sea around Sellafield, Barrow and Dublin.[149] Shortly afterwards, on 17 June, the three transport unions formally announced they were backing the NUS in blacking the movement of nuclear waste to port. The planned departure in July of the new dump ship, the *Atlantic Fisher*, adapted to allow the disposal of the barrels through an internal hatch to prevent interference, was thus threatened. The unions pointed out that the Prime Minister had signed a statement at the 1981 Commonwealth Conference in Melbourne strongly supporting a call from the South Pacific Forum on all countries not to store or dump nuclear wastes in the Pacific Ocean. Fears were expressed that the Royal Navy might be required to intervene and injunctions be brought against the unions and *Greenpeace*.

A protest armada from Britain and Spain was promised if the waste left port. The unions joint statement giving warning of an impending boycott said:

> regrettably, the government has decided to defy the LDC call and the United Kingdom Atomic Energy Authority is planning to dump 3,896 tonnes of nuclear waste 500 miles southwest of Land's End from July 11.[150]

Reinforcing in the public mind that they were deadly serious in their intent to stop the dump, another *Greenpeace* ship, the *Sirius*, interrupted the dumping of toxic wastes in the Channel and was arrested by the French navy as the joint union decision was being announced in London.[151] *Greenpeace* announced on 24 June that as a result of growing national and international pressure, Switzerland would cease dumping as of 1984.

During the summer of 1983 there were a series of protests in various countries against Britain's persistence with sea dumping. In Galicia, in northern Spain, Union Jacks were burned.[152] Villagers along Spain's northern coast tossed bottles into the Atlantic with messages such as, 'Let's leave a healthy planet for future generations'.[153] In the Hague, demonstrators in radiation suits petitioned the consular-general. In Copenhagen ten people chained themselves to three mock nuclear waste barrels outside the British Embassy. And in Auckland, New Zealand, the British consulate was blockaded by *Greenpeace* demonstrators. In London, the inquiry into the proposed PWR reactor at Sizewell was disrupted by protestors dressed as sea monsters and Neptune.[154] Despite the continued furore the UKAEA let it be known it had not abandoned hope of resurrecting the 1983 sea dump.[155] The owners of the *Atlantic Fisher* tried a direct plea to Jim Slater, leader of the NUS, which was rejected.[198] In Parliament William Waldegrave insisted it was still government policy to dump at sea. There was no evidence to show that present practices were harmful, he repeated, asserting there was 'a clear national interest' in ensuring difficulties were overcome.

Eventually, at the end of August 1983, it was reported that the Government had given up its dumping plans, as had the Belgians and Swiss.[156] It was not until nearly five years later in May 1988, that the Energy Secretary announced the Government had decided finally not to dump drummed nuclear waste at sea.[157] Even so the government did wish to keep open the option of sea disposal for 'large items arising from sea disposal operations'.

NO AL BASURERO
ATOMICO EN EL MAR
Organiza: MOVIMIENTO ECOLOGISTA DEL VALLE DE LA OROTAVA
AMIGOS DE LA TIERRA (M.E.V.O. - A.T.)
CANARIAS A LA
CONVENCION
DE LONDRES
Colaboran: EXCMO. AYUNTAMIENTO DEL PUERTO DE LA CRUZ • Excmo. Ayuntamiento de S. Cristóbal de La Laguna
Excmo. Cabildo Insular de Fuerteventura • Excmo. Cabildo Insular de El Hierro • Excmo. Ayuntamiento
de La Orotava • Excmo. Ayuntamiento de Icod de los Vinos • Excmo. Ayuntamiento de Los Realejos

The abandonment of sea dumping demonstrated the international politicisation of the nuclear waste issue. The publicity generated by *Greenpeace*, together with the action of the transport unions, had been decisive. For the British Government it represented a second retreat over waste management policy. With its options for HLW disposal and the disposal of radioactive waste at sea foreclosed, at least for the time being, the Government now turned to on-land disposal of ILW and LLW. In September 1983, a month after the sea dumping plans were abandoned, NIREX announced its plans for land burial. This is the subject of chapters 3 and 4.

CONCLUSION

The history of radioactive waste provides evidence which informs our understanding of contemporary conflicts. First, the tendency towards secrecy and centralisation of control over the nuclear industry derives from the military purpose of the earliest nuclear programmes. The atomic bomb projects were conducted under extremely secret conditions and the safe management of nuclear waste was a low priority. Whether through ignorance of the risks or neglect early workers on nuclear projects took practically no health precautions, nor was there any public pressure to isolate the wastes from local communities. In consequence thousands of tonnes of unlabelled and uncontrolled radioactive wastes were left on military reservations in the United States, and smaller amounts at sites in the United Kingdom.

By the time the civil nuclear programme was being developed in the 1950s the dangers of 'atomic ash' were recognised and discussed but little progress was made towards a long term management programme. The focus of attention throughout the sixties and early seventies remained predominantly on the expansion of nuclear energy. There was little political interest in the problem of nuclear waste and military nuclear waste remained outside the bounds of any public policy discussion. Nuclear waste management was perceived as largely a technical problem to be resolved by the nuclear industry under government direction. It was not a matter for public debate or political conflict.

It was only in the mid-seventies that specific anti-nuclear campaigns developed, mainly against reactor siting proposals. As comprehension of the problems of nuclear energy, including waste, began to spread among experts outside the nuclear industry so a challenge was presented to the nuclear establishment. This took the form of challenges to government policies (as at the Windscale Inquiry) and pressure for government to formulate longer term strategies for nuclear waste (as in the United States). The institutional centralisation of the industry, its obsessive secrecy and lack of openness, and the lack of a coherent nuclear waste strategy were proving obstacles to the public support necessary for the expansion of the nuclear industry. The industry's problems were compounded by periodic serious accidents such as those at Windscale, the Urals and Three Mile Island.

By the end of the 1970s nuclear waste had become a source of political conflict. At first, participation was limited to debate among experts as at the Windscale Inquiry. As nuclear waste shifted from being a generic issue to a practical matter of finding places for disposal, environmental groups and local communities began to participate through protest against the dumping policies. As this participation developed the balance of power began to shift perceptibly away from the nuclear industry and government. Although control over policy making remained centralised, political conflict ensured that waste management decision making was more responsive to public demands. The conflict between the nuclear industry and its opponents was engaged at the technical and political level.

Protestors achieved their first major success in the UK when local communities ensured the collapse of the HLW borehole drilling programme in 1981. Their second victory, the end of sea dumping, was brought about by the combination of *Greenpeace* and the transport unions. This left the politically available options as either some form of land disposal or extended storage. Nuclear waste was being perceived by the industry as a problem that had to be solved and by its opponents as a problem that could never be solved. Thus the political battles of the mid-1980s were framed.

NOTES

1. Kuroda, P. K., *The Origin of the Chemical Elements and the Oklo Phenomena* (Berlin, Springer Verlag, 1982) pp. 48–56.
2. Caufield, C., *Multiple Exposures: Chronicles of the Radiation Age* (London, Secker & Warburg, 1989). Clark, R. W., *The Greatest Power on Earth: the story of nuclear fission* (Book Club Associates, 1981).

3. Rhodes, R., *The Making of the Atomic Bomb* (Harmondsworth, Penguin Books, 1988).
4. Jungk, R., *Brighter than a Thousand Suns* (Pelican Books, 1964) p. 111.
5. Hiruo, E., *A Background Report for the Formerly Utilized Manhattan Engineer District/Atomic Energy Commission Sites Program* (US DOE, Washington, DC, September 1980) and *Nucleonics Week*, 22 September 1988, p. 12.
6. *Bulletin of Atomic Scientists*, January/February 1988, p. 13.
7. *US Department of Energy, Environment, Safety and Health Report for the Department of Energy Defense Complex*, Washington, DC, Government Printing Office, 1 July 1988.
8. *Science*, 24 June 1988.
9. *Bulletin of the Atomic Scientists*, op. cit., p. 36.
10. *Nature*, 16 June 1988, p. 591.
A US Nuclear Scandal: 'They Lied to Us', *Time*, No. 44, 31 October 1988, pp. 26–35.
11. *St. Louis Post – Dispatch* (Sunday Magazine) 18 January 1987.
12. Bernstein, B. J., 'Radiological Warfare: the path not taken,' *Bulletin of the Atomic Scientists*, August 1985, pp. 44–9.
13. Carter, L. J., *Nuclear Imperatives and Public Trust: Dealing with Radioactive Waste*, Washington, DC, Resources for the Future, 1987, p. 53.
14. Blackett, P. M. S., *Military and Political Consequences of Atomic Energy* (London, Turnstile Press, 1948).
15. Hanford Historical Documents, 1943–1957, US DOE released and declassified, 27 February 1987.
16. Steele, K. D., *Spokesman – Review* (Spokane) 'Hanford's bitter legacy', *Bulletin of the Atomic Scientists*, January/February 1988.
'The Hanford Legacy: milk, vegetation prime sources of potential exposure', *The Oregonian*, 11 May 1986.
17. Heinz, S., 'Emissions at Hanford: part of Cold War Plan', *The Oregonian*, 12 May 1986.
18. Schumacher, E., 'Hanford 1954: A deadly dose of Radiation', *Seattle Times*, 19 October 1986.
19. 'The Hanford Legacy', *The Oregonian*, 12 May 1986.
20. Interview by Dr. Lowry, New York City, at First Global Radiation Victims' Conference, September 1987.
21. Woodbury, D., *Atoms for Peace* (New York, Dodd, Mead & Co., 1955) pp. 137–40.
22. Ibid., p. 50.
23. Medvedev, Z., 'Two Decades of Dissidence', *New Scientist*, 4 November 1976, p. 264.
24. Contradi, P., 'Blast at Urals arms-factory "left nuclear wasteland"' *Daily Telegraph*, 17 June 1989. Rich, V., 'Thirty-year secret revealed', *Nature*, Vol. 339, 22 June 1989, p. 572. Smith, R. J., 'Soviets tell about Nuclear Plant Disaster', *Washington Post*, 10 July 1989.
25. Maxwell, A., 'The Accident at Windscale', *Contemporary Issues*, Vol. 9, No. 33, April–May 1958, pp. 1–49; Lowry, D., 'Disasters in the Dark', *Times Higher Education Supplement*, 5 September 1986, p. 13;

'The Windscale Fire 1957: a bibliography of publicly available material', *Atom*, October 1987, pp. 27–9.

26. Brown, J., 'Kyshtym Whitewash', *Undercurrents* No. 42, October–1 November 1980, pp. 24–6.
27. *Die Presse*, 17 March 1959.
28. Medvedev, Z., 'Facts behind the Soviet nuclear disaster', *New Scientist*, 30 June 1977, pp. 761–4.
29. Report No. CS–3/389, 4 March 1959.
30. Trabalka, J. R., Auerbach, S. I. and Eyman, L. D., 'Analysis of the 1957–58 Soviet Nuclear Accident', ORNL–5613, Publication, No. 1445, Environmental Sciences Division, Oakridge National Laboratory, December 1979. Soran, D. M. and Stillman, D. B., 'Analysis of the Alleged Kyshtym Disaster', LA–9217–MS, Los Alamos National Laboratory, January 1982. Soran and Stillman state (p. 3) that 'today, the Kyshtym Disaster is no longer merely an intriguing enigma. The allegations surrounding it bear heavily on the whole question of nuclear waste disposal in the US'. The report concluded that it was highly unlikely an accident at a nuclear waste facility was responsible. One of the authors of the earlier USDOE sponsored study, S. I. Auerbach, said of the Los Alamos report 'its lack of citations makes it difficult to differentiate between speculation and derived analysis' (the Soran–Stillman study was apparently based on classified CIA and other US government secret reports). Medvedev commented that he was disappointed that the Los Alamos report seemed to be more concerned to prove him wrong than to analyse all the scientific literature available. 'Urals disaster: explosion of just pollution?' *New Scientist*, 22 April 1982, p. 200.
31. Medvedev, Z., *Nuclear Disaster in the Urals* (London, Angus and Robertson, 1979).
32. For instance, correspondence in the *New Scientist* in 1976 (vol. 72, p. 692) and 1977 (vol. 76, pp. 352, 368, 547).
33. IAEA, Statute, p. 5, 1973 edition, (as amended), Vienna.
34. Scheinman, L., *The International Atomic Energy Agency and World Nuclear Order*, Resources for the Future, Inc. (Washington DC, 1987).
35. Ibid., p. 86.
36. *Radioactive Wastes, IAEA*, Vienna (undated).
37. *The Management of Radioactive Wastes, IAEA* (Vienna 1981).
38. International Radioactive Waste Management Advisory Committee (Revised terms of reference), IAEA, Rev. 2, 1988–12–12.
39. NEA 6th Activity Report, Paris, 1977.
40. 'Radioactive Waste Disposal Operation into the Atlantic 1967'. OECD/NEA Paris, 1968.
41. OECD reinforces international surveillance of sea dumping of radioactive waste. Press Release A (77) 31 Paris, 22 July 1977 Background note. In July 1977 the OECD established another framework programme entitled 'Multilateral Consultation and Surveillance Mechanism for the Sea Dumping of Radioactive Waste'. It spelled out standards, guidelines, recommended practices and procedures, including prior notification for sea disposal operators by OECD member states.

42. Treaty establishing the European Atomic Energy Community, Brussels, 1957.
43. Arnott, D. G., *Our Nuclear Adventure: its possibilities and perils* (London, Lawrence and Wishart, 1957).
44. Edwards, R., 'Hidden Caves Radioactive, says Scientist', *Guardian*, 22 February 1989.
45. Bertin, L. E., *Atom Harvest*, (London, Scientific Book Club, 1957) p. 230.
46. Mitchell N. T. and Shepherd J. G., 'The UK disposal of solid radioactive waste into the Atlantic ocean and its environmental impact'. Paper in symposium on Environmental Impact of Nuclear Power, BNES, London 1988, pp. 119–54.
47. 'Alderney Atomic Ammo Dump', *Peace News*, 15 June 1984, p. 3.
48. Gowing, M., *Independence and Deterrence: Britain and Atomic Energy, 1945–1952* (London, Macmillan, 1974) p. 92.
49. Hartley, H., 'Man's Use of Energy', *Bulletin of the Atomic Scientists*, November 1950, pp. 322–4.
50. Arnold, L. and Gowing, M., 'The Early Politics of Nuclear Safety', *New Scientist*, 5 December 1974, pp. 741–3.
51. McLean, A. S. and Marley, W. G., 'Health and Safety in a nuclear power Industry', *Journal of the British Nuclear Energy Conference* (hereinafter JBNEC) January 1956, pp. 26–27.
52. A Programme of Nuclear Power (Cmnd. 9389, 1955, p. 9).
53. Spence, R., 'The Role of Chemistry in a nuclear energy project', *JBNEC*, January 1957, p. 590.
54. Grover, J. R., 'Disposal of long-lived fission products', *JBNEC*, January 1958, pp. 80–5.
55. 'Mounting requirement for National Waste Disposal Service', *Nuclear Engineering*, December 1958, p. 514.
56. 'Britain solved the nuclear waste problem', *New Scientist*, 5 July 1979, p. 4.
57. Pocock, R. F., *Nuclear Power: its development in the United Kingdom* (Old Woking, Unwin Press, 1977), p. 226.
58. Hall, T., *Nuclear Politics* (Penguin, Harmondsworth, 1986) p. 90.
59. Jeffrey, J. W., 'The Collapse of Nuclear Economics', *The Ecologist*, Vol. 18, No. 1, January 1988.
60. Magnox sale dropped because of high cost, *The Times*, 25 July 1989. Donovan, P. and Travis, A., '£4.5bn climb down on Magnox', the *Guardian*, 25 July 1989.
61. 'Hazards of Past Low Level Radioactive Waste Ocean Dumping Have Been Over-emphasised'. Report by the US General Accounting Office (GAO) for Senator William J. Roth Jr, EMD–82–9, 21 October 1981.
62. Finn, D. P., 'Radioactive Waste Disposal in the Pacific Basin and International Co-operation on Nuclear Waste', Management: Undated paper.
63. Letter in *Bulletin of the Atomic Scientists*, September 1959.
64. Crick, B. and Robson, W. A. (eds) *Protest and Discontent* (Harmondsworth, Pelican, 1970).
65. Carter, L. J., op. cit., p. 61.

66. Ibid., p. 77.
67. O'Riordan, T., 'The Prodigal Technology: nuclear power and political controversy'. *The Political Quarterly*, Vol. 59, No. 2, April–June 1988, p. 164.
68. Slater, H. G., 'Public Opposition and the Nuclear Power Industry' in Conference Proceedings of *Environmental Aspects of Nuclear Power Stations*, IAEA, Vienna 1971, pp. 847–59.
69. Falk, J., *Global Fission; The Battle Over Nuclear Power* (Oxford, Oxford University Press, 1982).
70. Patterson, W., *The Plutonium Business and the Spread of the Bomb* (London, Paladin, 1984).
71. Abbotts, J., 'Radioactive Waste: a technical solution', *Bulletin of the Atomic Scientists*, Vol. 35, No. 8, October 1979, pp. 12–18.
72. Zinberg, D., 'The Public and Nuclear Waste Management', *Bulletin of the Atomic Scientists*, Vol. 35, No. 1, January 1979.
73. Abbotts, J., (1979), op. cit.
74. 'Radioactive Waste Policy is in Disarray', *Science*, Vol. 206, 19 October 1979, pp. 312–13.
75. Clelland, W., 'Solid Progress in Nuclear Waste Treatment', *New Scientist*, 26 September 1974, pp. 786–9.
75. 'Management of Radioactive Wastes from Fuel Reprocessing', OECD, Paris, 1972.
76. *ATOM* No. 218, December 1974, p. 204. Up to mid 1970s *ATOM* was a 'journal of record' for UK (and some foreign) nuclear issues. It was comprehensive in its coverage. Once the nuclear energy debate became politicised in the mid 1970s, *ATOM* was transformed into a promotional monthly magazine selling the nuclear industry. It therefore became less useful as an independent source of primary information.
77. 'Spotting words on atomic waste', *New Scientist*, 9 August 1973.
78. *ATOM*, No. 172, February 1971, p. 36.
79. Cutler, J. and Edwards, R., *Britain's Nuclear Nightmare* (London, Sphere, 1988) p. 46.
80. Patterson, W., *Going critical* (London, Paladin Grafton, 1985) p. 29.
81. 'The Battle to beat the dust of hell', *Financial Times*, 13 November 1974.
82. *New Scientist*, Energy File, 3 April 1975.
83. 'Sweden to dump atom waste on Britain', *The Observer*, 24 April 1975.
84. Silcock, B., 'Just a small nuclear accident, but . . .', *Sunday Times*, 25 May 1975.
85. 'Nuclear dustbin future denied', *The Guardian*, 22 October 1975.
86. 'Government considering whether to stop reprocessing of nuclear fuel from other nations'. *The Times*, 22 October 1975.
87. Pocock, op. cit., p. 225.
88. Hawkes, N., 'Muddled logic on nuclear dumping', 26 October 1975.
89. *Daily Telegraph*, 26 November 1975.
90. *Daily Telegraph*, 27 December 1975.
91. Reprocessing of nuclear waste in UK challenged, *The Times*, 16 January 1976.

92. *New Scientist*, 22 January 1976.
93. C. Conoy, 'What Choice Windscale', FOE/Con. Soc., 1978.
I. Breach, *Windscale Fallout* (Penguin, 1978). M. Stott and P. Taylor, *The Nuclear Controversy: a guide to the issue of the Windscale Inquiry*, TCPA/PERG, 1980. R. Williams, *The Nuclear Power Decisions* (esp. Chapter 11, The Windscale Decision (Croom Helm, 1980). D. Pearce, L. Edwards and G. Beuret, *Decision Making for Energy Futures: a case study of the Windscale Inquiry* (Macmillan/SSRC, 1979).
94. 'Nuclear dustbin full to capacity', *Guardian*, 6 March 1976.
95. *The Times*, 13 March 1976.
96. Letter to *Guardian*, 17 March 1976.
97. 'Atom bomb waste leaking in storage', *The Times*, 9 April 1976. *The Times*, 25 May 1976.
98. *Guardian*, 31 May 1976.
99. Williams, R., op. cit., 1980, p. 291.
100. e.g. 'Hazards of plutonium reactors alarm Royal Commission'; 'The politics of plutonium', *The Times*, 20 September 1976.
101. *The Times*, Letters, 5 November 1976.
102. Cook, C., 'Nuclear Waste sites sought in Countryside', *Guardian*, 9 November 1976.
103. Johnson, D., 'Cornwall out as atom dustbin', *Guardian*, 24 November 1976.
104. 'Dustbin politics', *Guardian*, 22 November 1976.
105. 'Another delay for BNFL', *Nature*, 2 December 1976.
106. 'How nuclear cover-up went wrong', *Sunday Times*, 12 December 1976.
107. *The Times*, 11 December 1976.
108. For instance, speech to Parliament on Nuclear Materials Offences Bill, Official Report 8 February 1983, cols. 953–955; and interviewed by Dr Lowry in London, June 1989.
109. Communication with D. G. Arnott, January 1989. The development of the opposition argument by one regional protest group, PANDORA, set up in mid-Wales in September 1979 may be traced in the ten PANDORA newsletters – called PANDORA'S BOX – published up to January 1982. The concluding edition, written by former IAEA nuclear research scientist, Dr D. G. Arnott, under the title 'The End of the Affair' summarised the politics and process of the local campaign. PANDORA stands for 'People Against Nuclear Dumping on Rural Areas'.
110. See, for example, D. G. Arnott, 'High Activity Waste: an alternative to Underground Disposal', PANDORA, April 1980.
111. *Poison in our Hills: the first inquiry on atomic waste burial* (Edinburgh, SCRAM, September 1980).
112. *The Nuclear Triangle. The Miners' Report on Waste Dumping in the Cheviot* (Newcastle, TUSIU, 1980).
113. DoE Press Notice, 309A. 24 July 1979.
114. *Glasgow Herald*, 8 September 1979.

115. *NOW!* 21–27 September 1979.
116. Tucker, A., 'Call to ease the control on nuclear Dumping', *Guardian*, 16 September 1979.
117. *Nature*, Vol. 281, 4 October 1979, pp. 332–4.
118. See for instance the letter by Professor Tolstoy to the *Dumfries and Galloway Advertiser*, 19 October 1979.
119. 'Government urged to create strategy for nuclear waste disposal', *The Times*, 31 December 1979.
120. *Guardian*, 31 October 1980.
121. Communication with D. G. Arnott, January 1989.
122. 'A-waste protestors get scent of victory', *Guardian*, 24 November 1980.
123. *New Scientist*, 24/31 December 1981.
124. 'Buried in Confusion', *New Scientist*, 24/31 December 1981.
125. Parliamentary reply, 15 June 1981.
126. Parliamentary reply, 10 December 1981.
127. Dunster, H. J., 'Waste Disposal in Coastal Waters', Conference Proceedings, Vol. 18, pp. 390–9.
128. 'Assessment of Radiological Protection Aspects of Disposal of High Level Waste on the Ocean Floor', NRPB–R–48. The NRPB report followed studies begun in the US in 1973 and research done by France, Japan & Canada under the OECD's Radioactive Waste Management Committee which held workshops in February 1976 and March 1977, jointly with the US. F. Feates and J. Lewis 'UK looks at geological and ocean disposal', *Nuclear Engineering International*, January 1978.
129. *Nature*, Vol. 267, 5 May 1977.
130. *The Times*, 19 July 1977.
131. These were formalised in another NEA study, published in 1979, recommending operational procedures for sea dumping.
132. *New Scientist*, 23 March 1978.
133. Robbie, D., *Eyes of Fire* (London, Ravette, 1986) p. 2.
134. Wilkinson, P., 'One dump or two?', SCRAM journal, February/March 1983.
135. *Guardian*, 25 July 1978.
136. Interview by Dr Lowry with Remi Parmentier, Greenpeace's first anti-dumping co-ordinator, London, 29 September 1988.
Wilkinson, P., 1983, ibid.
137. *Nuclear Energy*, Vol. 18, April 1979, pp. 85–100.
138. 'Atomic Waste dumpers foil saboteurs', *Daily Telegraph*, 12 July 1979.
139. For a review of developments see 'The Urban Transport of Irradiated Nuclear Fuel', edited by John Surrey (London, Macmillan Press, 1984).
140. 'Waste rail-roaded', *Guardian*, 9 July 1980.
141. A-Fraction (letter). *Guardian*, 19 July 1980.
142. Wilkinson, P., 1983, ibid.
143. *Time Out*, 6 December 1979, which published the Minutes, full copies of which were soon in 'underground' circulation.

144. *The Times*, 15 February 1983.
145. *The Times*, 18 February 1983.
146. *New Statesman*, 2 September 1985.
147. Roberts, L. E. J., *The Times*, 21 February 1983, 'UK Disposal of atomic waste'.
148. *Guardian*, 6 April 1983.
149. *Guardian*, 6 June 1983.
150. *Morning Star*, 18 June 1983.
151. *Daily Telegraph*, 18 June 1983.
152. *Guardian*, 7, 9 and 13 July 1983.
153. *Times*, 12 July 1983.
154. *Times*, 16 July 1983.
155. *Guardian*, 28 July 1983.
156. *The Observer*, 28 August 1983.
157. *Hansard*, 26 May 1988, col. 233.

3 The Battle of the Dumps

PART 1: THE CAUSES OF CONFLICT

On the 23 October 1983 *The Sunday Times* featured the impending announcement of two sites for the on-land disposal of intermediate wastes. One of these sites, the disused anhydrite mine at Billingham on Tees-side, had been anticipated (Figure 3.1). Already the Cleveland County Council had registered total opposition to the proposal[1] and over 800 people attended a public protest meeting in the week before the official announcement was made. But Billingham was not the only site. NIREX would be recommending a second site for radioactive waste disposal to the government. According to *The Sunday Times*,

> The second site will be a 600-acre area, north-west of London somewhere between Oxford and Cambridge. There, the nuclear waste will be buried in concrete-lined trenches. It will be convenient for the atomic establishment at Harwell. The site is believed not to be in private ownership.

It was suspected that the site was in Bedfordshire. The many abandoned pits left by the brickmaking industry, some of which were used for waste disposal, seemed the most obvious possibility though these were in private, not public ownership.

The following day the site was 'discovered' by BBC television researchers. It was the Elstow storage depot, a war-time ordnance depot owned by the Central Electricity Generating Board (CEGB). (For location of Elstow and other sites referred to in the text, see Figure 3.1.) The news was broadcast that same evening. Thus the issue which was to dominate the local politics of the Bedford community for almost the following four years was revealed to an unsuspecting public a mere 24 hours before the Secretary of State announced it officially in the House of Commons.

The selection of Elstow as a potential site for nuclear waste identifies the political dimensions of nuclear decision making identified in chapter 1 in the context of the UK. First, there is its secrecy, the tendency for political participation to be limited. Although Billingham had been alerted by a speculative article in *New Scientist*[2] in February and had begun to organise opposition, the people of

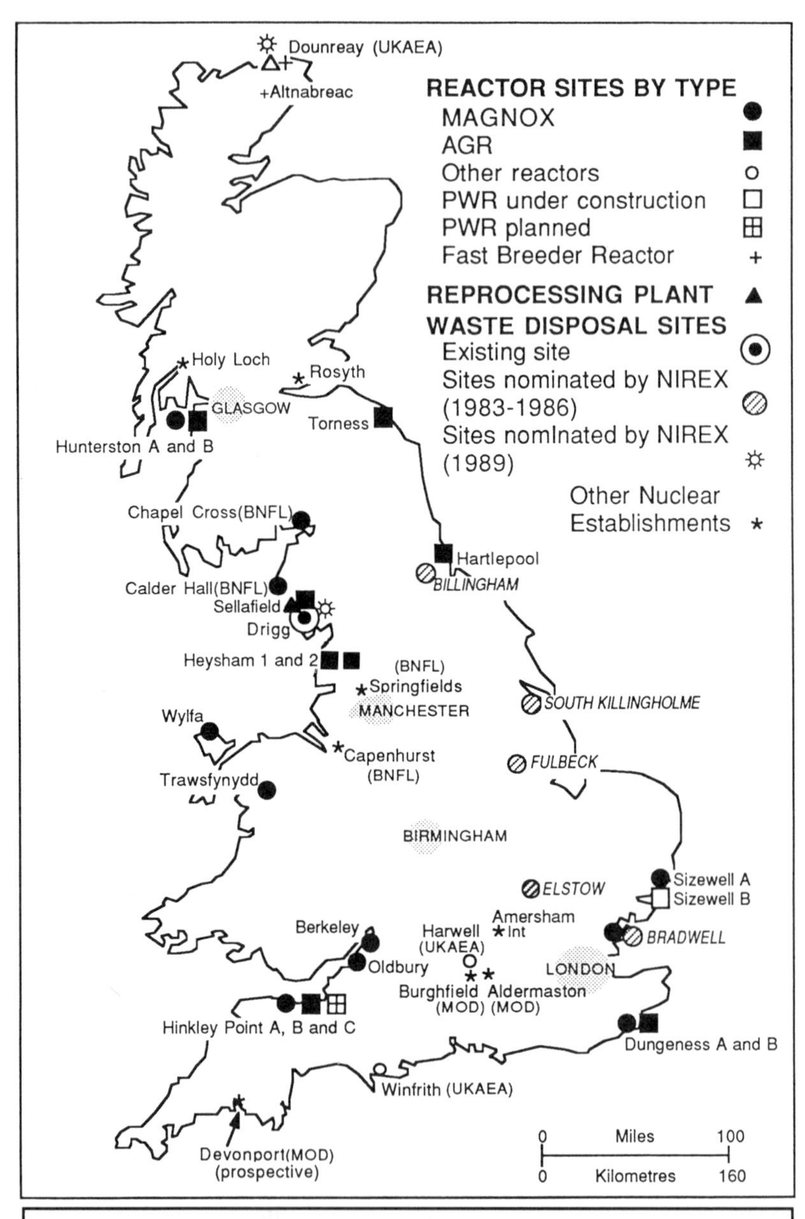

This map shows all *major* sites where radioactive waste is created and/or stored. In addition there are hundreds of other licensed sites using nuclear materials (such as hospitals) each producing small quantities of Low level Waste. There are also hundreds of other sites not requiring a site license under the 1986 Radioactive Substances (Substance of Low Activity) Exemption Order.

FIGURE 3.1 *Major nuclear facilities in Britain*

Bedford were caught unawares. It later transpired that one of the local MPs, Trevor Skeet (MP for North Bedfordshire), had been aware for two years that Elstow was likely to be chosen but had decided not to break the news to the local community on the grounds that 'Until the site is nominated it is not on the hit list. There was no point in my mind consulting anyone since Elstow might never have been chosen. People will have plenty of time to object'.[3] Until the press article two days before the announcement the local community had not been informed, let alone consulted about the proposal.

The second political dimension is the institutional framework for decision making. As we indicated in chapter 1 the UK's political system is centralised. Decisions about the type and location of nuclear facilities are taken by central government. Local councils are formally consulted and they may object but they do not have the power of veto. The local planning inquiry procedure is intended to ensure that national interests can be weighed against local interests. In the case of radioactive waste repositories it was made plain from the start that the planning applications would be 'called in' by the Secretary of State for the Environment so that he would take the decision on the basis of the evidence presented at the public inquiries.

Lack of a clearly defined and debated policy is the third dimension of nuclear decision making. Policy making tends to be characterised by incrementalism rather than a comprehensive strategy. The marathon Sizewell Inquiry into the country's first commercial Pressurised Water Reactor (PWR) became, in the absence of parliamentary debate, a forum for conflict over the direction of national energy policy.[4] The choice of Billingham and Elstow was not preceded by any government sponsored public debate about radioactive waste management strategy, let alone debate about possible sites. Two sites had been nominated before either alternative methods of managing wastes or criteria for selecting sites had been disclosed or debated. It was a case of premature legitimation.

These three features – secrecy, centralised power and premature legitimation – shaped the ensuing conflict. At the outset each feature seemed to favour the nuclear industry and the government. Secrecy was an essential prerequisite for making a choice without causing public alarm. Within the secret and centralised decision making the nuclear industry was able to gain privileged access to government, to seize the initiative and to deploy the formidable financial and technical resources at its disposal. At the outset the nuclear industry possessed considerable resources of power.

As time wore on these advantages waned and initiative was lost. The secrecy of the industry became a weapon in the hands of its opponents. To whom was the industry accountable? If there was nothing to hide why should not everything be revealed? Centralised power also made the industry vulnerable. Faced with massive resources of expertise and money funded by taxation, the local authorities, strapped for cash by government cuts and lacking the powers necessary to defend their interests could pose as Davids facing Goliath and achieve widespread public sympathy and media support. They could portray themselves as the defenders of democratic institutions. Finally, by picking two sites in advance of any public debate on alternative sites and strategies the nuclear authorities demonstrated a lack of foresight and political understanding. In the years ahead this political failure undermined the credibility of the proposals and ultimately destroyed them.

In this chapter we consider the causes of the conflict, the opposing arguments based on different evaluation of the evidence reflecting opposing technical and scientific arguments and political values. The conflict ranged over three broad areas. First there was the conflict over the decision making process, particularly its centralised nature and lack of consultation with local interests. Second was the conflict over policy brought about by the attempted preemption of debate by the government and the nuclear industry. And, thirdly, the centralisation, lack of consultation and inadequacy of policy fuelled the conflict over the specific sites chosen. The conflict was triggered by the announcement.

CONFLICT OVER PARTICIPATION

The announcement

Patrick Jenkin, the Secretary of State for the Environment made his much anticipated announcement on radioactive waste disposal on 25 October 1983. He began with the claim that 'effective disposal, in ways which have been shown to be safe, is well within the scope of modern technology'.[5] Consequently, 'there is no scientific reason for deferring the disposal'.

The statement introduced the main issues which were to dominate the conflict. On the question of decision making the statement was ambiguous. It was possible that site investigations 'may require planning permission' and if so, they would be called in for his decision. On the question of policy the minister stated that any debate over the 'merits of disposing wastes at the sites' would be reserved for a later public inquiry if NIREX applied for permission to develop 'at one of these sites, or at any other site'. As to the sites, two land disposal facilities were required for the disposal of low and intermediate-level waste. NIREX was announcing 'two sites which it considers sufficiently promising to justify further investigation' at Billingham and Elstow.

Some of these concerns were evident in the initial skirmishes in the Commons. Questioned on the integrity of the disposal method Mr Jenkin maintained that 'land-based disposal of intermediate wastes is the safest and best method provided that a site can be found with sufficient geological certainty and stability which will remain safe for the necessary period of time'.[6] MPs representing the local communities involved were quick to identify the arguments against the chosen sites. Frank Cook, Labour MP for Stockton North, which includes Billingham, made his points rhetorically asking the Secretary of State if he would 'ensure that the inquiry is made fully aware that 14 per cent – one sixth – of the registrable hazardous locations in this country are located in that area'. The two Conservative Bedfordshire MPs representing the Elstow area were more circumspect, seeking reassurances that no risks or blight would be imposed on the area. On the procedures the minister asserted that 'Consultations with the local authorities would not have been appropriate at this stage' and reaffirmed that the first inquiry 'if it is necessary, will be of comparatively limited scope'.[7] In the months ahead the strategy, the sites and the procedures would all be successfully challenged.

At the same time as the minister's announcement documents were published by the Department of the Environment (DoE) and by NIREX which provided further important background to the decision.

The DoE's Draft Principles

The DoE's document entitled 'Draft Principles for the Protection of the Human Environment' outlined the procedures and regulatory framework within which decisions on land disposal facilities would be taken. On the subject of disposal strategy it claimed that 'low-level wastes are already being disposed of safely'[8] a reference to the disposal site at Drigg in Cumbria near Sellafield, which was later generally accepted as inadequate and described as 'not an acceptable model for any future disposal site'.[9] The Principles further noted that the proposed disposal policy formed part of 'an agreed system for waste management',[10] a claim that was to be strongly contested by opponents of the proposals. Four types of land disposal facility were identified including an 'engineered trench at about 30m depth'[11] which appeared to correspond to the Elstow proposal. On the question of sites the document suggested that NIREX should follow a 'rational procedure' to ensure 'they have not ignored a clearly better option for minimising radiological risks or for satisfying town and country planning criteria and policies'.[12] Among the latter were such features as population distribution and the local economy.

In the following months the Draft Principles were subject to criticism especially and understandably from Bedfordshire and Cleveland County Councils.[13] Bedfordshire's response emphasised the absence of an agreed policy. Bedfordshire claimed that NIREX 'is acting prematurely in seeking sites before a national strategy on radioactive waste management has been developed'. In terms of siting Elstow did not qualify under the suggested criteria for demography, local economy and alternative land uses and should be ruled out on those grounds alone.

As a result of these and other criticisms the Draft Principles were revised and an interim version published for the Sizewell Inquiry.[14] This version omitted reference to the safety of existing low-level waste disposal routes and dropped the reference to types of land disposal facility. In the final version of the Principles published in December 1984 references to the agreed system and national strategy were omitted. The issue of siting criteria was stated as follows, 'In discussing alternative sites, the developer will be expected to show that he has followed a rational procedure for site identification. However, he will not be expected to show that his proposals represent the best choice from all conceivably possible sites'.[15]

The NIREX Proposals

The NIREX project statements published for Billingham and Elstow indicated the background to the proposals and consequently were used by both sides in the conflict. The design of the proposed repositories was outlined for each location. Elstow was to be the site for a shallow repository to take those low-level wastes (LLW) arising from power stations, research establishments, hospitals and industry, leaving the Drigg site with sufficient capacity to acommodate British Nuclear Fuel's (BNFL) reprocessing wastes from Sellafield for up to 50 years. In addition, Elstow would take short-lived intermediate level waste (that is, radioactive half-lives of 30 years or less and mainly beta and gamma emitters) consisting of sludges, resins and other contaminated items, some already existing in store at power stations, the rest expected to accumulate by the end of the century. The repository would consist of engineered trenches typically 6–15m deep, up to 20m wide and 150m long, lined with concrete, backfilled with clay, capped with a concrete intrusion shield and covered by further layers of clay or other materials to prevent water intrusion. By the year 2000 it was estimated there would be around 50 trenches covering 60 acres at the 450 acre site. The operational period would be about fifty years, after which it would take 'at most, two or three hundred years'[16] for the radioactivity to decline to very low levels.

Billingham had been selected from about a hundred existing cavities in the UK. An existing mine was to be preferred to a greenfield location 'in terms of cost and timing'.[17] It appeared to satisfy the siting criteria: 'The site is conveniently located and the geology is well understood'.[18] It would not pose an environmental problem in an area experiencing 'relatively high noise levels due to industrial operations which continue around the clock'.[19] The mine workings, owned by Imperial Chemical Industries (ICI) and abandoned since 1971, cover an area of 5km^2 immediately beneath the ICI works and parts of the town of Billingham. The workings have a capacity of 11Mm^3 at a depth of between 140 and 280 metres. The deep repository to be constructed at Billingham would accommodate the long-lived, mainly alpha emitting, intermediate-level wastes (ILW) consisting of fuel cladding, sludges, resins and concentrates predominantly from reprocessing operations. Existing stocks and future waste arisings to the year 2000 at Sellafield were expected to be 55 000m^3. The ILW wastes destined for Billingham were estimated to contain 80M curies(ci) of beta and gamma activity and IM ci of alpha in the short-lived ILW forecast for Elstow.[20] In both Elstow and Billingham safety considerations would be of 'paramount importance'.[21]

In both communities the NIREX proposals were summarily rejected. During the following years every aspect of the government's policy and the NIREX proposals were scrutinised by opponents representing the local communities. The conflict was, in part, a technical debate between experts on the justification for on-land disposal, the procedures for the selection of sites and the design and safety of the proposed repositories. But, it was also a deeply political matter in which technical issues became embedded in conflicts between national and local interests. In the rest of this chapter we look at the three major areas of dispute – over participation, policy and sites – by closely examining the arguments over Elstow. Unlike Billingham, which won an early victory, or the other shallow burial sites which came late into the conflict, Elstow was involved throughout the battle of the dumps.

Decisions and Local Democracy

On the issue of participation Bedfordshire County Council pointed out that local interests had been by-passed through the lack of any consultation before the Elstow announcement. This enabled the council to occupy the high moral ground against NIREX and the Government and thus gather public support. The council had two major procedural aims. One was to ensure that alternative sites to Elstow were named. The other was to ensure that the public inquiry process considered the widest possible range of issues rather than being restricted to site-specific issues, as had been indicated in the Secretary of State's announcement.

On the question of alternative sites Bedfordshire gained an early concession as the result of a delegation of councillors and officials who saw the Under Secretary of State for the Environment, William Waldegrave, in February 1984. He confirmed that NIREX, at the first public inquiry, would be required to list a number of alternative sites and say why they considered Elstow to be the most suitable. It would then be up to the Inspector to say which, if any, of the sites might be more suitable having heard the evidence. Further indications were given during the course of the Sizewell Inquiry at Snape Maltings on the Suffolk coast. The inquiry lasted for 340 working days from 11 January 1983 to 7 March 1985 and one of the subjects covered was radioactive waste disposal. From time to time the Inquiry provided interesting insights into the policy making (or lack of it) of NIREX, the CEGB and the DoE. In March 1984, Dr R. Flowers, the NIREX witness seconded from the CEGB, admitted a second site might be investigated. By October 1984 he confirmed the position when he said that 'We would really like to do field work on two or preferably three sites before making a formal application for construction.'[22] This would be essential in order to demonstrate that 'the site we have chosen is not obviously inferior to some other site'.[23] This decision had come as a surprise to the DoE whose witness at Sizewell declared, 'I was not aware of a firm intention on the part of NIREX in fact to do test drilling on three sites'.[24] But it was welcomed in Bedfordshire where the County Council had argued that Elstow should be compared with other named sites in a process they believed would demonstrate its unsuitability.

The question of the scope of the planning inquiries was complex. The County Council felt under severe time pressure for in the early days it appeared that decisions would be made quickly. The council initially expected a planning application for investigatory works in February 1984 followed by a public inquiry in the late spring or early summer. It would take about two years to undertake the drillings, analyse the findings, design a repository and prepare an environmental impact analysis. 'Thus it may be 1986 before any planning application for the repository itself is submitted.'[25] In fact it was not until the autumn of 1986 that the site investigations began.

Under such pressure of time the council had to be careful. If the terms of reference of an initial inquiry into investigatory works at Elstow were widened to embrace wider issues the County Council 'should be cautious instead of appearing early at a major inquiry unprepared and with a "hollow" case'.[26] As time went on and Bedfordshire was able to develop its case, so the need for caution vanished and the council was keen to test its arguments at a public inquiry at the earliest opportunity.

From the Government and NIREX's standpoint it was judged to be prudent to avoid an early exposure to public debate of the broader issues. It suited them better to have a first inquiry limited to the site works needed for drilling operations and to reserve the major debate until a second inquiry following an application to develop a repository. Better still would be to avoid a first inquiry altogether. At one time the DoE had considered amending the General Development Order (GDO) 'to make the planning process easier and simpler for NIREX'.[27] At the Sizewell Inquiry (Day 279, October 1984) the DoE's Mr Critchley indicated that a first inquiry might still be avoided through the use of a Special Development Order. Although overlooked at the time the use of this procedure was to provoke a most dramatic public reaction some two years later.

The time-scale of decision making became protracted as expected deadlines were missed and decisions deferred. The reasons for NIREX's delays may have been technical or political as they became enmeshed in conflict. Bedfordshire County Council regarded delay as a tactic designed to weaken their resolve as it became difficult to sustain a campaign over years rather than months. On the other hand delay provided time to organise and for the Council to build up its technical case. Delay proved to be the ally rather than the enemy of the local opposition.

One possible method of decision making was the use of a Planning Inquiry Commission, a statutory procedure which had lain dormant since its introduction in the Town and Country Planning Act of 1971. A Planning Inquiry Commission could be held if there were 'considerations of national or regional importance' and if 'the technical or scientific aspects of the proposed development are of so unfamiliar a character' and the question of whether development 'should instead be carried out at an alternative site' could be referred to the Commission.[28] The County Council held that a Commission 'would assist both NIREX, who need a site somewhere, and the Secretary of State, who would be able to make an informed decision, comparing a number of alternative sites, judged by the same criteria'.[29] A Commission would also, of course, benefit Bedfordshire and Billingham, by opening up the whole issue of radioactive waste management policy for debate.

The idea of using a Planning Inquiry Commission was one of the issues raised at the meeting with William Waldegrave on 15 February 1984. While acknowledging the case for a Commission he held that the strategy for radioactive waste management had been 'comprehensively debated' in the 1982 White Paper and 'it would merely be superfluous to rehearse the arguments again at a Planning Inquiry Commission'.[30]

The sometimes arcane discussion about planning procedures were fundamentally concerned with questions of democratic participation. The local councils had not been consulted about the plans before the announcement and they feared that a public inquiry into the need for exploratory investigations might be avoided by NIREX. The NIREX witness at Sizewell said as much, 'our love of public inquiries is not such that we would wish to hold them if there was another way'.[31] Once a site had been chosen and proved technically feasible, the chances of resisting it would be diminished, especially if the terms of reference of the inquiry excluded the broader political questions about radioactive waste management policy. Furthermore, the local community would be aggrieved if Elstow were to be the only site selected or alternative sites were investigated merely in order to provide comparative evidence that would help to legitimate the Elstow selection. As Mr Brooke, Counsel to the Sizewell Inquiry put it, 'they might think that as they have been chosen as the site for investigation the odds were stacked against them because there were no other sites being the subject of choice'.[32]

At the outset it was not yet certain whether alternative sites would be evaluated in order to justify the selection of Elstow or be put forward as options from which a final selection would be made. It was evident that neither NIREX nor the government had thought through the procedural implications of the Elstow announcement. There was uncertainty over the timing of the proposed development, the number and nature of the planning inquiries, and over the role of alternative site investigation. This air of uncertainty also pervaded the issue of the overall strategy for radioactive waste management.

CONFLICT OVER POLICY

In Search of a Strategy

Nobody seriously questioned the need for a policy for radioactive waste management. The nuclear industry regarded it as a necessary consequence of the benefits of nuclear power. 'No "justification" is necessary for waste disposal because the generation of electricity by nuclear power is considered a net benefit.'[33] Some opponents of the industry acknowledged the problem of existing accumulated wastes and regarded this as an opportunity to undermine the industry and either halt or slow down its further expansion. Thus, nuclear waste policy was inevitably a subject of profound disagreement since upon its resolution might hang the future of the nuclear industry.

By the time of the Elstow announcement the government had begun the 'preparation of an overall long-term strategy' with the publication of its White Paper, Cmnd 8607[34] in 1982. In its response to the Flowers Report in 1977 the government had foreshadowed 'the need for a national disposal facility'[35] for low and intermediate wastes. The White Paper (Cmnd 8607) on radioactive waste management represented government policy on nuclear waste. The government felt able to claim that 'it is feasible to manage and dispose of all the wastes currently envisaged in the UK, in acceptable ways' (para. 12). 'Waste management is not therefore a barrier to the further development of nuclear power as now foreseen' (para. 24).

There was 'no technical obstacle' (para. 53) to the disposal of LLW and ILW. This confidence was echoed in the Secretary of State's announcement as we have seen. At the Sizewell Inquiry in the summer of 1983 the government's confidence in its ability to manage wastes was clear. 'They can all be managed, held in storage and disposed of safely within the limits of present knowledge and present technology'.[36] Indeed, the probable method would be 'an engineered trench at a depth of about 20–30 metres, and a modified mine or purpose-built cavity at greater depth' (para. 37). This overall strategy provided the context for the specific proposals for Billingham and Elstow announced a year later.

To opponents the government strategy showed signs of expediency and pragmatism. The government had chosen disposal rather than storage on site; land burial rather than other disposal options; and had picked two sites without any public consideration of alternatives. 'In other words, the UK has moved from the identification of urgent need for a disposal option to the development stage with only a very short research stage allocated to what is a relatively complex technology with high performance requirements.'[37]

The exact status of the government's strategy was unclear. According to a DoE witness at Sizewell, Mr George Wedd, a strategy was in preparation and would be subject to continuous review. Asked if this meant there was no strategy at the present time he replied,

> There is indeed a strategy at the present time in existence in a simplified form appropriate to the amount of work that has been done on it, but we expect it to be developed, filled out and quite possibly corrected in the light of further work, further thought, further advice, as we receive them over the years to come.[38]

One issue was whether alternative management options such as storage or disposal on the seabed or under the continental shelf had been adequately considered. Certainly both NIREX and the government had accepted on-land disposal. NIREX had identified sites and the government clearly backed the solution. As the DoE's Mr P. Critchley stated at Sizewell in October 1984 'we are not disposed to be shaken in our pre-disposition in favour of trying to secure decisions on appropriate sites on dry land'.[39]

The issue of alternative disposal options was to play a decisive part in the later stages of the conflict. It began to emerge as a critical issue at Sizewell, in Bedfordshire and in Parliament. At Sizewell Mr G. Searle representing the Stop Sizewell 'B' Association emphasised the predetermination of policy during his cross-examination of Mr Critchley of the DoE,

> What I am suggesting to you is that you won't know who is going to win the race unless you tell the horses when you are going to start and if you run the race and you choose Elstow before you have given due consideration to the possible alternative, well it is a one horse race and one can predict the winner of that?[40]

Certainly the feeling that shallow land burial as the option and Elstow as the site had been decided upon prematurely was strongly reflected in Bedfordshire. One of the planks in the County Council's case against the proposal was that,

> it is premature for NIREX to apply for the investigation of the suitability of a specific site until the waste disposal technology which would be used to exploit the use of that site has been proven.[41]

The theme was also taken up in Parliament. Frank Cook, the Labour MP and former engineer who had worked on nuclear plant construction before entering Parliament, made the following claim in a debate in the House of Commons on May 3 1984,

> It appears that underground disposal was presumed the best from the outset. Its only shortlist has been a 'shortlist of one' – for high level intermediate waste at Billingham and one for low level intermediate waste at Elstow in Bedfordshire.[42]

Nicholas Lyell, Conservative MP for Mid-Bedfordshire suggested the Elstow site in his constitutency had been chosen 'it seems almost by means of a pin'.[43] Opponents of the proposals had developed the argument that NIREX, apparently supported by the government, had embarked upon an ill-considered and poorly researched strategy. The problem they faced was one of premature legitimation. And it was a problem compounded by the reckless decision to nominate specific sites with a minimum of warning and the absence of consultation.

Shallow Burial – Experience Elsewhere

NIREX sought to legitimate the land burial strategy by reference to experience elsewhere. Dr R. Flowers, one of their Directors, pointed out in 1982 that 'Shallow burial of ILW and LLW has been carried out routinely for a number of years at sites in the USA and in

France',[44] a point repeated in the NIREX project statements on Billingham and Elstow. In these statements NIREX claimed 'there is a wide consensus that the use of existing technology can lead to the disposal of radioactive wastes in a very safe manner'.[45] By the middle of 1984 the DoE was claiming that disposal of ILW 'has been validated as a safe option, by research and development, and by experimental or actual disposal and subsequent monitoring in several countries'.[46] In their second Annual Report in September 1984 NIREX asserted that 'France and the USA have similar programmes to the UK, actively using the shallow land burial option for low-level and short-lived wastes and looking for new sites for both shallow land burial and deep underground disposal'.[47]

These claims were rejected by opponents of shallow burial. First, they pointed out that the techniques used elsewhere were not comparable to those intended by NIREX. The French site at la Manche near Cherbourg (described in chapter 6) was essentially a surface disposal system sealed off from underlying rocks with water collected for analysis before discharge into the sea. By contrast the NIREX proposal was for an engineered repository below ground using the host geology to disperse and dilute radioactivity. The US disposal facilities (discussed in chapters 2 and 5), like the NIREX proposal, were based on shallow trench burial but without the containment barriers proposed by NIREX.

Secondly, experience elsewhere had not always been favourable. Although there were three shallow burial facilities operating in the United States, three had been closed down due to hydrogeological problems or structural failure leading to unsatisfactory levels of radionuclides being detected beyond the site boundaries (see chapter 5). In addition contemporary experience in the United States suggested a more cautious approach and considerable research investment. Bedfordshire's technical consultants concluded that, 'The scope and depth of this effort, prior to application for specific site investigation and development can be contrasted with the approach being taken by the DoE and NIREX in the UK'.[48]

A third reason to doubt the relevance of experience elsewhere was the fact that several countries (among them, Sweden, Finland, Belgium, Federal Republic of Germany, and Switzerland) had rejected shallow burial, opting instead for deep disposal. While France and the US among western nations had operating sites, the future trends in other countries suggested that the UK was 'in a considerable minority regarding international opinion on current acceptability for shallow burial, when other options are available'.[49] Considerably more research effort was being put into waste disposal strategies in other countries than in the UK.

There was a large LLW shallow disposal site operating in the UK at Drigg near Sellafield (see Figure 3.1). Indeed, one reason for the Elstow proposal was to provide an additional site so that Drigg's life could be extended and enable it to be used solely for the LLW from reprocessing at Sellafield. The Drigg site, opened in 1959, covered 300 acres and had accommodated 600 000m^3 by 1983. The wastes were brought to the site by rail and were disposed without prepackaging into shallow trenches 5–7m deep and covered with about a metre of excavated material.[50] Water draining from the trenches is monitored for radioactivity on its way to the Irish Sea.

Initially the Drigg disposal system was portrayed as unproblematic. At the Sizewell Inquiry the debate focused on future capacity at Drigg, its location relative to waste arisings and the timing of its replacement facility. There was little concern on the part of government about the safety of the Drigg operations. In any case the NIREX proposal for Elstow was conspicuously different – it was an engineered containment facility designed for LLW and some ILW. Drigg was excavated in clay without containment and accepted only low-level wastes. Nonetheless Bedfordshire considered the poor performance of the Drigg site could have an impact at a future inquiry. 'From published BNFL reports and from inside information, we know that not only have there been regular fires . . . occurring on the site which would give rise to small airborne releases of radioactive gases . . . but that considerable ground migration of radioactive leachate has occurred'.[51] To a degree the comparison between Drigg and the Elstow proposal was unfair since NIREX insisted that the Elstow repository would be a distinct improvement on Drigg. Nevertheless, opponents were prepared to use the problems of Drigg to underline their fears about safety in the industry.

Recourse to experience elsewhere was a weapon in the hands of both sides. While NIREX claimed that shallow burial elsewhere validated their proposal, their opponents pointed to the absence of comparability. But, Bedfordshire were not averse themselves to making unfair comparisons in the case of Drigg. In the first phase of the conflict the opposition was able to undermine the notion that the government strategy mirrored developments elsewhere. Later, the conditions at Drigg would be used to damage the credibility of the nuclear industry and lead to government action to improve the facility. Moreover, the NIREX plans would be compared unfavourably to the methods of disposal being introduced in other countries. The inadequacy of the policy of shallow land burial, once exposed, was to become a major aspect of the opposition's case.

Classification of Radioactive Wastes

The justification for shallow and deep burial relied on an accepted definition of the types of wastes destined for each facility. It was soon evident that there was confusion over this element of the policy also. The White Paper of 1982 had defined LLW as

> those wastes with a low level of radioactivity ... which can be safely disposed of by existing routes, the arrangements depending on the particular level and type of radioactivity

ILW were those

> which at present can be safely stored but for which disposal facilities are not at present available and which do not fall within iii [namely, the category of heat-generating wastes].[52]

By contrast, NIREX put forward a classification based on activity rather than disposal route. In his proof of evidence to Sizewell Dr Flowers asserted, 'The distinction between Intermediate (ILW) and Low Level Waste (LLW) is an arbitrary one based on the proportion of radioactive material to non-radioactive material in the waste.'[53]

This lack of agreement gave the DoE's George Wedd a difficult time during cross examination on Days 95 and 100 of Sizewell in July 1983. Wedd did not accept the 'arbitrary' definition of NIREX and maintained the definition for LLW was 'those for which there are disposal routes now'.[54] His subsequent efforts at clarification only served to obfuscate as the following exchange with Mr Brooke, counsel to the inquiry, illustrates,

> **Wedd** ...'I think one should be aware of the circularity of the language. Low level waste, by our definition is that of which one can dispose, so it would be equally true to say there are some grades of waste of very limited radioactivity which can now be disposed of and those are, for our convenience, referred to as low level wastes.'
> **Brooke**. 'Now, we are getting into an area where confusion is fast reaching chaos ... people are now more confused rather than less confused in an area in which ... it is essential for the public to understand what is being done.'[55]

The implication of the DoE's definition was that as new routes for disposal were brought into operation more waste would become classified as low-level and less as intermediate-level. The important question in Mr Wedd's view was not what the category of waste was but 'What are the characteristics of this material chemically and radiologically and what is the best way of disposing of it?'[56] Nonetheless the lack of clarity over definition was a problem as Mr Brooke was at pains to point out,

> the concern that some observers to the scene have expressed that the Department's method of categorising wastes involves a slurring from wastes of one form of radioactivity from one category to another, and possibly in the public eye a consequent diminution of the seriousness of the problem of disposing of them.[57]

Such concern was felt at Elstow since it appeared that the definition of the types of wastes coming into the site could alter over time. This lack of common definitions was further evidence that 'the DoE is in a considerable state of disarray over its strategy'[58] and could strengthen the case against the NIREX proposals. The DoE and NIREX were instructed to overcome their definitional differences by the Inspector at Sizewell and, accordingly, the DoE agreed to use the NIREX definitions. These were later published as definitions 'now in general use in the UK'[59] using levels of radioactivity to define the boundaries between low- and intermediate-level categories.

Although this definition overcame the confusion, the lack of coordination and agreement over this component of radioactive waste management strategy had been plain to see. There was even more confusion over the quantities and origins of the wastes destined for the proposed repositories, a point which further damaged the credibility of the government's strategy.

Volumes of Waste

It is important to remember that Elstow was intended both for LLW and some ILW. The case developed against the proposal by Bedfordshire County Council incorporated somewhat different arguments in respect of the two categories of waste. For ILW it was argued that the volumes of short-lived wastes were too small to justify separate disposal and that shallow burial was an untried and inherently unsafe method for these wastes. This became a crucial issue later in the conflict. The initial priority appeared to be to find alternative capacity for Drigg. In the case of LLW the existence of a disposal route at Drigg rebutted the charge that the technology was untried. If Drigg was unsatisfactory and unsafe then NIREX could argue that their proposed containment facility would prove much better. In any case since Drigg had been acceptable for many years why should another shallow burial facility be unacceptable? So long as Elstow was destined to receive both categories of waste, the relative weakness of the case presented by Bedfordshire against a low-level burial facility could be concealed by an emphasis on the fears about the safety of burying ILW.

The case against shallow disposal rested partly on the volumes, timings and origins of the wastes. Table 3.1 provides details of the estimates. It can be seen that NIREX in 1983 following their Proof of Evidence at Sizewell, using a high nuclear energy scenario (20GW capacity by 2000) calculated LLW arisings at 500 000m^3 and ILW at 80 000m^3 (of which about 25 000m^3 would be short-lived) up to year 2000.

The argument that additional capacity for LLW disposal was needed rested fundamentally on three points. One was the accuracy of the figures based as they were on particular assumptions about the size of the nuclear programme and waste packaging. Some experts believed the volumes could be considerably reduced by treatment including compaction and incineration. If that were so, then the life of Drigg would be lengthened. Furthermore, a second and related point, if the nuclear programme were to be altered the volumes might decline still further. This would particularly be the case if a future government abandoned reprocessing. As John Blake, representing the Town and Country Planning Association put it to the Sizewell Inquiry,

> Insofar as reprocessing accounts for almost four-fifths of the low level waste that is expected to be produced up to the year 2000 and would continue to account for a very large proportion of it after that time, clearly the strategy is very fundamentally dependent upon the policy with regard to reprocessing, is it not?[60]

Such a change in the nuclear programme would greatly extend the life of Drigg and render an additional site for LLW unnecessary in the immediate future. And, thirdly, there was the issue of the remaining capacity at Drigg. Originally, NIREX had assumed the remaining capacity to be of the order of 1M m^3 but later indicated that the figure had been revised downwards to 550 000m^3 on account of unsatisfactory

TABLE 3.1 *Estimated total of LLW and ILW to be produced by the year 2000*

1. Estimates prepared for Sizewell 'B' Inquiry
Assumes high nuclear energy scenario: 4GW(e) decommissioning of Magnox power station: 12GW(e) from PWRs by year 2000: total installed capacity of 20GW(e)

Total UK packaged waste volumes requiring disposal m³

	ILW	LLW
1981	30 000	—
2000	80 000	490 000

By 2000 51 500m³ ILW and 380 000m³ LLW assumed to arise from reprocessing.
Of the 80 000m³ ILW about 40 000m³ assumed suitable for shallow land burial.

SOURCE R. Flowers (NIREX) Proof of Evidence on: The Disposal of Low Level and Intermediate Level Solid Wastes, Sizewell 'B' Power Station Public Inquiry, November 1982, CEGB p. 21, p. 7, and tables 2, and 4 pp. 32, 34.

2. Estimates prepared for NIREX Preliminary Project Statements
Assumptions as above – waste conditioning reprocessing and 20GW(e)

Amounts of solid radioactive wastes expected to arise in UK by year 2000 in m³

ILW	LLW
25 000 predominantly short-lived[1]	500 000[3]
55 000 predominantly long-lived[2]	

[1]2M curies beta, gamma less than 1000 curies alpha

[2]80M curies beta, gamma 1M curies alpha

[3]About 10 000 curies beta, gamma. Negligible alpha
50 000m³ to be disposed of in shallow repository by 2000.
Remainder in Drigg

SOURCE NIREX (1983) 'The Disposal of Low and Intermediate Level Radioactive Wastes: The Elstow Storage Depot: A Preliminary Project Statement', October, Figure 6. Confirmed in NIREX (1984) Second Report, September, p. 22.

geology at the site.[61] There were clearly doubts and the future capacity of Drigg would make a considerable difference to the necessity and timing for a further LLW repository.

This area of uncertainty about the volumes of arisings and the capacity of Drigg provided yet another fruitful area for debate about the soundness of the government's strategy. Opponents pointed out that if different assumptions were made on waste compaction, the size of the nuclear programme (and possibly of reprocessing) and the capacity of Drigg the case for a further site for LLW (let alone for the much smaller volumes of short-lived ILW) was weak. If the assumptions proved correct and Elstow was found to be unacceptable then there was the possibility that Drigg would be filled up before an alternative disposal route was found. Again if, as NIREX suggested, the CEGB would not be embarrassed if Elstow were not opened when Sizewell started up,[62] then it was difficult to see why they were anxious to develop the site so quickly and why a site for LLW was the priority rather than a means of disposing of ILW for which there were no disposal routes. All these conundrums made it appear to opponents of the proposals that a nuclear programme was being pursued in the absence of a fully thought out strategy on waste management in which disposal routes had not been proven, let alone made publicly acceptable.

A final point was whether it was logical to use the shallow burial facility for ILW. Estimates of the volumes of ILW arisings varied, NIREX suggesting 80 000m^3 by 2000 (see Table 3.1) but others such as Duncan and Brown giving 50–60 000m^3,[63] with an estimate as low as 15 000m^3 from the NRPB.[64] The proportion of short-lived in these totals also varied from 10 per cent (NRPB) to 20 per cent (Duncan and Brown) to 31 per cent given by NIREX in its project statements or even about 50 per cent accepted by NIREX at the Sizewell Inquiry. Whatever the figures the volumes of short-lived ILW were small and the need for a separate facility for these wastes could be questioned. This point was noted in a confidential paper to Bedfordshire County Council which acknowledged there could be quite strong arguments for 'having only a single ILW disposal repository, which would be a suitable disused mine or deep engineered facility'.[65] At that stage such a revelation would have been imprudent since it would have threatened the united front developing between Billingham and Elstow against the proposals as a whole. But it was a point that would become critical later. Indeed, at Sizewell the DoE's witness Mr P. Critchley took the argument a stage further anticipating the future debate with the comment 'that the deep disposal site could also take an element of low level waste as well as intermediate level waste'.[66] It was the issue of co-disposal that provided the final twist in the debate nearly three years later.

Repository Design and Construction

The method of disposal proposed by NIREX was a multi-barrier engineered containment system (Figure 3.2). The concept was explained as follows,

> Containment of wastes disposed of to a land-based repository is ensured by a 'multi-barrier' concept, in which a number of barriers are interposed between the waste and man's environment, each of which can be shown to delay substantially any movement of radioactive materials away from the disposed waste. . . . One of these barriers can be formed by the natural properties of the geological strata, and therefore in identifying candidate sites, the geology is a prime consideration.[67]

The wastes would be contained by the mixture in which the waste was set, the metal container, a concrete container or concrete backfill and then the host rock. By the time the engineered barriers had been eroded the radioactivity would have decayed and the sorptive qualities of the host geology together with its very low groundwater flow would sufficiently retard migration to prevent any radioactivity from reaching the accessible environment 'before it has decayed to a very low level'.[68] At Sizewell Dr Flowers concluded that 'the risks from a repository . . . are very small indeed'.[69] Opponents were quick to point out that the proposed system lacked practical or even experimental validation.

The security of this system depended on four crucial variables – the half lives of the radionuclides in the waste; the integrity of the materials used, that is steel and concrete; the hydrogeology of the host rock; and the maintenance of stable conditions and lack of intrusion at the chosen site. The first two variables are common to any shallow repository and so will be dealt with here. The geological and hydrogeological variables are site specific and will be covered when we discuss the Elstow site itself.

NIREX estimated that the short-lived ILW would need to be contained for 200–300 years to enable radiation to decay to safe levels. This assumed very careful control over the waste streams entering the repository to ensure that any long-lived radionuclides were excluded. But monitoring systems 'are liable to errors'[70] and wastes above the specified limits could be disposed of as had happened in the case of telephone dials at Drigg in 1984.[71] Even accepting that careful monitoring could be sustained for such a long

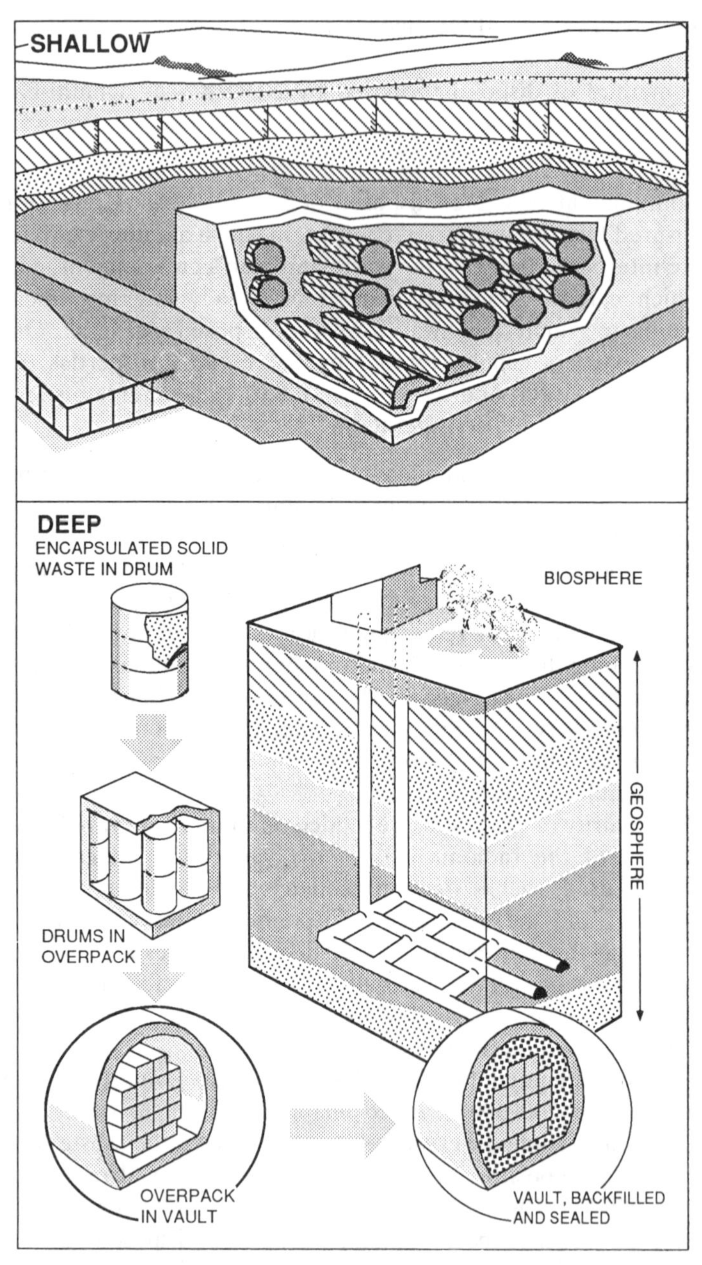

SOURCE Adapted from NIREX and UKAEA original illustrations
FIGURE 3.2 *Schematic representation of multi-barrier disposal sites*

period the prospect of engineering failure could not be entirely eliminated. In a confidential document, Bedfordshire County Council's technical consultants pointed out a number of possible scenarios. One was that such containment in saturated clays could lead to a concentration of contaminated trench water. This might result in the paradox that 'the adoption of an engineered containment system significantly increases the risks from radioactivity exposure to future generations if the trench containment system were to be disturbed in, say, 300 years time'.[72] If this seemed a remote possibility, the failure of the system before the point of safety was reached was more plausible. Trench failure resulting in the 'bathtub effect' in which the trench fills and overflows near the surface could be avoided by drainage systems excluding water from the trenches. 'However the fact is that to ensure this over a timescale of 200–300 years in a climate with a high rainfall is an extremely difficult engineering problem.'[73]

Considerable reliance had to be placed on the materials used. For LLW there was reliance on mild steel drums containing solidified wastes. These drums could be subject to damage during handling, and would corrode once in place. Although corrosion was an anticipated part of the design, the speed at which it occurred was disputed. NIREX assumed the life of the drums to be sufficient to contain radioactivity until it had fallen to safe levels. But, depending on the thickness of the steel, the coating materials, the amount of organic decay within the package and the rate of external microbiological corrosion, wastes might be leached out quite quickly. In the opinion of Bedfordshire County Council's consultants, 'on the information available from NIREX and analysis of local soil, the drums will not last beyond 20 to 40 years'.[74]

ILW would be further protected by concrete shielding and emplacement in concrete-lined trenches with the concrete acting as both a physical and chemical barrier (by preventing many radioactive elements dissolving in water). Opponents claimed that concrete would fail, citing examples of bridges, motorways and buildings. NIREX pointed to examples like the Colosseum, Hadrian's Wall, the Great Wall of China and the base of Salisbury Cathedral spire to substantiate their faith in the strength and longevity of concrete. Concrete structures 'can be designed which should not deteriorate significantly over a period of several hundred years in a repository environment and which will probably outlive the Greek and Roman concretes by a very large margin'.[75] A United Kingdom Atomic

Energy Authority (UKAEA) /NIREX archaeological expedition had discovered concrete in Yugoslavia aged 7600 years.[76] 'Recent research has indicated that specially prepared concretes can be made to last thousands of years under the right conditions' such as a moist environment, constant temperature and protection from wind and weather.[77] But, opponents argued, 'Most experience over any appreciable time-scale and structure size has been with the use of cementitious materials for bonding stone or brick, which cannot sensibly be used to predict the behaviour of mass or reinforced concrete'.[78] Concrete, like steel, corrodes and can suffer structural damage. The rate of degradation depends on the quality of the concreting materials and the potential for sulphate attack from microbial activity in the host rock.

There were various solutions to these problems such as using sulphate-resistant materials, protective coatings, or non-ferrous metals. NIREX had sponsored research into transport and packaging of wastes, and modelling of the behaviour of a repository. Opponents acknowledged that NIREX 'may potentially be able to evolve a system of almost perfect integrity'[79] but it could prove too costly and operationally unacceptable. In any event the method of emplacement was manifestly not proven and considerable research would be needed. Once built a repository would need 'continuous monitoring and policing'[80] for up to 300 years. The design and operation of a repository were necessarily at a conceptual stage. Opponents argued that the lack of research and evidence about long term integrity was yet another sign of a hastily conceived and premature strategy. Despite the assurances of NIREX, doubts about the safety of the proposed system were at the heart of public fears.

Justification and Safety

Safety was 'of paramount importance'.[81] The safety criteria employed are based on the three principles recommended by the International Commission on Radiological Protection (ICRP) in 1977.[82]

Two of these principles were open to interpretation. The first principle, that no practice shall be adopted unless it produces a positive net benefit, is taken as axiomatic by the Government. Radioactive waste management is the corollary of a nuclear energy programme. To the extent that radioactive waste exists and is being produced its management is necessary. But, it was also a policy objective that 'the creation of wastes from nuclear activity is minimised'.[83] This was interpreted by the nuclear industry as meaning 'that one should ensure that the creation of wastes from a given nuclear activity is minimised and not that the nuclear activity itself be minimised'.[84] It was automatically assumed that nuclear power was a positive benefit and that its further expansion would be ensured by successful waste management. Opponents argued that the development of nuclear waste facilities ought to be justified in terms of an overall social benefit sufficient to outweigh the specific social costs imposed upon populations surrounding the facilities.

The second principle, that 'all exposures should be kept as low as reasonably achievable' with economic and social factors being taken into account (the ALARA principle) was likewise open to differing interpretations. ALARA requires some assessment of cost-benefit, weighing levels of risk (radiation dose) against the safety achieved at certain levels of technology and cost. A more stringent approach, adopted in some countries is ALATA (As Low As Technically Achievable) which, 'while it cannot ignore economic factors, requires that a balance should be drawn heavily in favour of safety and environmental considerations'.[85] This approach was naturally favoured by opponents of the government's proposals.

Whichever approach was used it was clear that complex analysis would be needed 'to provide the proper guidance for the nuclear industry due to the longevity of radiation in the environment, the enormous capital investment required and long development lead times'.[86] The problem of interpretation became acute when ALARA was applied to specific proposals such as that for Elstow. While few people pretended to understand the mathematical principles involved it was a relatively simple matter for the opponents of the Elstow site to insist that by comparison with other possible sites or methods of disposal Elstow could not be regarded as the site where exposures were as low as reasonably achievable. But, if a site had already been chosen on geological, geographical and other grounds it was possible for proponents to argue that 'the role of ALARA is in determining the combination of burial facility design, waste types, and use-restriction period which is most appropriate for the site'.[87] Thus ALARA could be defined as a necessary criterion in site selection or as a constraint once a site had been identified. In such circumstances it was likely to generate confusion rather than clarification of the issues.

The third safety principle was to ensure that radiation doses were limited to levels recommended by the ICRP which were regarded as the minimum necessary to ensure safety. The National Radiological Protection Board (NRPB) advised tighter limits than those recommended by the ICRP.[88]

These principles were applied by the NRPB in a 'preliminary, generic assessment' of the safety of shallow land burial published in 1984.[89] The research evaluated the theoretical impact of different types of wastes disposed of in a minimum or a fully engineered facility, depending on the radioactivity in the waste streams involved. Several conservative assumptions were made in calculating the probabilities of exposure. Over time it was inevitable that barriers would be penetrated and radionuclides would migrate through the ground water. Exposure could occur either through human intrusion by excavation after closure or, in the longer term, by contamination of the food chain resulting from farming activity. A problem with engineered barriers was that they could only be effective if they lasted as long as the half-lives of the radionuclides they contained. The NRPB estimated that, for certain wastes in an engineered facility, the site might need to be restricted for up to 500 or 600 years, well beyond the 200 to 300 years assumed by NIREX.

A report by the same authors published only sixteen months earlier in 1982[90] had reached broadly similar conclusions on the likely radiological impacts, though the detailed calculations were different. It is instructive to compare the tone of the two documents. Both reports argued strongly the case for further generic and site-specific research and stressed the research should not be applied directly to make 'decisions on the acceptability of any specific burial site or facility design'.[91] The 1982 report concluded cautiously that 'emplacement in an engineered facility at shallow depth in clay strata could be a radiologically acceptable option' (p. 21). By contrast the 1984 report concluded, 'the results do indicate that shallow burial, in appropriate facilities, of reasonable quantities of any of the three waste types considered in the report is likely to be acceptable from a radiological protection viewpoint' (p. 29). Though still qualified the optimism is unmistakable.

There is little hint of qualification in the claim by NIREX that 'use of existing technology combined with sound management techniques can lead to the disposal of radioactive wastes in a very safe manner'.[92] By 1984 the DoE itself was proclaiming, 'there is no advantage from delay, and no requirement for lengthy research and development into methods of disposal'.[93] It must be said that the

NRPB reports did not provide the unequivocal evidence necessary for such sanguine pronouncements.

The safety case, as presented by the NRPB, was criticised as deficient in several crucial respects by the consultants employed by Bedfordshire County Council. One factor that had been ignored was the microbiological activity likely to dominate the degradation, solubility and transport processes of radionuclides. Another was the possibility of structural degradation. Certain radionuclides had not been included in the forecasts. The problem of disposing of contaminated trench water had been ignored. There was also the likelihood of methane gas production arising from decomposition, which was a major environmental hazard in a landfill near the Elstow site. The model used by the NRPB was generalised with simplifying assumptions, inadequate sensitivity analysis and had reached conclusions 'quite unsupported by the evidence'.[94] So, on the crucial question of safety there was considerable room for scientific disagreement and consequent political conflict.

The Need for a Strategy

The notion that there was a coherent strategy for radioactive waste management was clearly contestable. When each component of the strategy was examined there appeared to be neither technical nor political consensus. Shallow burial for LLW and short-lived ILW and deep disposal for long-lived ILW had been announced without any comparative evaluation of alternative management options. The argument that the methods had been validated by experience elsewhere was challenged. There was uncertainty over the classification, volumes and distribution of waste arisings. The safety case rested on controversial theoretical assumptions rather than empirical evidence. On each component there was considerable room for scientific, technical and political disagreement.

The lack of a coherent strategy led to some embarrassment for the NIREX witness at Sizewell when the following point was put to him in cross-examination by the Town and Country Planning Association's representative, Mr J. Blake in March 1984,

> There are a number, as it were, of uncertainties and the point I want to put to you is how can NIREX in fact proceed to its individual proposals for specific disposal routes in the context of the absence of a clearly defined plan or strategy for nuclear waste which shows, so far as it can, that these possible uncertainties will not arise?[95]

A major uncertainty was the classification of wastes and in an effort to cover the embarrassment the Department of the Environment subsequently produced for the inquiry a document described as 'The National Strategy'.[96] This provided an agreed classification and also attempted 'to set out the Government's current strategy for the long term in the management of radioactive wastes' (p. 2).

In practice the document attempted to rationalise the government's decision to go ahead with land disposal. Indefinite storage was rejected as an option since 'it would leave future generations with the burden of maintaining stores, and replacing them from time to time' (p. 11). Disposal at sea, though regarded by the Government as 'environmentally acceptable, and preferable for certain types of waste' (p. 16), was suspended. Land disposal could be 'relied on to provide at least an equal degree of safety to that provided by supervised storage' (p. 13). The problems were of 'a manageable nature' and 'the main need is to apply known technology in a consistent and coherent way in order to deal satisfactorily with each particular type of waste' (p. 27).

Critics described the document as 'cobbled together' without 'a sound technical scrutiny'.[97] The statement of a national strategy had been made nearly a year after the Commons announcement in which the Government had indicated both how and where it wanted to dispose of ILW and LLW. Bedfordshire's consultants commented on the haste with which the strategy had been put together; 'We have therefore in a matter of three years moved from a position of not having a nuclear waste disposal strategy to having potentially suitable sites identified and an assumption that shallow land burial of intermediate wastes is a proven technology'.[98]

The conflict over the government's policy was, at one level, technical and scientific. It was a debate among experts mediated through politicians or officials or, as at Sizewell, through legal procedures. The arguments over such issues as repository design, migration of radionuclides, volumes and types of waste, and dose and risk probabilities were inevitably and necessarily complex, sometimes esoteric and often inconclusive. The lack of any elite consensus on the issues was a factor in encouraging political conflict. The absence of a publicly-debated strategy evaluating possible waste management options was a powerful weapon for the government's opponents. The opposition's general case against the disposal strategy was reinforced by its case against the specific site selection.

CONFLICT OVER SITING

Elstow – Why Pick on Us?

The choice of Elstow gave specific focus to the criticisms of a centralised, closed and incremental approach to decision making for nuclear waste management. The claim that Elstow was a promising candidate proved hard to justify publicly on the basis of limited and general assertions. In their preliminary project statement[99] NIREX stated that Elstow had been chosen as 'one of the preferred sites' from among 'about 60 potentially suitable repository sites' (p. 1). There were few clues as to how the choice was made. The site was 'conveniently located' (p. 1) showing 'good potential' (p. 8). Among its favourable characteristics was the fact that local hydrogeological conditions were well known (p. 10) with about a 15m depth of Oxford Clay. Given the neighbouring brickmaking activity the repository was unlikely to 'significantly affect the environmental character of the area' (p. 11), where there was already 'a significant concentration of heavy goods traffic' (p. 14). Thus, the negative characteristics of the site were being promoted as a positive reason for its selection.

Four sets of factors had been identified in the site selection process; hydrogeology; ecology; land use and status; and socio-economic factors. The first factor was paramount since safety depended upon the security of the containment system. The initial step in site selection was to identify potential sites by size and geology. Sites with known problems were then eliminated. Comparisons were then made among remaining sites, which excluded all the existing shallow cavities leaving only 'greenfield sites'. These were assessed in terms of hydrogeology, site development and operation, and non-nuclear environmental considerations. From this analysis Elstow 'emerged as the preferred site' (p. 10).

The Sizewell Inquiry was unable to elicit any more information on the site selection process. According to CEGB witness Dr R. Flowers, Elstow and Billingham had been chosen on the basis of 'desk studies' to be examined further 'because they look promising on paper'.[100] The desk studies had taken account of many factors. 'It was a matter, for example, of identifying landowners who were willing to co-operate in the use of land for this purpose; for suitable geological and hydrogeological conditions; the geographical location . . . ; the question of other use of the land and whether it was earmarked or likely to be required for other purposes; disturbance of other resources; proximity of population'.[101] In their second Annual Report in September 1984 NIREX described a three-stage selection process – highlighting candidate sites; deciding which were worth investigating; and then selecting sites for field studies. Desk studies had been supported by field visits and final selection 'involved a detailed examination of each of the more promising locations embracing a review of all published material and available information'.[102]

Once it was clear that NIREX would have to identify sites other than Elstow they were under pressure to describe the selection process in more detail. Accordingly, in an edition of their news sheet *Plaintalk* issued in May 1985 they presented 'The meticulous five-step in the search for a site' (pp. 4–5). This showed how the available alternatives were gradually narrowed down by mapping appropriate geology, areas of highest accessibility to waste arisings, and avoiding areas of high population density or conservation needs. These factors defined the 'area of search within which potential sites could be identified'. The areas of search broadly coincided with the clay belts stretching across the middle of England from Yorkshire to the Severn, within which three of the four named sites duly appeared in 1986 (Figure 3.3).

At no stage were NIREX prepared to provide a list of the rejected sites. In this they were supported by the Under Secretary of State for the Environment, William Waldegrave, who told a Bedfordshire County Council delegation in February 1984 that naming of other sites 'would result in blight in those areas'. But, he did concede that a short-list of sites considered would be produced 'in outline form' at the first public inquiry.

The reluctance to name sites despite pressure to do so played into opponents' hands. They were able to argue that NIREX had something to hide. Either NIREX had failed to undertake a rational process of site selection, or worse had predetermined the selection of Elstow and Billingham and were using siting criteria as a method of post hoc rationalisation. In either case the lack of a comprehensive strategy for nuclear waste management was exposed.

Elstow, and indeed any other site selected, would have to be justified against the DoE's Principles for protecting the human environment and against the ALARA criterion. Critics believed NIREX would need to show that 'Elstow is at least as good as other potential sites, logistics and other socio-economic factors being taken into account'.[103] The failure to name other sites might make that task

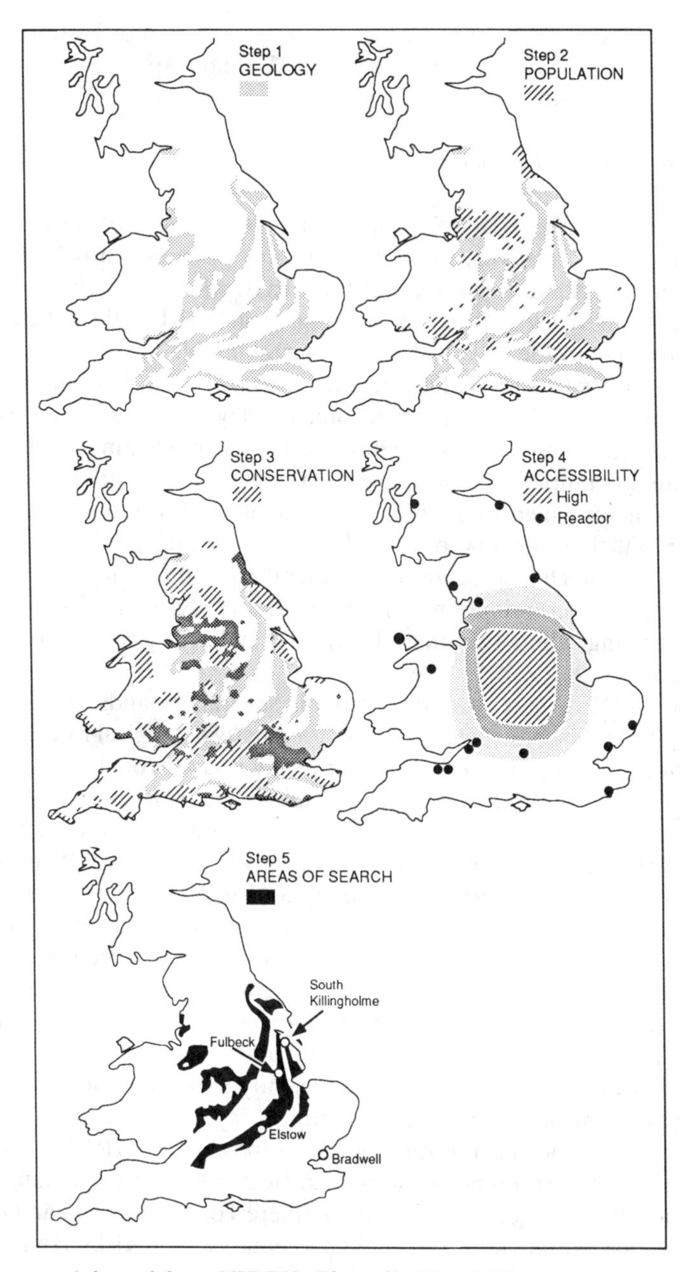

SOURCE Adapted from NIREX, *Plaintalk*, May 1985.
FIGURE 3.3 *The five-step search for a site*

more difficult. When specific factors in the selection of Elstow were examined the case put forward by NIREX appeared to critics well nigh indefensible.

Hydrogeological Factors

The claim in the NIREX project statement that 'hydrogeological conditions are relatively well known' (p. 10) was apparently based on the bore holes put down both for the CEGB who owned the site and for London Brick in the surrounding area. NIREX described the Oxford Clay in the area as 'a particularly suitable formation for the disposal of waste in shallow trenches' (p. 12). The hydraulic conductivities were low, the proportion of clay minerals that absorb radionuclides was high. The depth of clay was about 15m, though the saturated conditions would cause the trenches to fill with water over time, a factor which would have to be accounted for in the repository design. Further information would be needed on the geology, hydrogeology, geotechnical properties, mineralogy and radionuclide sorption properties, and seismic characteristics of the Oxford Clay and the overlying callow (weathered clay) and underlying Kellaway sands and clays.

The NIREX case was disputed on several grounds. First, the evidence of suitability was gathered 'for strictly geological rather than hydrogeological purposes'.[104] Oxford Clay was a strongly laminated clay-shale formation. Little was known about its hydrogeology though it was relatively impermeable with hydraulic gradients which would restrict the movement of radioactive leachate to 10 metres between 2000 and 200 000 years, quite long enough for radioactivity to decay to harmless levels. However, small faults had been observed in brick-pit excavations, cracks and fissures could be accentuated by weathering, sand pockets were also present, and excavation of clay might cause heave, all of which could increase the permeability at the site. There was little knowledge of the underlying aquifers, though the flows were anticipated to be sufficiently slow or static to inhibit the movement of water from the Oxford Clay. Overall, there was a need for site-specific information as proposed by NIREX. There were a number of unknowns which led Bedfordshire's consultants to conclude that on geological grounds 'there could be site characteristics which would indicate that unacceptably high risks could be associated with the use of Elstow as a radioactive waste disposal site'.[105]

The case against Elstow would be strengthened if it was discovered that the Oxford Clay was relatively thin at the site. This would increase the likelihood of contact with the underlying aquifers. There was considerable confusion over the depth of clay required to ensure the integrity of the disposal system. The 1982 White Paper suggested a depth of 20–30 metres[106] and in 1983 the DoE took its yardstick as 'about 30 metres depth' for an engineered trench in its initial draft of its general principles.[107] This appeared to conflict with the NIREX estimate of about 15 metres at Elstow. In an adjournment debate in the Commons in December 1983 William Waldegrave confirmed the NIREX figures as 10 to 15 feet of callow with 40 to 50 ft of unweathered Oxford Clay beneath.[108] Perhaps not surprisingly, no figure was given in the final version of the general principles published at the end of 1984.[109] When confronted with the fact that the depth of clay at Elstow was thinner than that originally envisaged by the DoE for a shallow repository, Dr Flowers told the Sizewell Inquiry, 'This is a question of fitting the repository design to the clay formation in question'.[110]

Not only was the clay thinner than that originally suggested by the DoE, the depth of unweathered clay at the site was found by Bedfordshire County Council's consultants to vary from a minimum of 11.43 metres to a maximum of 15.85, the depth quoted by NIREX. Furthermore, investigation of the Oxford Clay belt stretching from Dorset to Humberside revealed a diminishing thickness from 170 metres to 20 metres moving north. 'In this respect it seems less than fortuitous that the 16 metres remnant of Oxford Clay beneath the Elstow site should be among the thinnest developments of the unit in the Bedford area or even in the rest of England. Indeed one might safely claim that moving in any direction away from Elstow would find a thicker development of the Oxford Clay unit'.[111]

In addition to Oxford Clay there were several other clay formations that would be suitable for a shallow repository. These included the Triassic formations of the Midlands extending to 300m.; the Jurassic Lias clays varying from 30 to 300m., and Kimmeridge clay (450 to 40m); the Cretaceous Gault clay (up to 70m); and the Tertiary London Clay ranging from 30 to 180m (Figure 3.4). The Triassic and Liassic clays (which also include Oxford Clay) share the common characteristics of low water content, less weathering, low permeability and ability to absorb radionuclides. Given the wide distribution of apparently suitable formations it could be concluded that the properties of the Oxford Clay at Elstow 'are by no means unique'.[112] If

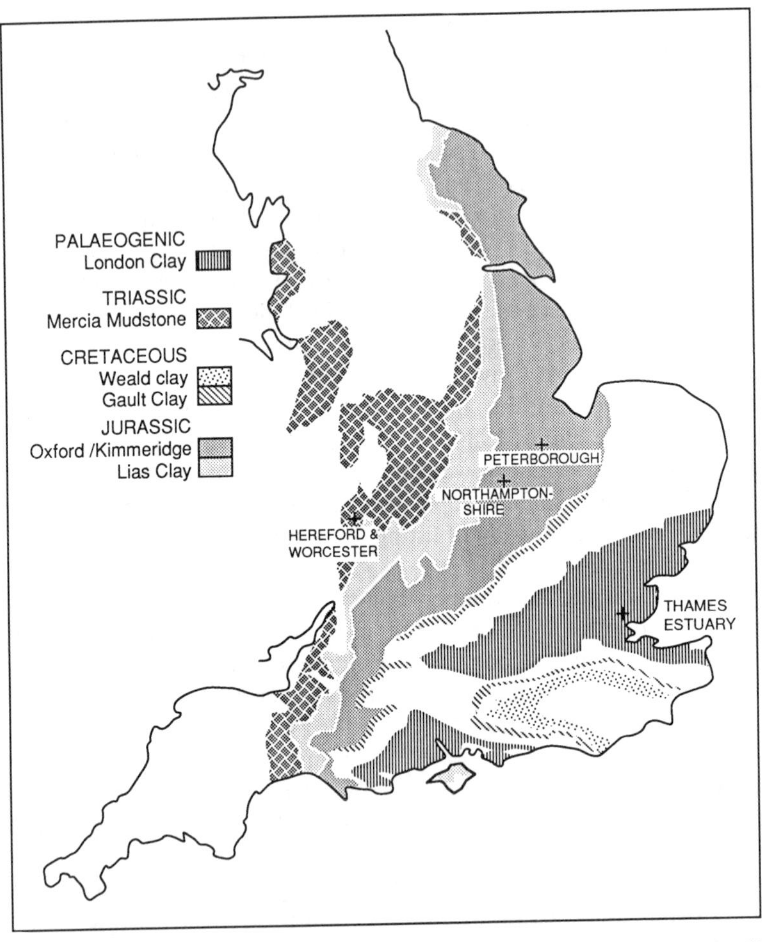

FIGURE 3.4 *Occurrence in England of clay mudrocks with potentially suitable geology/hydrogeology for LLW and ILW disposal*

safety was to be the major criterion in site selection then Elstow appeared to be only one among many available options.

Geographical Location

One of the factors in site selection was 'a more centrally located alternative site' than is presently available at Drigg in Cumbria for the receipt of non-Sellafield low level waste'[113] though this was a 'desirable' rather than an 'essential' feature. NIREX estimated that the volume of LLW from reprocessing by the year 2000 would be around 380 000m^3 out of a total of around 500 000m^3. Thus the bulk of the LLW was reprocessing waste still destined for Drigg leaving about 5000m^3 per annum to be sent to Elstow.

Assuming a central location were to be needed, it could be argued that Elstow was sub-optimal. Indeed, research commissioned by the CEGB had suggested as much, though, for perhaps obvious reasons, this finding was not publicised. Taking into account geology and the location of power stations and the need to minimise transport three areas appeared to be 'most geographically suitable'.[114] These were Worcestershire (Keuper marl); Northamptonshire/Peterborough (Oxford Clay); North of Thames estuary (London Clay) (Figure 3.4). If two sites were required then Worcester and the Thames estuary would be preferred; if only one site was needed Worcestershire was the best situated. In neither case was the Northamptonshire area, the obvious surrogate for nearby Bedfordshire, an optimal location. Although not a crucial issue in itself, taken with other more site-specific factors the location of Elstow relative to waste arisings suggested that the process of site selection might fail the DoE's test set out in its 'Principles for the Protection of the Human Environment', that a 'rational procedure' had been adopted.[115]

Economic, Environmental and Social Factors

NIREX regarded accessibility as a major advantage of the Elstow site. The Elstow storage depot covering 450 acres is about three miles from Bedford, close to two major roads and about seven miles from the M1 motorway, the main north–south artery. The main London–Sheffield railway passes close to the site and rail connection would be relatively easy to develop (Figure 3.5). The wartime depot buildings were used by a variety of businesses, altogether employing about 400 workers. NIREX would consider the needs of existing businesses and provide as far as possible alternative accommodation to avoid job losses. The Project Statement argued that development of the repository would create a 'substantial number of short-term jobs' (p. 15) and its operation would require about a hundred workers.

The surrounding area is a vale with rural villages and the large-scale excavations of the brickfields. The NIREX Project Statement claimed that 'adverse impacts can be minimised and kept within acceptable limits' (p. 14). There would be about two train loads or

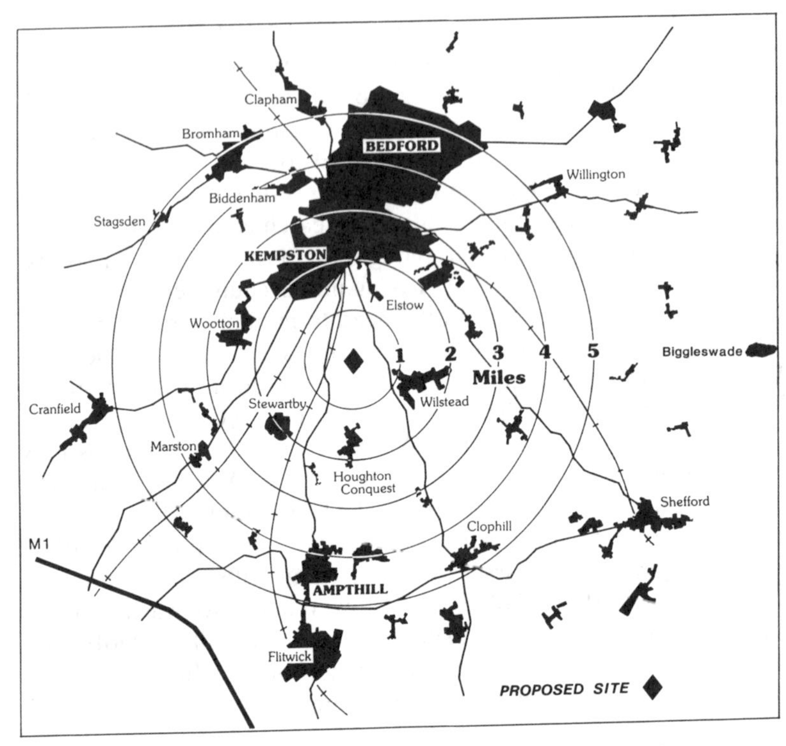

SOURCE Adapted from Bedfordshire County Council illustration.
FIGURE 3.5 *Location of the Elstow site*

30 lorry loads of waste per week, a small number compared to the 800 lorries already entering the site. Noise was unlikely to be a problem in an area with existing high noise levels. The scale of the operation 'would not be out of character with its existing use for storage purposes or with nearby brickmaking and landfill activities' (p. 14). It was even suggested the development could eventually enhance the local environment and in the long term return it to 'productive use' (p. 13).

From Bedfordshire's viewpoint the proposal represented a major economic and environmental disaster. The impacts were negative with no compensating benefits. Even if existing businesses were relocated there was likely to be economic blight caused by the possibility of what local people insisted was a radioactive waste 'dump'. Businesses might be deterred from investing in the area and house prices might fall, though it would be difficult to isolate the

"Thirty Lorryloads a Week"
...on a road near You?

- Sign the Petition against the dump
- Join Bedfordshire Against Nuclear Dumping (B.A.N.D.) tel: Bedford 214059
- Write to Patrick Jenkin, Secretary of State for the Environment
- Write to your MP
- Write to NIREX

NIREX wants to bring up to 30 lorryloads of nuclear waste a week to dump at Elstow. For the next 50 years.

Tell Them What You Think

Bedfordshire

SOURCE Bedfordshire Against Nuclear Dumping.

impact of the repository from other factors. Increased noise, disturbance and loss of amenity were further detractions that would accompany the proposed repository.

Transportation was another potential source of disturbance and risk. Transport of wastes involved some risk to workers and to the general public from continuous exposure along transport routes and from possible accidents, though the radiation levels and therefore the risks were low. However, the transport issue had potential political impact since wastes destined for Elstow would need to pass through neighbouring counties hitherto either unaware or unaffected by the transport of radioactive materials.

In proposing a site the DoE had indicated that the developer would need to consider the 'number and distribution of population in relevant areas'[116] and NIREX acknowledged they had assessed 'proximity to population and settlements'[117] in identifying Elstow. However, the site was in a densely populated area and close to a large town that was expected to expand on its southern side close to Elstow. There was a population of 120 000 within five miles of the site. It would seem likely that sites could be found elsewhere with much lower densities and that therefore 'the ALARA principle has not been satisfied at Elstow'[118] (Figure 3.5). The population near the equivalent facility in France was much lower, a point conceded by William Waldegrave in reply to local MP Trevor Skeet: 'it appears that a few hundred people live in the immediate neighbourhood of the Centre de la Manche and the town of Cherbourg – population 35 000 – is 15 miles away'.[119]

Moreover Elstow did not even appear to satisfy the population criterion employed by NIREX itself. Areas with low population density were to be preferred 'to minimise the risk, however small, of any adverse impact'.[120] NIREX quoted the Nuclear Installations Inspectorate's (NII) recommendation that there should be no settlement of over 10 000 within five miles of a nuclear power station and took two people per acre (ppa) as the maximum density. The equivalent density within five miles of Elstow was 2.4 ppa in 1981, which exceeded the NIREX maximum by 20 per cent. On this issue, just as in the case of the depth of clay at the site, NIREX appeared to contradict their own criteria.

A similar contradiction appeared over the issue of alternative land uses at the site. After its wartime use as an ordnance depot the site had remained in public ownership by the CEGB as a possible site for a future nuclear power station. Although Bedfordshire County

Council had opposed the idea in its first Structure Plan (a statutory statement of strategic land use policies), prepared during the late 1970s, the Structure Plan had recognised the possibility of a nuclear power station at Elstow and had proposed ' a general presumption against residential or industrial development within the area, ⅔ mile radius, centred on the potential power station site at Elstow'.[121] The relative absence of concern about the potential threat of a nuclear power station at that time contrasts markedly with the anxiety caused by a proposal for a nuclear waste repository a decade later.

In October 1983 the CEGB dropped their plans for a power station. Because of 'the lower requirements for new power stations in the medium term the Board has decided recently to relinquish its interest in the Depot'.[122] This left the way open for alternative uses for the site. One such use was brickmaking. The County Council had stated, in its Oxford Clay Plan, that it would 'protect workable deposits of Oxford Clay from unnecessary sterilisation'.[123] Moreover, the Plan specifically suggested that clay should be extracted for brickmaking from the Elstow site as an environmentally more suitable alternative to the nearby site at Houghton Conquest, which already had permission for clay extraction.[124] It was estimated that the volume of clay available at the Elstow site would be sufficient to produce enough bricks for 1.5M houses and could employ 300 people for about 46 years.

When the NIREX proposal was announced London Brick, actively encouraged by the County Council, duly applied for permission to extract clay from Elstow in exchange for agreeing to relinquish without compensation the permission at Houghton Conquest. From London Brick's viewpoint the deal was attractive since the Elstow site was far larger and could yield up to four times as much clay. From Bedfordshire County Council's perspective the application enabled it to claim that 'there is at least one viable alternative use for the Depot site that is of national importance; and that the NIREX proposal, if it went ahead, would sterilise a workable reserve of Oxford Clay'.[125] If the choice lay between brickmaking or a nuclear waste repository, 'The people of Bedfordshire, and their elected representatives are quite clear in their views about the issue – the development of the clay deposits which are a natural resource of national importance is seen as providing a far greater social good'.[126]

The DoE's Principles stated that 'a site must be selected where it is unlikely that future development of natural resources, or of the site, will disturb the facility'.[127] NIREX, somewhat ingenuously, suggested in their Project Statement that the clay extracted for the repository 'might be used at nearby brickworks or for land reclamation projects' (p. 13). Their Chief Executive, Maurice Ginniff, helped the opposition's case with the view that 'Repository sites, moreover, should not be close to valuable mineral resources.'[128] At the very least, Bedfordshire's technical consultants argued, the existence of clay used in the local brickmaking industry would 'force NIREX to prove that other sites have equally valuable alternative uses. This is likely to be very difficult for them to show'.[129]

The proximity to a large population and the existence of a nationally important alternative land use were regarded as important arguments against the Elstow site. Bedfordshire's consultants argued, 'If the site is unsuitable on local environment/economic impact terms ... then there is little point in NIREX carrying out investigatory drilling'.[130] Added to the uncertainties about the clay the case put forward by NIREX in favour of Elstow appeared ill-prepared, flawed and premature.

A Political, not a Technical Matter

When the selection of Elstow was related to the various criteria presented both by the DoE and NIREX it was difficult to perceive how this site had 'emerged as the preferred site'. On purely technical grounds there appeared to be alternatives that could equally or better meet the critieria. But rationality is not simply a technical matter; it also implies feasibility and that embraces political as well as technical issues.

Control of the site is an essential political precondition for an unpopular development such as a nuclear waste repository. The availability of sites was an integral part of the selection process and 'led to the exclusion of a number of potential sites which were committed to other forms of development or were otherwise unavailable'.[131] In the case of the Billingham mine the decision of the owners ICI to oppose the development was the primary reason for the abandonment of the project. The Elstow site was in the ownership of the CEGB, a partner in NIREX and the major source of the waste destined for the site. The coincidence of the release of the site by the CEGB and the NIREX announcement in October 1983 might be regarded as fortuitous or, more likely, premeditated. Even so there were other sites in friendly ownership available at the time.

Given the secrecy that shrouds the nuclear industry the other political criteria employed in arriving at Elstow are a matter of speculation. Among the possibilities are the known friendliness of local MP Trevor Skeet to the nuclear industry (he had even sponsored a private members' nuclear waste disposal bill in Parliament in 1980). Another factor may have been the existing pollution and dereliction from brickmaking in the Marston Vale. The County Council welcomed the importation of waste, including some toxic materials, into its abandoned brick pits as a means of restoring derelict land. NIREX may have perceived the area as a 'pollution haven' where an already deteriorated environment could readily accommodate a relatively innocuous radioactive waste disposal facility.

Bedfordshire proved far from being a politically soft option. Initially, the centralised and closed system of nuclear decision making appeared to provide advantage to the government's proposals. But, the lack of a coherent strategy was a major weakness that was effectively exposed by opponents. In the months and years following the announcement, the local community and NIREX were locked in a conflict which eventually assumed national significance and engulfed the government's radioactive waste policy. In this chapter we have reviewed the issues at the heart of the conflict; the lack of consultation and the absence of a coherent policy. They formed the substance of the political conflict providing the protagonists with scientific, technical and procedural arguments. But success depended on the power of the various interests involved and their ability to deploy political resources at crucial points in the conflict. As the conflict evolved so the institutional, political and policy context changed in response to the power exercised by the participants. It is to this political dimension of power that we must now turn.

NOTES

1. Cleveland County Council, Minutes of the Industrial and Land Committee, 22 September 1983.
2. Pearce, F., 'Britain plans new nuclear dumps', *New Scientist*, 24 February 1983.
3. *Bedfordshire Times*, 10 November 1983.
4. Armstrong, J., 'Democracy and the Sizewell Inquiry', In Blowers, A. and Pepper, D. (eds), *Nuclear Power in Crisis*, (London, Croom Helm) 1987, Chapter 4.
5. Hansard, 25 October 1983, col. 156.
6. Ibid., col. 161.

7. Ibid., p. 158.
8. Department of the Environment (DoE), 'Disposal Facilities on Land for Low and Intermediate-level Radioactive Wastes: Principles for the Protection of the Human Environment', Draft, October 1983, para. 1.1.
9. House of Commons Environment Committee (1986), *Radioactive Waste*, London, HMSO March, para. 75.
10. DoE, op. cit., para. 1.3.
11. Ibid., 1.5.
12. Ibid., para. 5.4.
13. Bedfordshire County Council, Comments on 'Disposal Facilities. . .', April 1984. Cleveland County Council, Response to 'Disposal Facilities. . .', March, 1984.
14. Department of the Environment, interim version submitted to the Sizewell Inquiry.
15. Department of the Environment, 'Disposal Facilities on Land for Low and Intermediate-level Radioactive Wastes: Principles for the Protection of the Human Environment', London, HMSO, December 1984, para. 5.3.
16. NIREX, 'The disposal of low and intermediate-level radioactive wastes: the Elstow storage depot', A Preliminary Project Statement, October, 1983, p. 15.
17. NIREX, 'The disposal of low and intermediate-level radioactive wastes: the Billingham anhydrite mine', A preliminary project statement, October 1983, p. 10.
18. Ibid., p. 1.
19. Ibid., p. 14.
20. The types of ionising radiation are described in Chapter 1. The strength of a radioactive source is traditionally measured in curies (ci) but under the metric system the curie is replaced by becquerels (1 curie $= 3.7 \times 10^{10}$ becquerels).
21. NIREX, Billingham, p. 17.
22. Sizewell Inquiry, Transcript of Proceedings, Day 272, 11 October 1984, p. 52.
23. Ibid., p. 48.
24. Sizewell Inquiry, Transcript of Proceedings, Evidence of P. Critchley, DoE, Day 279, 24 October 1984, p. 20.
25. Beds. C.C. Oxford Clay sub-committee, December 16, 1983, agenda item 6, para. 22.
26. Counsel's briefing for councillors for meeting with William Waldegrave, 15 February 1984.
27. Notes of meeting with William Waldegrave, 15 February 1984.
28. Town and Country Planning Act, 1971, statute 48.
29. Briefing notes for councillors for meeting with William Waldegrave, 15 February 1984, p. 6.
30. Notes of the meeting with William Waldegrave, 15 February 1984. The White Paper referred to is, *Radioactive Waste Management*, Cmnd 8607, London, HMSO, July.

31. Sizewell Inquiry, Day 272, 11 October 1984, p. 53, evidence of Dr R. H. Flowers.
32. Ibid., p. 77, Mr Brooke's role as Counsel to the inquiry was to help the Inspector, Sir Frank Layfield, to address and clarify issues that might otherwise not have been raised.
33. CEGB, 'Feasibility study of shallow land burial of specified intermediate and low level radioactive wastes', Study by Pollution Prevention Consultants Ltd., April 1982, p. 39 (submitted to Sizewell Inquiry).
34. HMSO, *Radioactive Waste Management*, Cmnd 8607, 1982.
35. HMSO, *Nuclear Power and the Environment*, The Government's Response to the Sixth Report of the Royal Commission on Environmental Pollution (Cmnd. 6618), Cmnd. 6820, para. 8, 1977.
36. Sizewell Inquiry, Evidence of Mr G. Wedd, witness for the DoE, Day 100, 25 July 1983, p. 10.
37. Environmental Resources Ltd. (ERL) (1984) 'A review and preliminary technical evaluation of the case against the NIREX proposal for a radioactive waste repository at Elstow' June, p. 4. (ERL were consultants to Beds. C.C.)
38. Sizewell Inquiry, Day 95, 15 July 1983, p. 13.
39. Sizewell Inquiry, Day 279, 24 October 1984, p. 36.
40. Ibid., pp. 36–7.
41. ERL op cit., see footnote 37, p. 4.
42. Hansard, 3 May 1984, p. 636.
43. Ibid., p. 638.
44. Sizewell Inquiry, Proof of Evidence on the Disposal of Low Level and Intermediate Level Solid Wastes, by Dr R. Flowers of the CEGB, November, 1982.
45. NIREX, 'The Disposal of Low and Intermediate-Level Radioactive Wastes: the Elstow Storage Depot', October 1983, p. 3.
46. DoE, *Radioactive Waste Management: the National Strategy*, July 1984, para. 28.
47. NIREX, Second Report, September 1984, p. 8.
48. Environmental Resources Ltd., 'The status of research and policies on intermediate waste management in the UK and other OECD countries', Working Paper No.1, 1984.
49. Ibid., p. 5.
50. Environmental Resources Ltd., 'Review of the capacity of the LLW disposal site at Drigg' Working Paper No. 9, February 1986.
51. ERL, op. cit.; see footnote 37, June 1984, p. 22.
52. HMSO, *Radioactive Waste Management*, Cmnd 8607, para 5.
53. Sizewell Inquiry, Proof of Evidence by R. Flowers (NIREX) on The Disposal of Low Level and Intermediate Level Solid Wastes, November, p. 3.
54. Sizewell Inquiry, Day 95, 15 July 1983, p. 9.
55. Sizewell Inquiry, Day 100, 25 July 1983, pp. 12–13.
56. Ibid., p. 15.
57. Ibid., p. 17.
58. ERL, op. cit.; see footnote 37, June 1984, p. 19.

59. DoE, *Radioactive Waste: the National Strategy*, July 1984.
60. Sizewell Inquiry, Day 201, 27 March 1984, p. 41.
61. The downward revision was revealed by Dr Avery on Day 275 of the Sizewell Inquiry.
62. Evidence of Dr R. H. Flowers given at Sizewell Inquiry, Day 201, 27 March 1984, p. 67.
63. Duncan, A. B. and Brown, S. R. A., 'Quantities of waste and a strategy for treatment and disposal', *Nuclear Energy*, Vol. 21, No. 3, June 1982, pp. 161–66.
64. Hill, M. D., Mobbs, S. F. and White, I. F., 'An assessment of the radiological consequences of intermediate level wastes in argillaceous rock formations', National Radiological Protection Board, 1981.
65. ERL, op. cit.; see footnote 37, June 1984, p. 6.
66. Sizewell Inquiry, Day 279, 24 October 1984, p. 29.
67. NIREX, Second Report, September 1984, p. 15.
69. NIREX, Fact Sheet 4, 'What is a radioactive waste repository?'
69. Sizewell Inquiry, Day 272, 11 October 1984, p. 61.
70. Environmental Resources Ltd., 'Integrity of containment systems for low and intermediate level radioactive wastes', Working Paper No. 5, November 1985, p. 11.
71. *The Times*, 29 August 1984.
72. Environmental Resources Ltd. 'The future disposal of low and intermediate level nuclear wastes in the UK', November 1984, p. 5.7.
73. ERL, op. cit.; see footnote 37, June 1984, p. 13.
74. Environmental Resources Ltd. 'Integrity of containment systems for low and intermediate level radioactive wastes', Working Paper No. 5, November 1985, p. 11.
75. NIREX, Fact Sheet, 'How long can concrete last?'
76. Reported at BNES Conference on Radioactive Waste Management 26–29 November 1984.
77. NIREX, Article in *Plaintalk*, September 1985.
78. ERL, op. cit.; see footnote 70, November, p. 16.
79. Ibid., p. 20.
80. Environmental Resources Ltd., op. cit.; see footnote 37, June 1984, p. 14.
81. NIREX, 'Preliminary Project Statement', October 1983, p. 17.
82. ICRP, Recommendations of the ICRP, Publication 1, 1977.
83. DoE. Proof of Evidence, Sizewell Inquiry, P/20, p. 15.
84. Sizewell Inquiry, Day 271, Evidence of Dr R. H. Flowers (CEGB), 10 October 1984, p. 83.
85. House of Commons, Environment Committee, *Radioactive Waste* (1986); Minutes of Evidence, Bedfordshire County Council, 26 June 1985, p. 467.
86. Environmental Resources Ltd., 'The future disposal of low and intermediate level nuclear wastes in the UK', November 1984, p. 5.8.
87. NRPB, 'An assessment of the radiological protection aspects of shallow land burial of radioactive wastes', prepared by Hemming, C. R., Hill, M. D., and Pinner, A. V., Didcot, NRPB, April 1984, pp. 25–6.

88. The criterion used is *dose equivalent*, the quantity obtained by multiplying the absorbed dose by a factor which takes into account the differences of various types of radiation in terms of the harm they produce in tissue. The *effective dose equivalent*, is the quantity obtained by multiplying the dose equivalents to various tissues and organs by risk weighting factors appropriate to each and summing the products. It is expressed in sieverts. The ICRP advised that the effective dose equivalent from all sources, excluding background, should not exceed 5 millisieverts (mSv) to a critical group in any one year. With the application of ALARA the NRPB considered that the average dose to a critical group would be less than 1mSv per year. See DoE (1984) Principles for the Protection of the Human Environment', p. 13.
89. NRPB (1984), op. cit.
90. NRPB, 'Radiological protection aspects of shallow land burial of PWR operating wastes', prepared by Hill, M. D., and Pinner, A. V. Didcot, NRPB, October 1982.
91. NRPB (1982), op.cit., p. 21; NRPB (1984), op. cit., p. 29.
92. NIREX (1983), Preliminary Project Statement, p. 3.
93. DoE, *Radioactive Waste Management: the National Strategy*, July 1984, para. 21.
94. ERL (1984), 'Comments on the NRPB Report'.
95. Sizewell Inquiry, Day 201, 27 March 1984, p. 42.
96. Department of the Environment, *Radioactive Waste Management: the National Strategy*, July 1984.
97. ERL, 'Comments on DoE's radioactive waste management – the national strategy' (undated).
98. ERL, op. cit.; see footnote 86, November 1984, p. 5.5.
99. NIREX, 'The disposal of low and intermediate-level radioactive wastes: the Elstow storage depot', October 1983.
100. Sizewell Inquiry, Day 201, 27 March 1984, p. 38.
101. Sizewell Inquiry, Day 271, 10 October 1984, p. 91.
102. NIREX (1984), Second Report, p. 15.
103. ERL (1984), 'Suitability of the Elstow site for radioactive waste disposal compared with other options in the UK', Working Paper No. 3, p. 1.
104. ERL (1984), 'Preliminary hydrogeological appraisal of Elstow as a disposal site for radioactive wastes', Working Paper no. 4, p. 1.
105. ERL, op. cit.; see footnote 37, June 1984, p. 11.
106. HMSO, *Radioactive Waste Management*, Cmnd. 8607, 1982, p. 13.
107. DoE, Draft Principles, op. cit., October 1983, para 1.5.
108. Hansard, 5 December 1983, col. 16.
109. DoE, 'Principles for the Protection of the Human Environment', op. cit. December 1984.
110. Sizewell Inquiry, Day 201, 27 March 1984, p. 69.
111. ERL (1984), 'Suitability of the Elstow site for radioactive waste disposal compared with other options in the UK', Working Paper No. 3, p. 6.
112. Ibid., p. 8.

113. Sizewell Inquiry, Day 271, 10 October 1984, p. 88.
114. Pollution Prevention (Consultants) Ltd., 'Feasibility study of shallow land burial of specified intermediate and low level radioactive wastes', Study carried out for CEGB, April, 1982, p. 16.
115. DoE, op. cit., December 1984, p. 19.
116. Ibid., p. 20.
117. NIREX (1984), Second Report, p. 15.
118. ERL, op. cit.; see footnote 37, June 1984, p. 8.
119. Hansard, 5 December 1983, p. 15.
120. *Plaintalk*, May 1985, p. 4.
121. Bedfordshire County council (1980), County Structure Plan, Policy 90, p. 62.
122. CEGB Statement, October 1983.
123. Bedfordshire County Council (1984), 'Oxford Clay Subject Plan', Policy 1, p. 20.
124. Ibid., Policy 4, p. 24.
125. Bedfordshire County Council, Oxford Clay sub-committee, February 24, 1984.
126. Bedfordshire County Council's draft comments in response to DoE's Draft principles for the Protection of the Human Environment, April, p. 4.1c.
127. DoE, op. cit.; see footnote 15, December 1984, p. 16.
128. Ginniff, M. E., 'Implementation of UK policy on radioactive waste disposal', Paper presented to BNES Conference, London, 26–9 November 1984.
129. ERL, Review and preliminary technical evaluation, June 1984, p. 22.
130. Ibid., p. 17.
131. NIREX, Second Report, September 1984, p. 14.

4 The Battle of the Dumps

PART 2: THE COURSE OF THE CONFLICT

Until 1983, Elstow's main claim to fame was as the home of John Bunyan, the radical preacher, who had written *The Pilgrim's Progress* three hundred years earlier. In that allegory he described 'a very miry slough, that was in the midst of the plain . . . The name of the slough was Despond'.[1] The location of this slough is traditionally thought to be near Elstow and in the wake of the announcement of a nuclear dump Bunyan's vision seemed to have become reality for the people nearby. But initial despondency and bewilderment were quickly replaced by a mood of resentment and determination to prevent the proposal. In a conflict lasting nearly four years the local community eventually achieved total victory. In this chapter we shall examine the course of the conflict and the reasons for the outcome.

Elstow is chosen as the main focus of the story for two reasons. First, it was the one site which remained on the agenda throughout the conflict and thus provides an overall perspective. The second reason is pragmatic, the fact that two of the authors were closely in touch with events at Elstow. The conflict can be conveniently divided into three phases. Phase 1 began with the announcement of sites in October 1983 and ended with the abandonment of the Billingham proposal in January 1985. After initial enthusiasm the local protest movement faded. This phase of the conflict was characterised by much technical and procedural debate within local councils and Parliament, among experts and at Sizewell. The second phase beginning in January 1985 ended in February 1986 with the announcement of three other potential shallow disposal sites. During this phase Bedfordshire was on its own as the only named site. This was a period when the local campaign was reinvigorated and the broader issues debated as various reports were published. The final phase lasted from February 1986 until the Government abandoned its plans in May 1987. With four sites now identified a united front was established among the four communities using parliamentary pressure and protest to secure victory.

THE OPENING PHASE: OCTOBER 1983 TO JANUARY 1985

Elstow's Reaction

A victory seemed improbable in October 1983. Indeed, even the leader of the newly formed protest group, Bedfordshire Against Nuclear Dumping (BAND), former police officer Ray Selden, was not sanguine about the prospects,

> I have the feeling that this Government has been thwarted in other parts of the country to dump this waste and is now determined to find somewhere. It is the Secretary of State who will eventually decide. Given that we have nuclear waste and the stuff has to be disposed of he will probably choose Elstow. That sounds pessimistic but it would not surprise me.[2]

By the time the Elstow story opens the government had already suffered two reversals discussed in Chapter 2, first over high level waste disposal sites (in 1981) and second over sea dumping just two months before the Elstow announcement. No doubt defeat had hardened the resolve of government and the nuclear industry, but equally it demonstrated to opponents that radioactive waste plans were vulnerable.

Certainly the 25 October 1983 announcement of the Elstow site came at an unpropitious moment for the nuclear industry. A few days later, on 1 November, Yorkshire TV transmitted to the nation its documentary called 'Windscale – the nuclear laundry'. This struck at the heart of the nuclear industry by suggesting a link between major nuclear installations such as Sellafield (formerly called Windscale) and local leukaemia clusters. The programme provoked a controversy which was to reverberate in the following years and led directly to the setting up of an inquiry under Sir Douglas Black, an experienced medical researcher. Shortly afterwards, in mid-November, divers acting for *Greenpeace* trying to block the discharge pipe from Sellafield into the Irish sea were contaminated by radioactive 'crud'. Some of the slick was washed onto the shore leading British Nuclear Fuels plc (BNFL), the operators of the plant, to close 200 yards of beach for 24 hours. The incident provoked considerable criticism of the plant's operation and its safety record. Levels of radioactivity found in seaweed along the Cumbrian beaches led to warnings to the public not to use them for 20 miles in each direction from the plant. *Greenpeace* was prosecuted and fined for its actions but had the satisfaction of exposing the problems at Sellafield. A report by the Nuclear Installations Inspectorate revealed that the discharge was deliberate and arose as a result of poor management, inadequate monitoring, and deficient safety procedures. As a consequence the company was prosecuted, convicted and fined in July 1985.

The leukaemia scare and the Sellafield discharge coming within a month of the announcement of sites were an unhappy coincidence for the embattled NIREX at Elstow. Its spokesman Dr Ian Blair ruefully confided, 'We are seriously wondering whether we are the victims of an orchestrated campaign... They do us a tremendous amount of damage and do not help our case.'[3] In the weeks after NIREX dropped what the *Bedfordshire Times* called its 'nuclear bombshell',[4] media coverage in the area was intense and opposition determined and virtually unanimous (Figure 4.1). The nuclear dump (as it was universally called) succeeded in uniting councils, communities and local groups and political parties.

Within days of the announcement of the Elstow site a steering group of environmentalists organised a public meeting which was held in Bedford and attended by nearly a thousand people. Speakers included campaigners from Billingham who were already well organised. The meeting demonstrated the level of public support and agreed that the steering group called BAND (Bedfordshire Against Nuclear Dumping) should spearhead the public campaign.

The local press gave the issue prominent coverage. The county newspaper, the *Bedfordshire Times*, announced initially that it would look at all the facts and then decide. Within six weeks it declared 'We believe the Government announcement ... was nothing short of an unmitigated disaster for Bedfordshire'.[5] *Bedfordshire on Sunday*, a freesheet with a local reputation for its investigative political coverage, proclaimed it was placing itself 'unequivocally on the side of those who oppose the dumping'.[6] This strong support was a crucial factor in the later stages of the conflict. The local Press sustained a debate throughout the conflict.

The County Council was a 'hung authority' but political differences were overcome in its unanimous resolution to oppose the dump. The Council called for support from other councils in neighbouring areas; agreed to appoint expert consultants to support the Council's opposition and informed the Secretary of State for the Environment that 'they consider the way this matter has been approached by his Department and NIREX to be unacceptable'.[7] All these points were pursued in the following months as the County Council took a lead in

Bedfordshire Times

First published in 1845 | THURSDAY, OCTOBER 27, 1983 | Price 17p

County stunned as Elstow named for radioactive dump

Nuclear bombshell

NUCLEAR WASTE DEBATE

An aerial shot of the proposed dump site at Elstow Storage Depot.

FULL public inquiries will be held before any radioactive waste is dumped at Elstow.

That was the message this week from Environment Secretary Mr Patrick Jenkin after the shock Government announcement that the village is one of two sites in Britain earmarked for the dumping of nuclear waste.

The dump is on the 454 acre Elstow storage depot which is owned by the Central Electricity Generating Board.

Waste classed as 'low level' and 'short lived intermediate' will be brought to Elstow by road and possibly rail from nuclear installations all over Britain.

The 'low level' waste includes such items as contaminated clothing and will be packed in sealed containers such as barrels.

More serious is the 'short lived intermediate' waste which covers radioactive items like pipes and filters from nuclear power stations.

It will come to Elstow in concrete containers probably by lorry. The waste will continue to be radioactive for at least 300 years.

Both types of waste will be buried in concrete filled trenches about 30 to 40 feet in depth covered with more concrete and earth. The whole process will use several million tons of concrete.

The trenches will stretch across the 454 acre site and officials of NIREX, the nuclear industry's radioactive waste body which singled out Elstow as a potential dump, believe the area will be used a a nuclear dump for at leat 50 years starting in the late 1980's.

About 30 lorry loads of waste would be transported to the site every week and the total tonnage of radioactive items stored at the dump could be in the region of 50 to 60,000 tons.

News of the plan to dump nuclear waste at Elstow leaked out at the weekend and was confirmed in a House of Commons statement by Mr Jenkin on Tuesday afternoon.

In his statement the Min-

● Continued on page 3

United against dump

THE nuclear waste row has united the three political parties on the county council in a common fight against the dumping proposal.

The Tory group's deputy leader, Coun Ian Dixon, says his party will oppose the decision and added: "We will certainly oppose this proposal and demand answers to important questions.

"We must have positive assurances on safety and the environmental impact of such proposals. If we don't get these assurances we will fight the plan to dump nuclear waste at Elstow."

ARROGANT

Labour group leader, Coun John Tizard, said he hoped the council would use "all means available" to resist the proposal. "We regard this as an appalling decision and we will fight it with everything we have. Although I will not endorse a programme of civil disobedience at this stage I will not rule out any action in the long run."

Liberals have also condemned the proposal and group leader Coun Brian Gibbons has pledged full support for a protest campaign.

Ian Dixon ... answers.

John Tizard ... appalling

Brian Gibbons ... arrogant.

Message from the minister

A TOP Government minister this week issued a personal message to the people of Bedfordshire — trust us.

Speaking in the House of Commons Mr Patrick Jenkin, Environment Secretary, said the Government was fully aware of the safety of present and future generations.

He was replying to a question from Mid Beds MP Mr Nick Lyell who had asked for

● Continued on page 3

Facts of the matter

IF anyone tipped a load of rubbish on your doorstep how would you feel? Upset? Angry? Who would dare do such a thing without consulting you?

That is how we feel about the Government's decision to dump radioactive waste on our doorstep in Elstow without prior consultation.

But that is history now. The announcement has been made. Now is the time to look at all the issues involved with a clear head.

It is easy to over react to the nuclear issue. But before emotion gets the better of good sense people must look at the facts. That is exactly what we will be doing in the coming months.

COMMENT

The Beds Times will obtain all the information it can about the storage of radioactive waste. We will give you the facts and then, and only then, will we decide whether to wholeheartedly resist the plan or reluctantly support it.

As the leading newspaper in the community we have a responsibility to reflect public opinion. to do so we need to know your feelings.

Write to us now, Letters, Bedfordshire Times Group, Caxton Road, Bedford, or telephone your views to our Newsdesk on Bedford 63101.

Report by Andy Diprose and Malcolm Stevenson

SOURCE *Bedfordshire Times*, 27 October 1983.
FIGURE 4.1 *Bedfordshire Times 'Bombshell'*

the opposition to NIREX with the full support of Mid Bedfordshire and North Bedfordshire District Councils whose boundaries crossed the Elstow Storage Depot, the site of the proposed repository.

NIREX with a publicity budget of £70 000 for the Elstow proposal distributed the first issue of their news sheet, *Plaintalk*, to every household in the area in November 1983. An advertising campaign had begun emphasising the safety of the project, claiming for instance, that low level waste 'typically gives off radiation comparable to an old luminous bedside alarm clock'. The risks were minimal,

> Each year you've got a one in 7000 chance of being killed in a road accident. A one in six million chance of being fatally struck by lightning. And only a fraction of *that* chance of dying from the nuclear industry's waste disposal.[8]

NIREX also opened up an information centre on two days a week in Bedford town centre and began a series of local public meetings in an effort to allay public fears. The first of these meetings, held in Bedford in November, attracted an all-ticket audience of about 450 who gave a very hostile reception to the panel of directors from the AEA, NIREX, CEGB and BNFL.

Within a few weeks of the announcement the major protagonists, the County Council, BAND, and NIREX, had established their positions. The public meetings reflected the air of bewilderment and confusion that had settled on the area. There was speculation as to why Elstow had been picked. NIREX's Chief Executive Maurice Ginniff proclaimed that 'the Elstow area was the most suitable site in the UK for dumping low and intermediate level waste'[9] though he also acknowledged 'I can't see any council that hasn't had previous experience of radioactive operations wanting to accept a radioactive disposal facility'.[10]

Opponents seized on the possible political reasons for the choice. The Friends of the Earth (FoE's) Energy campaigner, Renée Chudleigh, saw a clear political motive since similar sites in Kent or the salt mines in the Cheshire stockbroker belt had been looked at 'but these may have been rejected because they are more politically damaging'.[11] Similarly, Peter Wilkinson of *Greenpeace* argued that Bedfordshire had been chosen because it was 'politically safe and likely to offer the line of least resistance'.[12] It was an area 'unlikely to offer the sort of protest that has beaten the Government's plans for nuclear dumping in the past'. But, the strength of feeling in the area was aptly captured in the phrase, 'No one is going to ride roughshod over the people of Bedfordshire'.[13]

It was also suggested that one of the reasons for the choice of Elstow 'was the cynical calculation that one of the MPs concerned would be hamstrung by his past support for nuclear power'.[14] This was a reference to Trevor Skeet, Conservative MP for North Bedfordshire, which borders the Elstow site, who confessed he had known of the proposal for two years. He was adopting a 'rational approach' and would 'look in depth at the whole matter' before making up his mind.[15] At the Bedford public meeting he declared, 'It is perfectly natural that you should have anxieties if you are not knowledgeable on the subject.' This equivocal and disdainful stance was in marked contrast to Nicholas Lyell, Conservative MP for Mid-Bedfordshire who told NIREX, 'I tell you frankly your proposals are not welcome to us.'[16] Skeet maintained that panic was unnecessary, 'The event may never happen. . . It could be that the blue Oxford clay is faulted with strata of sand and gravel common in the area or that the water-table constitutes a grave risk. . . Obviously the time is not ripe for decision.'[17] He maintained this position in the face of critics who felt he was undermining the campaign to stop the proposal.

The Problem Nobody Wants

After the initial flurry of excitement a period of waiting began. In the early months an application from NIREX for permission to undertake investigatory drilling was anticipated, first early in 1984, then later, until it became clear that the conflict might last years rather than months. The discussion at Sizewell during 1984 indicated a lengthening time-scale of decision. For the opponents of the proposal this created the problem of sustaining public interest over a long period during which nothing seemed to be happening, although it also provided time for an effective case to be developed against the plans. The initiative remained with NIREX who were able to dictate the timing of the application.

NIREX continued to present their scheme in an elaborate public relations exercise. Fourteen village meetings were held with attendances varying from 15 to around 200 in the village of Elstow itself. Everywhere they met predictable hostility. Five of the local parishes formed an action committee who conducted a 90 per cent household survey which revealed 97 per cent opposed to the dump.[18] NIREX officials conducted a ritual defence of their case at meetings and in the press.

Public opinion as expressed in the media and at meetings was a combination of incredulity and resentment that Elstow should have been chosen. The following comments on NIREX from John Greenway, the vicar of a local parish, sum up their feelings,

> It came across to me that they were in deadly earnest. They were not pretending. They truly believed they would be providers of boon and benefit to mankind if they were allowed to go ahead.

SOURCE Bedfordshire Against Nuclear Dumping

> Above all it seemed to me that they regarded anyone who doubted this as either unenlightened or perverse, and not having a viewpoint which should be listened to with respect.[19]

NIREX officials betrayed an air of injured innocence and scientific disdain in their responses to such attacks. But they maintained their willingness to debate the issues,

> The nuclear industry does make considerable efforts to both discuss and explain its technology. For example, if John Greenway, or anyone else as an individual or in a group, would like to discuss the subject matter with ourselves they have only to ask. Furthermore, should they feel that our scientific understanding may be biased and therefore wish at the same time to involve others who might take an opposing view then we would welcome such an exchange of opinion.[20]

At first NIREX had seemed unwilling to share a public platform with opponents. The early meetings became a confrontation between NIREX and the audience. The first public debate between the sides occurred in April 1984, and in September, Central TV transmitted a nationally networked programme called 'The Problem Nobody Wants' featuring NIREX and anti-nuclear experts. In it Dr Lewis Roberts, Chairman of NIREX, claimed, 'People have jumped to the conclusion that we are hell bent on coming here, but that is not the case' thus conceding that alternative sites would have to be taken into account.

Early Skirmishes

During the first year after the announcement the opposition had made some inroads into the NIREX case. In February 1984 London Brick, encouraged by the County council, applied for planning permission to use the site for brickmaking (see Chapter 3). The site had been requisitioned during the war and London Brick claimed 'Our rights in the land still exist.'[21] 'NIREX's intention to use it for the storage of nuclear waste could render the Elstow area useless for brickmaking purposes... We believe that no decision should be taken on the NIREX proposal without the importance of this area to brickmaking being taken into account.'[22] Both uses could be given planning permission and the ultimate use would depend on ownership, something the CEGB were unlikely to surrender. But, at least, an alternative use of national importance had been demonstrated.

NIREX were also challenged on the depth of clay at the site. Using borehold findings from Cambridge University showing a depth of clay at the site of 15m a local district councillor, Mike Dewar, argued that this was less than the depth quoted by NIREX for a shallow repository. NIREX official Dr Ian Blair replied, 'We are now considering burial at a depth of 9 metres–12 metres covered by a landscaped mound of earth.'[23] But 15m was consistent with the NIREX project statement and, as the County Council pointed out, various depths could be suggested or assumed. It was later discovered that 15m was a maximum at the site (see Chapter 3). Although the issue was debatable NIREX displayed some uncertainty which intensified suspicions and damaged their credibility.

The County Council made early headway on the question of alternative sites. In February 1984 a delegation of county and district councillors and officials met the Under Secretary of State for the Environment, William Waldegrave, in London. He conceded that, at the first public inquiry, NIREX 'would be required to name a number of alternative sites and say why they considered Elstow to be the most suitable'.[24] He was prepared to consider whether work should be undertaken at the other sites so that, if Elstow failed, there would be other possibilities. It was also agreed that the first inquiry should consider the technical feasibility of the proposal and also 'whether it was worth having a major inquiry in two or three years' time'.

The County Council also decided to add a final policy to the County Structure plan which provides the statutory strategic policies governing development. The proposed policy stated that 'The County Council will oppose any proposals for the location of nuclear waste disposal facilities within the County, including proposals for the exploration of any sites for the possible disposal of nuclear waste; furthermore the County Council will oppose the transportation of nuclear wastes within or through the county.'[25] Explaining its position, the Council argued that since there was no coherent strategy, proposals for specific disposal sites were premature and, in any case, sites remote from human settlements could not be found in a small, urbanised county such as Bedfordshire. Public consultation predictably showed overwhelming public support for the policy. The policy had no chance of approval by the Secretary of State; rather it was part of a ritual demonstrating public opposition in a variety of ways.

Another part of this ritual was the legal challenge presented by the Council to NIREX in the summer of 1984. NIREX had indicated an intention to put down three 2 inch boreholes at 60ft depth at the site in order to develop measurement techniques and to obtain samples for a study of the soil chemistry. The County Council served NIREX with a temporary injunction to prevent the operation on the gounds that it was a form of development and would, therefore, require planning permission. At a high court hearing the judge concluded that 'it is a very trifling matter indeed'[26] and declined to grant the injunction. The Council appealed and the appeal judges confirmed that 'all that is wanted is to bore three – true deep but tiny – holes in a 400-acre site'.[27] While the judges acknowledged there was an interesting point of law on the question of whether planning permission was required they did not feel the appeal should be upheld.

Having failed to win an injunction the County Council resorted to serving Stop and Enforcement Notices on NIREX the moment they entered the site on 4 July. A Stop Notice could only take effect after three days. Although the contractors encountered difficulties with their equipment and were unable to complete the drillings they were able to gather sufficient information for their purposes. This episode, perhaps unimportant in itself, demonstrated the determination of the local authorities to use every avenue to stop NIREX.

The unity of the opposition, embracing all political parties and including those who favoured nuclear energy, depended on the narrow focus of opposing the NIREX proposal which locally nobody wanted. In March the County Council, together with the four Bedfordshire District Councils, declared a 'united front'. The protest group BAND also emphasised it was not a party political organisation, 'It is simply concerned with the proposal by NIREX to investigate radioactive waste dumping at Elstow'.[28] The ability to mobilise unanimous public and party support was an enduring asset throughout the conflict. The one dissident voice remained that of the local MP Trevor Skeet. He argued that the County Council 'don't know anything about it'. He was 'taking the pragmatic approach. I have no opinion . . . I am waiting for evidence as to whether it should be thrown out completely or considered further'.[29]

During this early phase the County Council perceived its role as a largely technical one, gathering evidence with which to defend its case at the imminent public inquiry. Accordingly, the Council appointed as technical consultants Environmental Resources Ltd (ERL) who demonstrated a comprehensive knowledge of the issues, perceived the weaknesses in the NIREX case and showed a commitment to defeating the proposals. The two consultants assigned to the NIREX proposal were Richard Johnson and Tim Roberts who combined scientific authority with a determination to win the political battle with NIREX. Alan Wheeler, a county planning official, was assigned to the NIREX project and for the next three and a half years effectively coordinated the county's technical and political case.

Apart from such events as the delegation to the minister and the injunction, the councillors, officials and consultants held their discussions in secret in the belief that they would soon be called upon to defend their case at a public inquiry. The Council was content to let BAND and the media maintain a high public profile. Within a few weeks of the announcement, however, there were signs that BAND's public campaign was beginning to falter.

After the initial publicity, organisation of the campaign was left to about forty activists. BAND established a bureaucratic structure with an Executive Committee to which four working groups – technical, campaign, fund raising and publicity – reported. Its objective was limited to opposing NIREX, not taking on broader nuclear issues. It would use 'parliamentary democracy' and 'grass roots democracy' 'to mobilise public opinion'.[30] By February 1984 meetings had been held with other groups such as Billingham Against Nuclear Dumping, two issues of a newsletter *Band Wagon* had appeared, and a petition had been launched aiming to canvass all 150 000 electors in North and Mid Bedfordshire. Thereafter, as it became clear that a long campaign was in prospect, energy began to evaporate. By May Ray Selden, the original inspiration behind BAND and its Chairman, confessed,

> we look as though we're running out of steam . . . some people may think the campaign has foundered. I suppose it was inevitable . . . The difficulty is keeping the debate going while there appears to be little or nothing going on.[31]

Two months later Ray Selden resigned and was succeeded by Dr Jerry Fitch, a sociology lecturer, who was to become a prominent and effective leader throughout the remaining years of the conflict. Under Fitch BAND abandoned its formalised structure and redirected the campaign to attracting more public attention since 'the public were in danger of becoming too complacent'.[32]

All Over by Christmas?

A year after the announcement complacency and apathy were the problems facing opponents of the Elstow site. In a four-page feature on the first anniversary of the announcement the *Bedfordshire Times*, concerned that 'the message is not getting across sufficiently loudly and clearly', declared its 'total opposition to the plan'.[33] *Bedfordshire on Sunday* likewise warned 'We must not let ourselves be lulled into a false sense of security.'[34] The government had promised a decision on whether to proceed with the Billingham and Elstow proposals by around Christmas and it was time for a 'big push to convince the Government that it will be better to look elsewhere'. The *Bedfordshire Times*, in cooperation with BAND, began a regular printing of petition forms and by January 1985 the petition containing 5000 signatures was delivered to Downing Street.

By the end of 1984 the County Council, too, was concerned that the campaign was lacking impact,

> In the past year the local authorities have maintained a very low profile, refusing to deal with NIREX in public or private . . . and making no public pronouncements other than a few short press releases. As a result, the field has been left very much to NIREX.[35]

It was agreed that the Council should take a more prominent role by meeting NIREX, linking with potential allies, coordinating opposition, raising the issues at a national level and providing practical assistance to BAND in the effort to encourage local opposition.

The attempts by BAND, the local media and the County Council to breathe new life into the flagging campaign came too late to influence the Government's decision. On 24 January 1985 the Secretary of State for the Environment, Patrick Jenkin, announced that he had asked NIREX 'to select and announce two further sites for investigation in addition to Elstow'.[36] Thus, Elstow remained as the only named site for a shallow repository. All that had been achieved was the prospect of other sites being nominated. Mr Jenkin also announced, 'To avoid a further period of uncertainty at Billingham, I have invited NIREX, and it has agreed, not to proceed further with that site'. Billingham's total victory meant that, until further sites were named, Elstow would have to fight on alone.

Bedford Mourns while Billingham Celebrates

The success of the Billingham campaign had been envied in Bedford. Billingham had benefited from prior knowledge of the proposal. An article in *New Scientist* in February 1983[37] had stirred rumours in the area and in March Imperial Chemical Industries (ICI), the owners of the abandoned anhydrite mine, confirmed they had been approached by NIREX. Cleveland County Council reacted stating that 'in no circumstances will it countenance the deposit of nuclear waste in any part of Cleveland' and conveyed its view to ICI and other bodies.[38] By September NIREX revealed that the Billingham mine was one of the shortlisted sites. In the weeks before the government announcement the local Press were already campaigning against the project. On 20 October, five days before the proposal was confirmed in Parliament, a packed protest meeting was held in Billingham Technical College at which a petition was launched.

Once the announcement was made the campaign intensified. Within three days Billingham Against Nuclear Dumping (BAND) was formed. Its objectives were to prevent disposal of radioactive wastes in Cleveland; to campaign for a national policy of safe storage for wastes; and to aid any other community threatened by the disposal of radioactive wastes. Its leadership included the Rev. Peter Hirst of Billingham, the 'father-figure' of the campaign; a full-time Secretary, Ian Wilson; Cllr Maureen Taylor, the chair of the Cleveland planning committee; and Cllr Jimmy Vaughan of Stockton Borough Council, deputed to liaise with the Elstow campaign. The newly elected Labour MP for Stockton North, Frank Cook, whose constituency included Billingham, provided informed, uncompromising and determined leadership both locally and in Parliament. On 1 November NIREX officials were shouted down during a debate before an audience of 500 at the Billingham Forum with 600 listening outside the hall.

The scale of the opposition was impressive. The district councils of Stockton, Hartlepool, Middlesbrough and Langbaurgh followed Cleveland's lead in opposing the proposal. Stockton declared the decision, 'demonstrates a monumental mistake in backstairs decision making to the exclusion of the normal decision making process'.[39] The local councils gave financial support to BAND and an office was provided in Billingham. The opposition embraced MPs and councillors from all parties on Tees-side. The MPs, led by Frank Cook, met ministers, put down motions and called debates in the Commons. In Billingham the public turned out in large numbers at public meetings and 3000 joined a four-mile protest march around the ICI plant in November. Among other events organised by BAND were a protest drive to Sellafield, picketing of York railway museum where a waste flask was displayed, and lobbying at the Conservative Party Conference. Press coverage in the Tees-side newspapers was extensive.

Around 40 000 letters were delivered to Patrick Jenkin and a petition of 83 000 signatures was presented to the Prime Minister in May 1984. Campaigners displayed an 83 ft banner, one foot for every thousand signatures, outside Parliament.

Billingham's campaign had an emotional energy generated by its leadership and reciprocated in the public response. Unlike Bedfordshire where there was a dual leadership between BAND and the County Council, Billingham BAND worked closely with the local councils who played a more peripheral role. The emphasis was on mobilising public opposition rather than the preparation of detailed technical arguments. The feelings of Billingham were summed up in the *Northern Echo*,

> The opposition may be irrational, it may be unscientific, but it must be heeded because the people of Billingham have first claim on their town. They have invested their lives in it.
> It is not Billingham which is unloved and unloveable. That description best fits the grotesque plan to turn it into an atomic age wasteland.[40]

The contrast between Billingham's dynamic campaign and the lack-lustre efforts of Bedfordshire reflected their different political cultures. Billingham was part of the industrial North, dominated by Labour politics with heavy industries, high unemployment and close-knit communities providing a solidarity against threats from the outside such as NIREX. Elstow was in the South in a relatively prosperous, politically conservative, area with an increasingly mobile, middle-class population lacking deep identity with the area. Ian Wilson, Billingham BAND's secretary summed up the contrast, 'It's the difference between the *Guardian* readers of Bedford and the *Sun* readers here'.

There were other differences too. The proposed repository already existed in the form of the deep (430 to 920 ft) anhydrite mine formerly the basis of the Billingham chemical plant. Therefore, 'no major underground excavation work is envisaged'.[41] Existing rail and road links could be used. Moreover, the entrance to the mine was in the middle of a conurbation of half a million people rather than on the rural fringes of a country town. It was underneath the town and the fear of an explosion was great. 'I do not doubt that many people in Billingham thought that there was a proposal to put a bomb underneath their houses' said Ian Wrigglesworth, MP for Stockton South[42]; though it must be said the people had lived close to the danger of a major chemical plant for many years. The wastes destined for Billingham were long-lived intermediate alpha-emitting wastes, altogether more dangerous than those intended for Elstow.

There were some political similarities between the two proposed sites. Tees-side, like the Marston Vale in Bedfordshire, was an area of environmental disadvantage. It was one of the worst pollution blackspots in Britain. It had 36 'major hazards', places where dangerous substances were stored. The decline of its industrial base in steel and chemicals had made Cleveland Britain's unemployment blackspot too. But, for NIREX, the socio-economic character of the area was an advantage. 'It believed that the area's existing concentration of industry, and the history of mining at Billingham, would help to make its proposal acceptable to the local community!'[43] If that were truly so, the naïveté of NIREX was, indeed, palpable.

There was a crucial, and as it proved decisive, difference in the ownership of the two sites. The CEGB owner of Elstow, was a partner in NIREX; ICI, owner of the Billingham mine, was a multinational company with a substantial stake in Tees-side. ICI had long connections with the nuclear industry. The company had provided some of the key scientists who planned Britain's atomic energy project, including Sir Christopher (later Lord) Hinton who developed Windscale in the late 1940s. It had strong links with government and Sir Robin Ibbs, representing Billingham's agricultural division on the ICI Board, had been chairman of the Government's Think Tank. NIREX expected, and initially received, cooperation from the company. Once the government's announcement had been made and the full force of opposition unleashed the company equivocated. While it had 'no wish to pursue this proposal which is unrelated to its traditional business interests . . . ICI would be willing in the national interest to co-operate with NIREX to work at an arrangement to make the mine available'.[44] The pressure on the company intensified with threats of strikes over the proposal and worsening relationships between the company and the community. By March 1984 the company had made up its mind: 'Having carefully considered the implications that the proposal would have for its business, ICI has concluded that it would not be in the company's best interests and is therefore opposed to it'. The announcement was first made at a public meeting attended by the bewildered NIREX officials.[45] The company would oppose the plan at a public inquiry, prevent the use of the mine for the project unless forced to do so, and, of greatest immediate importance, refuse NIREX entry for purposes of inspecting the mine's suitability.

ICI's opposition was crucial since NIREX had no powers of compulsory purchase or access and with the company and community ranged against them would find it extremely difficult to pursue the proposal. NIREX acknowledged that without access it would be 'very difficult' for them. The Billingham campaign continued, culminating in a visit to the area in October by William Waldegrave who was clearly impressed by the scale of opposition and indicated a decision would be made by Christmas. When the Government's capitulation duly came in January 1985 the Secretary of State, Patrick Jenkin, argued the decision was to avoid uncertainty while studies on conditioning and containing ILW were undertaken. He claimed, too, that, 'There is not the same urgency about the identification and construction of a deep facility.'[46] However the decision was presented it was clear it was a political triumph for the people of the Billingham area. Although it was a Labour-dominated area its powerful industrial interests and political clout had clearly alarmed the Government. With the earlier abandonment of plans for high level waste disposal and sea dumping (see Chapter 2), Billingham was the third defeat for the Government's nuclear waste policy. William Waldegrave did not minimise the scale of the reversal,

> I have never met a scientist interested in this subject who did not say to me that Billingham was a most magnificent site in terms of scale, depth, geology and so forth.[47]

Neither Victory, nor Defeat

There was dismay in Bedfordshire that outright victory had not been achieved as in Billingham. The Bedfordshire campaign had, so far, failed to capture the public's imagination locally, let alone nationally. The focus of attention nationally had been on events at Sellafield and Sizewell. Sellafield was rarely out of the news. In July 1984 the Black Report concluded that the link between high leukaemia levels and the Sellafield plant was 'not one which can be categorically dismissed, nor on the other hand, is it easy to prove'.[48] Predictably such a 'qualified reassurance' failed to quell public alarm. The Sizewell inquiry continued throughout 1984, and, on several occasions, took evidence on radioactive waste. In August a cargo ship, the *Mont Louis* carrying uranium hexafluoride from France to the Soviet Union collided with a ferry and sank off Belgium raising fears of contamination of the English Channel.

While these events fuelled the controversy surrounding the nuclear industry in general, they had little impact in Bedfordshire. The local campaign against the nuclear dump refrained from controversies likely to disrupt its fragile unity. Conversely, the Elstow issue was unable to compete with Sellafield or even Billingham for national attention. It was occasionally mentioned in the national press and the Central TV programme gave it a fleeting national prominence. But, in general, the energies of the national media and of the national environmental lobbies were absorbed in the developing dramas at Sizewell and Sellafield.

Despite the failure to make a national impression, Bedfordshire's campaign made some progress during this opening phase. There was cross-party opposition to the dump (with the singular exception of Trevor Skeet); BAND had been established; the local media had backed the campaign. But NIREX and the government retained the initiative. The County Council, working in secret for tactical reasons, had begun to develop, with its consultants, an impressive technical case for use in the impending inquiry. Persistent pressure had ensured that other sites in addition to Elstow would eventually be investigated.

Thus, as the second phase of the conflict began in January 1985, the foundations of opposition had been laid. So far the opposition to the Elstow proposal had made little obvious impact on government policy and participation in decision making had been restricted. Hitherto Bedfordshire's campaign had focused on NIREX. But, Billingham had shown that a well-organised campaign, combined with propitious circumstances, could yield victory. It was possible to tilt the balance of power away from NIREX through political action which focused more directly on Parliament. That was a lesson which was not to be lost on Bedfordshire.

THE MIDDLE PHASE: JANUARY 1985 TO FEBRUARY 1986

A Battle We Must Win

Patrick Jenkin's statement to the House of Commons on 24 January 1985 contained four issues which were to determine the course of the conflict in the following years. Of most immediate importance was the abandonment of Billingham and its implications for Elstow. Second, was the decision to investigate other sites which later broadened the opposition to the Government's plans. Thirdly, the Secretary of State announced that he had decided to revise the planning procedures and proposed to ask Parliament to give planning permission for geological investigation 'by way of a special development order'.[49] Opposition to this proposal would eventually provoke the climax of the conflict and ensure that national attention was focused on the issue. Finally, Mr Jenkin indicated that further research would be undertaken including a review of the technical options for storing and disposing of intermediate wastes. The options for ILW disposal became a major factor in the denouement in 1987.

In Bedfordshire the analysis of the state of the campaign was cathartic. In the opinion of the local press it had been a 'tactical error' to believe that NIREX 'would only be beaten by scientific facts, not petitions and emotions'.[50] The dual roles played by the County Council operating largely in secret, and BAND providing a sluggish momentum to the public campaign, had not worked. A mere 5000 signatures to a petition 'would be interpreted as a sign of lack of opposition'. The failure of Trevor Skeet to commit himself to the local campaign had undermined the unity of the opposition. It was the 'ferocious and superbly organised campaign of opposition'[51] which defeated the Billingham project despite its ostensibly attractive technical properties.

Billingham's victory was the spur for a rejuvenated campaign in Bedfordshire. Jerry Fitch commented, 'We shall take heart from this and strengthen our resolve. A united opposition can stop NIREX and the Government'.[52] The immediate impact was for the four local newspapers to declare 'an unprecedented display of unity'[53] to defeat the dump and jointly encourage the public to sign the petition. The County Council took a public lead by circulating 250 000 leaflets to every household in the county urging people to 'Say No!' and setting out the reasons for opposing the dump and what could be done to prevent it. A coordinator, Richard Lemon, was appointed to liaise with BAND and to organise the County's Council's high profile campaign. Financial and practical assistance was given to BAND and advertisements to counter those from NIREX were prepared for the Press. By the summer an old mobile library had been converted into a campaign vehicle carrying publicity materials to villages and towns throughout the country. Most conspicuous of all were the banners proclaiming NO TO NIREX NUCLEAR DUMP carried on the Town and County Halls in Bedford.

The renewed enthusiasm of the campaign leadership soon

transmitted itself to the general public. There was a surge in the numbers signing the petition and it passed the 50 000 mark in April. A packed meeting of 500 people in the Corn Exchange, Bedford, in February was addressed by representatives from Billingham, national environmental groups *Friends of the Earth* and *Greenpeace* and local community leaders. Billingham BAND offered advice and support. The strength of opposition in Bedfordshire was given statistical confirmation in a Marplan Poll which revealed 91 per cent against the dump.[54] The poll also found that 53 per cent felt more unhappy about the proposed dump as a result of seeing the adverts and leaflets put out by NIREX. On a more political note 52 per cent, said they were less likely to vote Conservative at the next General Election if the dump were to go ahead and 86 per cent felt local MP Trevor Skeet should do more to fight the Elstow plans.

Mr Skeet's position remained enigmatic. In the Parliamentary debate following the January announcement he asked for assurances that 'the best site will be selected . . . the prime consideration will be public safety ... and ... there will be further research into the behaviour of radionuclides ... to ensure that they do not enter the food chain'.[55] The minister had little difficulty in agreeing. In March Mr Skeet called on the government to consider using a Planning Inquiry Commission. His speech in the House of Commons revealed his comprehensive knowledge of the history of the radioactive waste issue and the potential options available but also displayed his irritation at the campaign against the dump in Bedford,

> It is bound to bore even the most persistent sympathisers. More extreme stories have to be discovered in rapid succession to maintain local interest, and the facts that there are have become immersed in emotion and alarm. I have consistently adopted the policy of waiting for the facts before reaching judgement rather than prejudging the case without the evidence.[56]

The debate provided the other Bedfordshire MPs with a platform to urge the case for other options and to criticise the role of the Department of the Environment. Replying for the Government William Waldegrave predictably turned down the idea of a Planning Inquiry Commission, and denied any conflict of roles. He expressed the Government's dilemma over alternative options,

> we may sometimes have to tell people that we will not deny a rumour that a particular option is being considered. We will tell the

truth . . . Such an attitude may create anxieties in the country, but if we take the easy way out and rule out all anxieties early on, we may find that we have ruled out investigations of options that might, perhaps, be feasible.[57]

It might equally be said that the creation of public anxiety led directly to the ruling out of the options of sea dumping and the Billingham anhydrite mine.

Experience had demonstrated that radioactive waste was foremost a political issue. This point was conceded by NIREX when they confronted the County Council, local MPs, BAND, and environmental groups at a meeting in Bedford in March. John Baker, a member of the Boards of both the CEGB and NIREX, told the meeting, 'The final choice will depend on social questions and judgements of political acceptability.' Dr Lewis Roberts, the Chairman of NIREX, confirmed that Elstow would be judged on political criteria, 'On the information we have it is a site worthy of close examination, but it is not my decision. We have to adopt what they (the Government) think.'[58]

Up to this point NIREX had been the major focus of attention by protesters. As William Waldegrave observed, 'It is easy to blame the messenger when one does not like the message.'[59] The County Council and BAND recognised that the campaign should now be directed to Parliament and the government where the crucial political decisions would be taken. BAND switched its attack, beginning with a letter to all MPs and the County Council, prepared to take its case to the minister and to the House of Commons Environment Committee which had begun a study of radioactive waste. By the summer of 1985 the reinvigorated and united campaign in Bedfordshire appeared to have wrested the initiative from NIREX and was opening up its offensive at Westminster.

Not Here, Not Anywhere, Not Yet

Bedfordshire's dismay at the proposed use of a Special Development Order (SDO) to give planning permission for investigatory works, thus avoiding an early local planning inquiry, was the main issue when William Waldegrave once again met a deputation of councillors in June. When the Elstow site was first announced the Secretary of State had indicated planning permission might be required for investigatory works and, if so, he would call in the application for determination after a public inquiry. At this first meeting with Bedfordshire's representatives in February 1984 William Waldegrave had discussed the scope of this first inquiry (see Chapter 3). Now it was proposed to by-pass an inquiry which the County Council agreed 'denies the local authorities and the community a chance to debate the suitability of the site . . .'.[60]

In response, Mr Waldegrave adroitly claimed that the change in procedure had occurred as a result of their previous meeting. He had taken the point about the need for 'real comparison between one site and another' and felt that the anxiety of blighted communities would be reduced if the decision time were shortened and there were one inquiry. He indicated there would be no preferred site when the shortlist was announced and 'there will be no preference for the Inspector or the Department until a major inquiry had been held'.[61]

Among the other issues raised with the minister was the absence of a fully debated strategy for radioactive waste management. This was the main argument put to the members of the House of Commons Environment Committee by a Bedfordshire delegation led by Councillor Andrew Blowers on 26 June. The delegation was anxious to impress their view that specific sites had been chosen in advance of any debate on general principles. Cllr Blowers argued that the stage had not yet been reached 'when any community should be saying, "No, not here" because it seems to us that the overall options have not been debated'.[62] Cllr Janice Lennon used the analogy of a gypsy caravan site to demonstrate how general criteria are laid down before a specific site is selected,

> whilst the public at large might not agree to the gypsy site being at the end of their garden, at least they can see some logical process whereby you have arrived at the decision to put the gypsy site at the end of their garden instead of someone else's garden.[63]

No national criteria had been set out before shallow burial and the specific site at Elstow had been identified.

In its submitted evidence to the Committee the County Council pointed out that Billingham had been dropped to avoid further uncertainty while research was undertaken into conditioning and containment of intermediate wastes: 'No mention was made of the fact that intermediate wastes are also destined for Elstow or that the necessary research would give rise to a further period of uncertainty in that community.'[64] Furthermore, since most ILW came from Sellafield it seemed that cost and environmental factors 'suggest that a single deep mined facility, or possibly burial under the sea bed'[65] would be more practicable disposal options than shallow burial at Elstow. In a subsequent submission the Council emphasised the dearth of basic research in such areas as repository design, waste packaging, radionuclide migration, ecological impacts and safety assessment, and underlined the problems associated with existing shallow disposal methods. It was essential to pursue deep burial both for the greater geological protection it afforded and 'also since public confidence is likely to be immediately greater in such a development programme as opposed to shallow burial'.[66]

The Best Practicable Environmental Option

The County Council had stressed the case for alternative radioactive waste management options. Various alternatives were being canvassed by those involved in the campaign. Nick Lyell, Elstow's MP and a supporter of nuclear power, had identified a case for an isolated coastal site. At an inland site radionuclides 'could find a pathway to man, without the protection of dispersal by the tide'.[67] BAND, along with national environmental groups such as *Friends of the Earth* and *Greenpeace* favoured surface storage at nuclear sites where wastes could be monitored and retrieved if necessary. Among the advantages of storage were that 'The waste can be regularly monitored, and it is readily accessible should any problems with the repository arise. Furthermore, any delay in final disposal results in a decay in the level of radioactivity in the waste . . .'[68] All these options, opponents argued, were to be preferred to shallow burial, an untried, uncertain and unnecessary technology.

The government and its advisers were still committed to shallow burial. The Radioactive Waste Management Advisory Committee (RWMAC) had, in September 1984, urged the adoption of Special Development Order (SDO) procedures and, in its 6th Annual Report, supported the case for developing a further LLW disposal site as a matter of urgency and rejected indefinite storage as an acceptable management strategy.[69] Government policy on radioactive waste secured predictable endorsement from the Department of the Environment (DoE). Arising from the abandonment of sea dumping a review of policy had been initiated in December 1983 by a Government–Trades Union Congress (TUC) Committee under Professor Fred Holliday, a geologist from Durham University. The Report, published in 1984, noted that 'So far there has been no serious attempt in the UK to compare sea dumping and land options' and as 'a matter of urgency' recommended the government to undertake 'a comparative assessment of all disposal and storage options with a view to establishing the "Best Practicable Environmental Options"' (BPEOs).[70]

The DoE's BPEO assessment was released in draft form in November 1985 and published in March 1986.[71] Produced internally by the Radioactive Waste (Professional) Division, the Report was described by some of its critics as, 'in parts almost unintelligble'.[72] But there was the danger that 'its very complexity and opacity may begin to confer an authority and permanency upon it which is neither justifiable or desirable'.[73] The Report assumed different nuclear power scenarios although all included reprocessing of Magnox fuel. An assessment was then made of the BPEO for over 70 different types of ILW and LLW using a multi-attribute decision analysis procedure which 'must be considered as somewhat experimental'.[74] The method described in the Report involved applying different weightings to emphasise the relative priorities given to reducing radiological risk or to limiting economic costs. The weightings were based on different judgements reflecting 'different social and ethical preferences' (p. 27). The Report stated that disposal routes could be chosen for each type of waste to ensure that all five disposal options considered (sea disposal, shallow burial, deep burial, off-shore disposal from bore holes and indefinite storage) are 'practicable and ... financially realistic' (p. 51).

The study concluded that storage was the least attractive option unless the need to retrieve wastes was overriding. The assessment demonstrated that 'near-surface disposal is the BPEO for over 80 per cent [90 per cent in draft report] by volume of all the waste considered' (p. 51). Therefore, 'The BPEO for most LLW and some short-lived ILW is near-surface disposal, as soon as practicable, in appropriately designed trenches' (p. 3), precisely what was, at that time, being proposed by NIREX and the government. Bedfordshire's technical consultants, ERL, pointed to the contestable assumptions such as the existence of 'reliable data' on shallow burial from 'practical experience'. The methodology was flawed in that shallow burial and sea disposal were included in each of the options being compared. The weightings were not explained and relied on the DoE's perceptions of society's views. But, from the government's point of view, the Report achieved its purpose. ERL concluded, 'Its raison d'être might appear to be a means of confirming policy choices rather than identifying BPEOs as envisaged by the Holliday committee'.[75]

Bedfordshire's local opposition to shallow burial was reinforced by growing criticism at the national level. In a hostile reaction to the BPEO report the TUC argued that 'Radioactive waste should be accessible and open to human control at all times, to allow present management policies to be reversed, if necessary, at some future date'.[76]

At the end of 1985, the findings of the House of Commons Environment Committee were leaked to *The Times* (16 December). The Committee were highly critical of the nuclear industry's lack of coherent policy, its 'defensive secretiveness', lack of public accountability and the primitive state of existing disposal sites. There were signs that Bedfordshire's reservations were widely shared.

In Search of Fresh Sites

From the point of view of government and NIREX the BPEO Report offered a lifeline in their effort to legitimate the shallow burial option. Opponents had succeeded in opening up the debate, thus preventing a narrow focus on the merits of specific sites. As confidence rose BAND geared up for the offensive. By October 1985 over 90 000 people had signed the petition. The final total when the petition was closed in January 1986 was 94 025. New personalities appeared on the scene. One, Mieke Hinchcliffe-Wood, a Biggleswade housewife with experience of Dutch protest movements, provided the organisational flair and commitment which would sustain the campaign through its final victorious phase and was soon visible in a Fun Day held on Elstow Green. Another, Jim Eldridge, a writer and author of radio plays, took charge of the non-violent direct action campaign. It began by distributing leaflets outside the DoE's headquarters in Marsham Street and was followed by dumping 45 gallon barrels marked 'Radioactive' but containing soil on the DoE's doorstep. Eldridge told a DoE official, 'We want you to accept these five barrels as a present in return for the thousands of barrels of lethal waste you are offering us'. At the end of 1985 *Friends of the Earth* pulled a giant Christmas cracker containing four barrels outside Marsham Street, and in, January 1986, a group from BAND managed to lock four of the five main entrances into Marsham Street.

BAND's efforts were not limited to protests and local consciousness raising. Early in 1986 BAND published its own evidence against shallow burial.[77] This claimed that the designation 'low level wastes' was misleading since among the wastes destined for Elstow were long-lived isotopes such as plutonium. The Report produced evidence to demonstrate the problems of concrete corrosion to refute NIREX's confidence in containment. Finally, the Report considered the evidence for linkage between nuclear facilities and leukaemia which, it was claimed, had been ignored in the Black Report. Although BAND's report was dismissed as inaccurate by Trevor Skeet and NIREX criticised it for introducing 'fanciful allegations',[78] it indicated that BAND intended to present its own technical case at any future inquiry.

There were changes, too, in the approach of the other groups engaged in the conflict. Within a few months Bedfordshire County Council lost the services of consultants Richard Johnson (through untimely death) and Tim Roberts (through resignation) both of whom had provided the technical expertise and advice behind the campaign. ERL's replacement team included Peter Floyd, whose extroverted anti-nuclear commitment suited the shift from technical to political emphasis that characterised the final phase of the conflict. The replacement of Patrick Jenkin by Kenneth Baker as Secretary of State for the Environment in September 1985 provided the opposition with a new target. 'If the campaign wanes now Mr Baker could be almost oblivious to the fact that it existed in the first place.'[79] Mr Baker was bombarded with 4000 postcards with the message 'No to NIREX' and, along with the Prime Minister received Christmas wishes from the County Council with the request, 'Make ours a Happy New Year by saying NO to the NIREX proposal to dump radioactive waste at Elstow!' At the end of the year Tom McInerney was appointed as Chief Executive of NIREX and subsequently proved more skilful at public relations than his predecessor Maurice Ginniff.

In the national and local press there was speculation about the alternative sites for shallow burial. Among the possible candidates suggested were places as far apart as Teesdale and Weardale in the North East and Lyme Regis in Dorset. There was considerable interest aroused by a list of sites provided by *Friends of the Earth* which included Arncott and Harwell (home of NIREX) in Oxfordshire, Newport Pagnell in Buckinghamshire and South Killingholme on South Humberside. As the time of the announcement drew near there were rumours of conflicts in Cabinet as proposals for sites in Oxfordshire were resisted by powerful Conservative interests in that county.

When the announcement finally came on 25 February 1986 it contained some surprises. In addition to Elstow and South Killingholme which had been confidently predicted, sites at Fulbeck in Lincolnshire and Bradwell in Essex were also to be investigated (Figure 3.1). Mr Baker confirmed the use of an SDO to enable test drillings and soil sampling at each site, expected to take 12 to 18 months, leading to a public inquiry beginning in 1988. If an inland site was selected a small coastal site might also be needed for decommissioned reactors from nuclear submarines. NIREX would also 'be examining the feasibility of deep-mined cavities ... possibly under the seabed'[80] for long-lived ILW. Mr Baker also announced measures in hand to improve the management of the Drigg facility.

With the announcement Bedfordshire was no longer alone. The decision to evaluate four sites rather than pursue Elstow alone was the fourth defeat for the Government and NIREX's policy. After the victory of Billingham, Bedfordshire, led by the County Council, BAND and the local Press had rejuvenated its campaign and shifted its attention from NIREX to Parliament. Government policy was under close technical scrutiny. Political pressure was beginning to open up the closed decision making process. With the naming of three further sites there was now the prospect of a dramatic shift in the balance of power as the conflict entered its decisive phase.

THE FINAL PHASE: FEBRUARY 1986 TO MAY 1987

Potentially Suitable Sites

With four sites under consideration the government and NIREX might have hoped that one site would emerge as clearly preferable. Some sites could be ruled out on safety grounds after geological investigation. The four had been chosen after exhaustive site identification procedures. Each site had 'a variety of features which make it potentially suitable for repository development'[81] and, unless a totally cynical view is taken, it was improbable that sites would have been selected which were likely to fail NIREX's safety criteria.

If all sites remained technically feasible then the eventual choice would rest on political criteria. Yet each site was in a safe Conservative seat 'unshakeable in a landslide and full of conservative people unlikely to take to the streets'.[82] The difference in the strength of opposition could be a decisive factor measured in terms of local unity

"Look, I keep telling you, this stuff is absolutely safe."

SOURCE *Daily Mirror*, 19 August 1986, p. 6

and political influence. It was possible that each community in its anxiety to avoid the dump would emphasise reasons for the dump being sited elsewhere. In such circumstances of implicit political competition among the candidate sites, the government and NIREX might hope to select the one which appeared the political 'soft option'. In the weeks following the announcement of sites this was a distinct possibility as opposition was mobilised within each of the three additional communities where sites had been identified. In each area different objections to the proposal were emphasised and each community had a distinctive political culture.

Fulbeck was the only site of the four which appeared on the list of 15 sites identified as meeting area and geological criteria by the DoE's consultants in 1981.[83] The site is a disused airfield (600 acres) owned by the Ministry of Defence on the 120m thick Lower Lias clays. Fulbeck lies in a farming area and the nearest towns are Newark (8 miles) and Grantham. Access is by country road to the trunk road A17 and the East Coast main railway line is about two miles away.

Almost a year previously, in April 1985, there were rumours of a possible coastal repository in Lincolnshire.[84] But the choice of Fulbeck was unexpected and local media reaction was strident. 'Not an hour should go by without the people of Lincolnshire and South Humberside telling their neighbours, their vicars, their doctors, their councillors, their MPs that they do not want this filth on their doorsteps.'[85] The local MP Douglas Hogg was not present in Parliament when the announcement was first made but made clear his outright opposition in a debate on March 13. He argued that Fulbeck should be ruled out on grounds of hydrogeology and poor access but emphasised the general point that shallow burial 'is not a proper, safe or necessary way of disposing of intermediate level waste'.[86]

Such an intensely conservative area was unaccustomed to the idea of protest against a Conservative Government's proposal and there were fears its resistance would be weak relative to the other areas; 'Only in Fulbeck have things been somewhat low key.'[87] But the local community demonstrated its opposition when around 3000 marched in protest from Fulbeck village to the airfield. A protest group, Lincolnshire and Nottinghamshire Against Nuclear Dumping (LAND), was established. Its Chairman Julian Fane, local landowner with impeccable Tory credentials, personified the restrained but resolute method of campaigning in this rural area. His was the image of the country squire, a traditional local leader but with waning political influence,

> To count, he has to become a protester. And to do this Fane wants to shun his 'aristo' image, harnessing all the support he can. And if it may appear slightly comic for the ruddy faced pinstriped 47 year-old to forge allegiance with the duffel coated Friends of the Earth activists, so be it.[88]

The County Council, too, was Conservative-controlled but unanimously opposed to the dump. The Fulbeck community was prepared to offer uncompromising resistance to NIREX.

South Killingholme is close to the River Humber about four miles from the deep water port of Immingham. The 320 acre site on boulder clay (about 25m) overlying chalk was bought by the CEGB for an oil-fired power station project which was subsequently shelved. It is in an area of major industrial investment including oil refining and port development and, within a radius of 12 miles, it was estimated there were half a million people including the large cities of Grimsby and, across the Humber, Hull.

As in the case of Billingham, Humberside had anticipated the announcement forming Humberside Against Nuclear Dumping (HAND) after a meeting attended by 350 people at Immingham. The meeting was addressed by both BANDs of Billingham and Bedford, evidence of the solidarity that was to play such a significant part in the conflict. The local Conservative MP for Brigg and Cleethorpes, Michael Brown, who had previously dismissed rumours as scare-mongering said, 'NIREX and the Government have indulged in gross deception and told outright lies'.[89]

Like Bedfordshire, Humberside County Council was 'hung' but there was all-party opposition to the proposal, expressed in an additional policy in the Structure Plan.[90] Joint action was agreed with the local District Councils. By January 1986 the Council had established a campaign costing £240 000 a year with the slogan, 'Say No to a Shallow Grave.' A major argument against the dump was the danger and blight it would bring to a developing industrial area. 'Any suggestion of a nuclear waste dump in the centre of this vitally important array of modern industry – and within a stone's throw of Conoco, Lindsey Oil Refinery, and Calor Gas storage caverns – defies belief. It is the last place on earth for such a forbidding land use.'[91]

This theme was echoed by the local media which featured the issue and backed the petition being circulated to every home in the area. The three local MPs, Michael Brown and fellow Conservative Richard Hickmet (Glanford and Scunthorpe) and Labour MP Austin Mitchell (Great Grimsby), declared their opposition. When the site was announced Mr Brown threatened 'a guerilla war against the unelected and unaccountable body NIREX'[92] and promised to resign his seat and fight a by-election should South Killingholme be chosen. Mr Hickmet made the prophetic point that, unless public fears were allayed, 'the Government have no alternative but to accept the forth-right opposition of members of Parliament who are affected by the decisions of NIREX'.[93] A discordant note was provided by the MP for Gainsborough, Edward Leigh, who suggested it was 'sheer hypocrisy and humbug to say that it should go anywhere but on one's own doorstep'.[94]

The campaign in Humberside resembled that in Billingham. It was a northern, industrial area able to mobilise an aggressive and extroverted campaign by the time the dump was announced. It was ready to cooperate with other areas in a united opposition to shallow burial. The third area in which a dump was proposed, Essex, had an altogether different political culture.

Bradwell, on the Blackwater estuary in Essex, was the only one of the four sites near an existing nuclear facility. The site lies in flat, coastal marshes dominated by the nuclear power station which dwarfs the previous landmark, the simple seventh century church of St Peter a mile or so away. The Bradwell nuclear power station, commissioned in 1962, was one of the first commercial Magnox stations. It is interesting to compare the relative lack of concern caused by the proposal for a nuclear power station there in the 1950s with the furore created by the NIREX proposal thirty years later. The public inquiry into the Bradwell power station lasted five days compared to the 340 days for Sizewell B. Among the major concerns at the time was the aesthetic damage that would be done to a peaceful, relatively remote coastline and the potential loss of the oyster breeding grounds. Anxieties about safety were also raised. 'Apart from any danger to personnel, radioactive material might be released to such an extent that it might be necessary to evacuate a large city, to abandon a major watershed and sterilise the site for years to come.'[95]

Such fears were overwhelmed by the confident reassurances of the nuclear industry. The power station was 'inherently safe, it was only in deference to public opinion that the first two stations . . . would be built away from large centres of population'.[96] The relaxed way in which the potential consequences of nuclear catastrophe were discussed is chilling to the modern ear: 'Would an atomic explosion once every generation, or perhaps less, be acceptable as the price to pay for this new power?'.[97] Nuclear power was needed to meet demand and 'They in Bradwell were the guinea pigs for England and they might feel proud to be selected for this honour.'[98] There was a similar complacency about the problem of radioactive wastes: 'Methods of dealing with radioactive waste matter were so well-known in all the countries of the world that there was no hazard attached to the disposal of them.'[99] A generation later the contrast in attitude to the problem is arresting. Having lived in the shadow of an ageing nuclear power station with apparent unconcern, the local population was aghast at the selection of Bradwell for what, by contrast, was a relatively innocuous nuclear waste repository.

Although the site is in a thinly populated area the towns of Colchester, Clacton and Maldon lie within a radius of ten miles and within twenty miles come Southend and Chelmsford, bringing the total population to over 700 000. The site was in the constituency of John Wakeham, the Government's Chief Whip and member of the Cabinet. He took the unusual step, for a member of the Government, of issuing a statement supporting government policy on nuclear power but stating that Bradwell would be found 'wholly unsuitable' once investigations have been completed: 'It is my personal view . . . that Bradwell does not represent a solution to the long-term problem of radioactive waste disposal.'[100]

Wakeham's dilemma in supporting government policy but opposing its implications for Bradwell was echoed in the ambivalence of Conservative-controlled Essex County Council who, while totally opposed to the Bradwell proposal, were reluctant to join forces with the other areas in a united front against the concept of shallow burial. The County Council urged more emphasis on research into alternative options especially under the seabed. In particular Essex was trying to reduce the importation of hazardous wastes into the county, and, for that reason, believed that 'the case against Bradwell is materially different and more likely to succeed if presented separately'.[101]

The local community displayed no inhibitions in its campaign against the dump. In the remote and scattered villages near Bradwell there was intense opposition, the subject of a nationally televised current affairs 'World in Action' programme. About a hundred protesters walked out of a meeting with John Wakeham after he declared his intention of voting for the SDO while maintaining his opposition to the Bradwell proposal. A locally based Bradwell Action Group was formed and Essex Against Nuclear Dumping (EAND) carried the campaign to the wider area. As in the other areas there was all-party opposition to the plan, a petition and wide Press coverage.

The populous parts of the county lay some distance away across river estuaries and distance created an indifference that made mobilisation of the wider community difficult. The confidence that the proposal could be prevented by political influence initially seemed to isolate Essex from the other threatened communities. But, when eventually put to the test, the resistance in Essex proved as strong as that elsewhere.

Unite and Fight

While attention was inevitably drawn to the three new sites, Bedfordshire was dismayed it was still on the list. The County Council restated its opposition and for a third time, a deputation visited William Waldegrave in London to press the county's objections.

When Waldegrave visited Elstow in April 1986 he gained an impression of the opposition the proposal was facing and conceded, 'It is certainly possible to stop things.'[102] A party of councillors visited Drigg just before conditions at the dump were severely criticised by the House of Commons Environment Committee. BAND too, were very active with the local campaign against Elstow culminating in April when eight coachloads and the County Council's campaign bus formed a convoy to deliver the 95 000 petition to the House of Commons.

Bedfordshire County Council announced it would offer an alliance with other authorities to oppose the whole concept of shallow burial of radioactive waste. In March representatives from the local authorities in the newly selected areas were invited to a seminar at County Hall, Bedford. As a result three County Councils (Bedfordshire, Humberside and Lincolnshire) formed a Coalition of County Councils supported by their District Councils. The Coalition declared it was not prepared to accept shallow burial until a coherent national policy had been devised and approved; the quantities of waste had been rigorously defined; the safest option had been independently established with a proper design brief; the criteria for acceptance of waste had been defined; proper compensation had been made; and proper procedures for final site selection had been approved by Parliament.[103] This formulation ensured that the inherently diverse political attitudes in the three authorities could be reconciled in a common objective. This unity was weakened by the decision of Essex that its interests would best be served 'by presenting our own case'.[104]

There were no defectors among the various pressure groups. BAND and Bedfordshire County Council hosted a National Conference on Radioactive Waste Dumping in Bedford in early April 1986 attended by protest groups, environmentalists, media and local authorities from the four areas and Billingham. The speakers included prominent local and national anti-nuclear campaigners covering a spectrum of subjects from health effects, policy and procedures to organising opposition. In his summing up Frank Cook, Billingham's MP, commented on his visits to Bedford over the previous two years.

> The first time we met in someone's living room and there were seven of us. The look on the eyes of the people there was one of outright shock. They were shattered and bemused.
>
> But look at the people from Bedford now – see how they've

> learned. They've run an event here that started out as perfection and got steadily better. It's been bloody marvellous! It's been a privilege to be here.[105]

The outcome of the conference was the formation of Britain Opposed to Nuclear Dumping (BOND) comprising the four protest groups plus BAND of Billingham. BOND's aims were to halt site investigations; to advocate above ground dry storage of waste; to demand a publicly acceptable nuclear waste policy; to alert the public to health risks from low level radiation and 'to demand complete openness, consistency, and public accountability in all matters relating to nuclear waste disposal'.[106]

The government's decision to pursue shallow disposal as its strategy and to announce four potential sites had resulted in a fourfold increase in opposition. And the opposition had developed an organisational capability and political resource that could be deployed to fight its common cause.

A Live and Growing Issue

By the time the Coalition and BOND had been formed, radioactive waste management policy had become 'a live and growing issue'.[107] Opponents of potential sites had demonstrated a capacity for exploiting political circumstances. In the spring of 1986 the Coalition and BOND were presented with unexpected political opportunities that yielded an immediate pay-off and shaped the future course of the conflict.

The first of these was the publication on 12 March of the House of Commons Environment Committee's Report on Radioactive Waste.[108] Written before the announcement of further sites and partly leaked earlier, it was broadly acknowledged as the most wide-ranging and comprehensive survey of the subject so far produced in the UK. Its political impact rested on the fact that it was an agreed statement by a group drawn from the three major political parties. The Report echoed Bedfordshire's criticism of a lack of clear policy. It stated that 'the UK is still only feeling its way towards a coherent policy' (para. 3). Unless there was public acceptability for safe disposal 'the continuing production of such waste and, consequently, the nuclear power policy will be put into question' (para. 5).

The Report was scathing about the method of site selection. 'On the one hand, bold announcements about prospective new disposal sites are issued. They are then withdrawn, left hanging in the air, or modified ad hoc' (para 3). There was a lack of openness, public involvement or research into options. Instead the industry and government had rushed pell-mell into site proposals. 'The kind of urgency suggested by NIREX's haphazard activities with regard to Billingham and Elstow is very unhelpful and can only fuel the public's anxiety and distrust. In addition, NIREX's rush risks the loss of what might be geologically prime sites' (para. 58). The price paid for 'the premature and uncoordinated approach' was the loss of Billingham. 'Ultimately the penalty for the largely uncoordinated approach may be the loss of Elstow too' (para. 251).

Existing methods (as used at Drigg) of classification of wastes failed to relate type of waste to disposal route so that dangerous long-lived toxic radionuclides could enter the waste stream. 'With nothing labelled, nothing recorded, pinpointing the offending waste would be virtually impossible' (para. 71). Coming to the question of shallow disposal, the Committee considered there were two possible objections. One was the predilection for deep disposal for ILW in many other countries, the other the problem of public anxiety. The Report recommended that shallow burial was 'only acceptable for short-lived low-level wastes' (para. 99) but emphasised the need for greater research and development into seabed options, especially tunnels under the seabed from land.

The Report went on to emphasise the gap between the industry and the public measured both in opinion polls and in the strength of opposition to Elstow and Billingham. 'Public anxiety is significant and deep-rooted' (para. 221) and compounded by the impression of 'error and misjudgement' (para. 223) conveyed by the industry. But its opponents, too, could damage their credibility by 'distortions or untruths' (para. 229). The gap could only be bridged by greater openness, public involvement in the policy process and site selection and through adequate compensation. Above all, the industry should abandon its 'shoe-string' approach and adopt 'Rolls-Royce' solutions whose 'very small extra cost would be a price worth paying in the end to win public confidence' (para. 244).

Bedfordshire's arguments had clearly found favour with the Committee. 'We were impressed by the detailed critique prepared by Bedfordshire County Council' (para. 87). 'We agree wholeheartedly with witnesses for Bedfordshire County Council who stressed the importance of having an agreed policy or strategy before attempting to carry out executive functions' (para. 252). Naturally, Bedfordshire

Council Council was delighted with the Report: 'It is gratifying to find such an overwhelming weight of opinion coming round to our way of thinking'.[109] The Coalition drew heavily on the Report in preparing their initial statement on radioactive waste disposal and employed professional lobbyists (Parliamentary Monitoring Services) to organise support in Parliament. The chances of success were greatly increased by the event which, more than any other in the history of nuclear energy, had a transforming impact on the nuclear debate.

The Edge of the Abyss

The disaster at Chernobyl in the Ukraine on the 26 April 1986 caused a radiation cloud which drifted over Europe leaving its print both in the increased radiation levels and on the minds of people. For a moment the world had stared at the abyss. Anxieties were registered in the high level of opposition to nuclear power in the opinion polls and in the increasing sensitivity of politicians to nuclear issues. As in the autumn of 1983 so in the spring of 1986 the coincidence of a nuclear accident with the debate over radioactive waste left the nuclear industry and the Government on the defensive and its opponents confident of public support.

The immediate issue being faced by the Government in the aftermath of Chernobyl was the opposition to its shallow burial policy. At the very least, opponents argued, the Government should make a response to the House of Commons Report before it put the SDO before Parliament. The Government yielded to the pressure and issued an interim response to the Report on 2 May which confirmed its support for both a shallow and a deep disposal facility.[110] However, there was one significant concession. The response underlined that the distinction between ILW and LLW 'was arbitrary and intended to be so' (para 17) and not intended to prescribe particular disposal routes. Turning to shallow burial it went on,

> Nevertheless, this is an area where it has proved particularly difficult to bridge the gap between scientists' assessment of risks and the honestly-held perceptions of the local communities. The Government takes seriously the distinction drawn by many between the acceptability of LLW and ILW in such a site and recognising that many people would be reassured if this site were used only for LLW. In these circumstances, the Government has

> decided that NIREX should proceed on the basis that a near-surface facility will only be authorised for the disposal of what is broadly described as low level waste (para 32).

The government's full response was published as a White Paper in July.[111] This indicated there would be no other significant concessions and in particular reprocessing, the major producer of ILW and LLW, would continue.

The decision to drop ILW from the shallow burial proposal was initially hailed as giving 'strategic ground' to the opposition.[112] A more considered judgement was highly critical suggesting that the concept of 'broadly described' LLW was arbitrary and 'could well result in the co-disposal of currently defined short-lived intermediate waste'.[113] Privately the Coalition were worried that the government's decision would undermine their case since much of their technical argument rested on the dangers of hitherto untested shallow burial of ILW. The government could claim to have accepted the arguments and to be putting forward a method 'based on current knowledge and experience' (Cmnd 9852, para. 43). While Drigg was not an acceptable model, at least there was operational experience in the UK and elsewhere of shallow burial of LLW.

On the other hand, the government had accepted 'there are no technical reasons why both LLW and short-lived ILW should not be placed in a near-surface facility' (para. 20). Presumably there were equally no reasons why they should not be disposed of together in the same deep facility. The government also acknowledged that 'Public understanding is needed. Public confidence is required.' Confidence would be difficult to build in the emotional aftermath of Chernobyl. The withdrawal of plans for shallow burial of ILW marked a fifth retreat on nuclear waste policy leaving only the question of disposal of low-level wastes on the immediate political agenda. The final stages of the conflict would test the government's will to resist the political pressure mounted by the Coalition and the protesters.

There were some indications of the political problems facing the government when the Commons debated nuclear energy on 13 May. It was an umbrella debate ranging widely over a range of nuclear issues prompted by Chernobyl and providing a platform for MPs from the four communities to press their opposition to shallow burial. Douglas Hogg, representing Fulbeck, urged deep disposal for all wastes; Nicholas Lyell, for Elstow, argued for remote disposal; and Richard Hickmet, on behalf of South Killingholme, warned that without public acceptability 'the Government run the risk of losing the nuclear industry argument altogether'.[114] In contrast, Sir Trevor Skeet from Bedfordshire judged the most acceptable site would be 'on the coast' (that is, Bradwell) or 'on property owned by the Ministry of Defence' (that is, Fulbeck). Shallow burial for LLW was not only government policy but was also supported by Labour's spokesman on the Environment, Dr John Cunningham whose Copeland constituency included Sellafield and Drigg. But he was opposed to the use of a Special Development Order, the immediate target of the opponents of shallow burial.

A week later, on the same day as Kenneth Baker was replaced by Nicholas Ridley as Secretary of State for the Environment, the arguments were rerun in the debate on the Special Development Order (SDO). The strength of opposition was measured in the petition of 150 000 from Humberside and in the hyperbole of Austin Mitchell, MP for Great Grimsby,

> I am fighting for my area ... I would rather build Jerusalem in Humberside's green and pleasant land and I certainly will not allow it to become a dump. We are not allowed to build Jerusalem under this Government, but we can fight to keep out the detritus of their nuclear policy.[115]

For the government William Waldegrave reaffirmed the belief in shallow disposal and the necessity for an SDO in order 'carefully and steadily, to go ahead'.[116] Seven Conservatives voted against the Government, including four – Douglas Hogg, Michael Brown, Richard Hickmet and Edward Leigh – representing the areas around South Killingholme and Fulbeck. Despite their opposition to the proposals Nicholas Lyell and the Chief Whip, John Wakeham, stayed loyal to the Government in which they both held office. Mr Lyell argued, 'I must support the Government or resign. I would be happy to resign if it gained something for Bedfordshire'.[117] A few weeks later there was a polite and restrained debate on the SDO in the House of Lords, but the matter was not pressed to a vote. The focus of conflict now shifted away from Parliament and back to the four sites where NIREX faced the protesters as they waited for the SDO to come into operation on 7 July.

The Enemy is in Our Midst

NIREX had been carefully planning for the time when drilling would begin on each of the four sites. For each site consulting engineers had been appointed to design, supervise and interpret the geological investigations.[118] The overall cost of the programme was around £10M and drilling would take about six months enabling NIREX to assess the reports and make a choice of site by the end of 1987. The drilling programmes varied. For instance, that for Elstow included 70 boreholes to test the geology and water movements and an extensive excavation to examine the fabric of the clay and its mechanical and hydraulic response to 'a large stress change'.[119]

NIREX hoped to engage in 'open consultation ... in a professional, non-combative forum'.[120] This would help to provide relevant information on environmental impact; to 'enable the local authorities to influence NIREX thinking' and to 'help NIREX choose which one site to submit ... for permission to develop a repository' (ibid). In this way it hoped to 'close the gap between public and technical perceptions of the state of the art' (ibid). There were certain mutual advantages in cooperation. While NIREX might get necessary information for the environmental impact assessment which had to be submitted as part of the planning application, the local authorities could benefit from gaining full access to the information from the drilling programme. But, if NIREX really thought their site investigations would be unchallenged and unimpeded, they were soon disabused.

BOND and the local protest groups had drawn up their strategy at a meeting in June. In July a demonstration at each site by families pinning photographs to the fencing signalled the style of peaceful protest and coordinated opposition that NIREX might expect. But, it gave little indication of the scale and commitment of the opposition when the battle of dumps was engaged in the middle of August.

On Monday morning, 19 August, the contractors tried to enter Elstow, Fulbeck and South Killingholme but at each site were prevented by the protest groups. At Elstow the convoy was met by 'a wall of protesters forming a human blockade'[121] and were advised to turn back by police. Jim Eldridge of BAND, one of the organisers, stated that 'everyone on this blockade is prepared to be arrested and is prepared to go to gaol'.[122] At Fulbeck the contractors were repulsed twice by about 300 demonstrators blocking the single road entrance to the site. 'We are going to continue along the same lines, silent, passive and non-violent' said their spokesman Trevor Cartwright.[123] At South Killingholme about a hundred people turned back the contractors. Each of the blockades received coverage on national television throughout the opening day.

GUARDIAN

and Portsmouth Tuesday August 19 1986 25p

FAST.FORWARD.
METROPOLITAN
WIGAN
WORKING HARD WITH INDUSTRY

Villagers foil nuclear dump tests

DITOR

any dumping of

Daily

Nuclear dumping

From Mr Julian Fane

Sir, Your second leader, "Green but not pleasant" (August 19) suggests that our case has no meri but deserves a hearing.

NEWS

N-dump tests bl by human walls

when Nirex monitorin county

Killingholme
Fulbeck
Elstow
Bradwell

win round one

By Staff Reporters

August 21, 1986

blockade

have finally agreed arrangements with the council; when the Government have answered our letter of

PROTEST lages opening round to prevent low-grade waste being dumped near their homes.

and her children outside

had taken demonstration airfield with men, children of every age and

This fact alone should give Government more cause for co than any other protest. Th question that needs asking is wha a diverse group o

For the second day running protestors successfully turned away Government contractors yesterday from the proposed site for a nuclear waste dump in Fulbeck, Lincs.

THE TIMES

TUESDAY AUGUST 19 1986

Protesters nuclear

hicles turned back when confronted by the protesters. The demonstrations were peaceful and police made no attempt to intervene. Nirex, the Government's waste agency, has test programme

Continued from page 1

Essex, where drilling is scheduled to start next month. A spokesman for Nirex last night that the organization had no immediate intention of taking legal steps to access to the sites, al "we realize we might resort to it eventually".

At Elstow, 75 pe

SOURCE Montage from Press, August 1986, p. 1

NIREX were prepared to make further attempts to gain access in the hope that 'now the opposition had had some publicity they may rethink their position';[124] 'We want to make sure we have exhausted every reasonable avenue'.[125] But, the blockades remained. At Fulbeck a 24-hour vigil was maintained and a 'telephone tree' warning system enabled large numbers of protesters to be gathered to thwart every attempt by the contractors to gain entry. Similarly, South Killingholme sustained a 24-hour presence at the site. At Elstow, Tom McInerney, NIREX's Chief Executive, attempted to pacify the protesters during a 70-minute open meeting outside the site concluding 'It went as I expected it to go, really. But it's always worth while. I think the odds were somewhat against me, but I did my best'.[126] At each site several attempts were made to gain entry, each one obstructed by protesters blocking the gates.

In Essex there were fears that 'Bradwell might be put in the shade by protesters at the other three possible sites'.[127] But, when the contractors arrived on 1 September, they were met by a mass protest (variously estimated at 300 to 1000) which had begun at dawn sitting down in the road to the site. Here, too, a 24-hour vigil was mounted in order to mobilise the blockade whenever entry was attempted by the contractors. Protesters from other sites joined those at Bradwell to demonstrate the solidarity of the protest among the four sites.

The blockades were backed by the local media using language of Churchillian belligerence. 'D-Day at Fulbeck',[128] 'Battle lines at Bradwell'[129] and 'Flare-up at Elstow'[130] described an unyielding confrontation. The protesters were portrayed as heroic, 'the small group of people who have been defending OUR rights at great personal sacrifice to themselves'.[131] Some protesters became well-known, like Mieke Hinchcliffe-Wood, who organised the Elstow blockade and was described as 'Mother to media star'.[132] She and other members of BAND, Jerry Fitch, Richard Lemon and Jim Eldridge were featured in the national *Sunday Times Colour Supplement* on 24 August. The blockades were given considerable coverage in the national press. The *Guardian* described the protest as 'Prejudiced, maybe; irrational, no' (23 August). *The Times* considered the protesters to be 'Middle-class, middle-aged hooligans from middle-England' (19 August), a charge rejected as 'a slur on the people of this country'.[133]

The blockades had succeeded in presenting a united front among four politically diverse communities and in achieving considerable – and in the main – sympathetic publicity which could yield political dividends in future. At Fulbeck the Chairman of LAND, Julian Fane, intimated the political significance of the protests, 'I would warn the powers that be that there will be an election before any decision is made here'.[134] The Bradwell protest was directed at their local MP John Wakeham with the slogan, 'Wake-up Wakeham, Dump Us and We'll Dump You'. In each of the four areas local MPs

SOURCE *The Times*, 2 September 1986.

were fully aware of the political sanctions that might be applied if the policy of shallow burial was maintained.

The protesters realistically could not hope to keep out the contractors indefinitely. At Elstow the contractors gained access when the pickets had left on the evening of 1 September. Two weeks later on 16 September at Bradwell they slipped past the handful of protesters at 4 a.m. Two days afterwards they gained access to South Killingholme in another dawn 'raid'. By that time injunctions had been gained by NIREX to ensure entry. This was sufficient for the Fulbeck blockade, the last survivor, to capitulate as around fifty police secured entry for the contractors on 22 September. In a final gesture of defiance, one protester warned Nicholas Ridley the Environment Secretary, 'The Conservative party is going to lose so many votes, he won't know what's hit him'.[135] While NIREX had gained access to the sites the protesters hoped it would prove a pyrrhic victory.

Light Years Ahead

With the contractors on site the intensity of the demonstrations subsided. At Elstow a token picket was maintained and there was some harassment of the engineers by women monitoring workmen's activities and persistently inconveniencing their work. There were intermittent demonstrations, a candle-lit mock funeral procession at Elstow, a mock burial ceremony and a tractor procession at Fulbeck, a night raid by saboteurs (unconnected with the protest groups) on the Bradwell site. The official launch of BOND in October was marked by a demonstration at the House of Commons featuring an 18 ft high grey polystyrene monolith resembling the kind of marker proposed to warn future generations of the presence of nuclear waste sites.

NIREX had used blanket injunctions at all four sites covering named individuals and all those who were 'affiliated to' or 'associated with' the protest groups. LAND and HAND fought a successful high court action to get the blanket injunction lifted and the bulk of the named persons removed. In the case of Elstow NIREX served a writ on Mieke Hinchcliffe-Wood for breach of the injunction accusing her of interfering with and obstructing test drilling work and organising others to do the same, of creating a black list of contractors working for NIREX and of intimidating workmen on the site. 'Can women intimidate 40 burly contractors?' she asked.[136] The High Court judge was unimpressed by the NIREX case, refusing to fine or imprison her though making her pay the costs of the action. The NIREX tactic provoked a predictable reaction. 'If they want to make a martyr out of someone they just might find some more volunteers'[137] was a typical response. In Parliament Austin Mitchell described it as 'legal bullying . . . employed in a dubious cause which was decided prematurely without proper consideration and without proper thought'.[138]

In an effort to achieve greater public credibility the Secretary of State appointed three independent directors to the Board of NIREX. They were biochemist Professor Sir Hans Kornberg, Master of Christ's College, Cambridge, chairman of the British National Committee on Problems of the Environment and a former chairman of the Royal Commission on Environmental Pollution 1976–81; Ray

Buckton, the General Secretary of the train drivers' union ASLEF and chairman of the TUC group on radioactive wastes; and Angela Rippon, journalist and broadcaster. This last appointment was criticised as a means of attracting favourable publicity through a well-known TV celebrity. But Angela Rippon, along with her new colleagues, proclaimed her independence and the need for careful consideration of the problem. Meanwhile, the drilling proceeded at all four sites with the contractors adopting a studied disinterest in the political conflicts surrounding their work.

As the activity at the sites diminished, attention shifted once more to Parliament. The County Councils Coalition undertook a tour of European radioactive waste facilities to obtain first-hand evidence of alternatives to the NIREX proposals. A party of eighteen councillors and officials chartered an aircraft for a four-day visit to Sweden, West Germany and France at the end of October. In Sweden they saw the nearly completed first phase of the purpose-built repository constructed beneath the Baltic coast, near the nuclear power station at Forsmark, about 100 miles north of Stockholm. In West Germany they toured the abandoned deep iron ore mine proposed for nuclear waste disposal at Konrad near Brunswick. In France they visited Centre de la Manche, the operational above-ground engineered facility adjacent to the reprocessing works at the tip of the Cotentin peninsular 15 miles from Cherbourg. They were clearly impressed by what they saw describing the Swedish and German approach as 'light years ahead of this country in research into the disposal of nuclear waste'.[139] They felt there was no comparison between the NIREX plans and what was proposed or completed elsewhere.

The Coalition presented their findings in a statement and report by their consultants, Environmental Resources Ltd (ERL), presented at a seminar for MPs in January 1987.[140] There were three main findings. The first concerned the technique of disposal. In each country there was a different technical solution but all were 'far more advanced than the UK in the development of policies and practices for the disposal of radioactive waste' (p. 1). Although classification of wastes varied, in all three countries LLW and ILW were to be disposed of in the same facility. Incineration and compaction could considerably reduce the volumes of LLW enabling co-disposal in the UK. The Coalition recommended that NIREX and the Government 'should reconsider the current ad hoc policy and prepared comprehensive plans for the disposal of both LLW and ILW in a single deep repository' (p. 15).

The second finding was that the costs of alternatives might be greater than shallow disposal, whereas, if co-disposal were adopted, costs would be significantly lower than are assumed by the BPEO Report for the separate disposal of LLW and ILW. The BPEO Report had calculated the costs of a deep cavity at £2600 per m^3 of long-lived ILW. But, if short-lived ILW and LLW were added the Swedish and German evidence suggested a unit cost of around £1000 per m^3. The Coalition recommended that NIREX provide cost estimates of both LLW and ILW disposal, 'These could then be used to estimate the likely additional, if any, costs of a more sophisticated repository for the combined disposal of both LLW and ILW' (p. 16). In any case, ERL concluded, the total costs would be less than 1 per cent of the costs of nuclear electricity generation, a small price to pay for public acceptance.

The need for public acceptability constituted the third finding of the Coalition. The Swedish approach was especially attractive, 'the Swedes have spared neither effort nor expense in coming to terms with the problem' (Foreword). In West Germany where nuclear power was deeply controversial nuclear waste was regarded as a political matter and any attempt by the Federal Government to overrule the Lower Saxony *lande* over the development of Konrad would be 'political suicide' (p. 10). Even in France with its heavy nuclear commitment and relative lack of opposition 'there had been considerable efforts to ensure local, at least, public acceptance for Centre de la Manche' (p. 16). The Coalition concluded that 'If public acceptance is to be gained, a truly convincing solution to the disposal of low and intermediate level waste needs to be found' (p. 1).

The Coalition's report was the subject of a Commons debate led by Michael Brown, MP for Brigg and Cleethorpes, on 9 February 1987. He emphasised the main findings of the Report and the way in which the politicians in the Coalition 'have set aside the party political divide and worked together wonderfully'.[141] Despite the solidarity among the three counties Sir Trevor Skeet persisted in his view that 'the two sites which should be carefully scrutinised as possible locations are Bradwell and Fulbeck'.[142] There were no signs that the policy would be altered. Even Sir Hugh Rossi, Chairman of the Environment Committee whose report had been extensively quoted by the Coalition in support of their case, emphasised the acceptability of shallow disposal of low-level waste. In summing up William Waldegrave seemed to dismiss the case for alternative means of disposal,

> The scientists would want to know what we were doing fiddling around with permeable stone just below the seabed when we have a more Rolls-Royce type solution in the clay in this country. It is that crescent of clay which is, above all, the determinant of why the four sites have been chosen for investigation.[143]

He was convinced there would be no consensus: 'I fear that any proposal will find opposition'.[144] The Coalition's case for co-disposal in a deep repository as the means of achieving public acceptability appeared to have been denied.

Predictably NIREX maintained in a point by point rebuttal of the Coalition's arguments that shallow burial in an engineered facility was a proven, safe and appropriate technology for the UK. They indicated that work was being done, however, on the potential of different geologies for 'a sub sea-bed repository, accessed either by shafts and tunnels from a suitable land-based site or from a sea-based rig or even artificial island'.[145] In February 1987 it was not yet clear whether such straws in the wind were to be clutched by NIREX's opponents as a forlorn hope of change or whether they presaged a more dramatic transformation.

The Burial of Shallow Burial

There was little reason in the early spring of 1987 to expect any significant policy changes. The publication of the long-awaited Layfield Report at the end of January on the Sizewell 'B' Inquiry had little impact on the debate over nuclear waste.[146] Indeed, if anything Layfield's opaque pronouncements on nuclear waste tended to confirm the Government's postulated 'clear strategy' (p. 41.51) and the NIREX proposals. Layfield affirmed that the provision of new sites 'is a matter of urgency' (p. 41.53) and asserted that the feasibility of safe disposal was 'not disputed' (p. 41.52). While arguing that 'public disquiet about waste disposal should be substantially reduced' (p. 41.55) the Report concluded that 'There are no reasons connected with the management of radioactive waste why consent for Sizewell B should be refused' (p. 41.56).

There were signs, too, that the opposition to the proposals was beginning to lose coherence. BAND in Bedfordshire was riven by a major split over an accommodation reached with NIREX. In return for an undertaking by BAND not to do anything that breached the terms of the injunction NIREX removed the injunction from all but six people. The six included Jim Eldridge and Mieke Hinchcliffe-Wood who led a breakaway group of activists prepared to break the law to stop the dump. At Fulbeck there were rumours (later proved false) that a secret DoE document showed the site to be favourite on the grounds of good access and the impregnable majority of the sitting MP.

It was the visits by the three directors that provided some tangible intimation of a shift in the wind. At each visit they stressed their independence. Ray Buckton declared he was against shallow burial;[147] Hans Kornberg confessed he was pressing NIREX to consider co-disposal of ILW and LLW;[148] and Angela Rippon said, 'We are independent and try and represent the views of the waste receivers.'[149]

Despite the scepticism towards shallow burial shown by the three independent NIREX directors none of the four communities had expected the sudden abandonment of shallow burial announced in Parliament on 1 May by the Secretary of State, Nicholas Ridley. The decision confirmed the arguments about cost, alternatives and public acceptability that had been pressed by the Coalition and the protesters. NIREX had commissioned consultants to undertake a comparative costing of co-disposal of ILW and LLW at a deep site with deep disposal of ILW and shallow burial of LLW. In a letter to the Secretary of State, dated 30 April, John Baker, Chairman of NIREX indicated that there was no reason to doubt that a repository meeting the safety criteria could be engineered at any of the four sites. However, the exclusion of ILW, and improvements in design, had increased the costs of shallow disposal, reflecting 'the need to respond to public perception rather than any technical requirements'. He estimated the cost of disposal of LLW in a shallow repository in the range of £500–£1000 per m^3 of waste compared with a marginal cost of around £750–£1200 per m^3 if it were disposed of in a deep burial facility which was necessary for ILW. 'So whereas a year or so ago when the policy of investigating the current four sites was endorsed, the ratio of cost per metre cubed of disposing of low-level waste deep compared to shallow was probably four to one, I now consider it more likely that the costs are broadly similar if low-level waste is regarded as being "piggy-backed" into a deep repository on the back of intermediate level wastes.' Therefore, he advised that NIREX efforts be switched to deep disposal options for both ILW and LLW. This would 'reassure the public that we are not following a predetermined course'. He concluded,

> I do not believe that ready public acceptance of radioactive waste disposal is likely to be achieved in the short-term. I would hope, however, that if you are able to agree to the changes I am proposing in this letter, that will go some way towards providing reassurance that not only are we all determined to find good solutions to the disposal of wastes, but that we are indeed responsive to changing circumstances.

Nicholas Ridley replied the following day, 1 May, accepting the recommendation as 'the responsible course of action'.

The cause of the sudden *volte-face* was perceived as a means of disposing of 'an intermediate level political embarrassment in advance of an imminent general election'.[150] Indeed, a General Election was called a week later and all four MPs representing the sites were comfortably re-elected. The speed of the announcement was clearly prompted by immediate political expediency. Ridley's quick reaction was said to be 'entirely typical of the man' with a proclivity to abandon schemes 'directly he was convinced of their even short-term political impracticability'.[151]

The impending election may have affected the timing of the decision but not its necessity. There is no reason to doubt the conclusions about costs reached by NIREX. Once short-lived ILW had been removed, a Rolls Royce shallow repository for LLW alone became an expensive option. The need for deep burial for ILW meant that a potentially safer alternative for LLW too was available.

The underlying cause of the change in policy was a technical reassessment prompted by political calculations. The opponents had articulated a convincing technical case against shallow disposal based on detailed research on alternatives and it was a case that could well have stood up to the rigorous examination of a public inquiry. John Baker apprehended the political message. 'One lesson from our work is surely that the public does not like feeling pressurised to accept imposed solutions in this area and I would think time spent now in considering all the issues ... would be time well spent.'[152] The jubilation with which the decision was greeted in the four communities was made sweeter by the knowledge that their arguments had been accepted and heeded by the Government. This sixth retreat on nuclear waste policy was a comprehensive victory for the four communities.

THE WAY FORWARD

A Politically Inevitable Choice

The conflict over shallow burial which we have described in these last two chapters played a crucial role in transforming the politics of nuclear waste in the UK. First, the political participation by local communities had prised the issue free from hidden decision making into the political arena. This was not the first time that nuclear waste had aroused opposition. As we saw in chapter 2, communities in remote parts of the country had successfully resisted exploration for potential high level waste disposal sites. But, the sustained campaign by, first, Billingham, then Elstow and later the other three communities had given the issue a public visibility that made any reversion to secretive decision making unlikely. Furthermore, opponents of the proposals had demonstrated a capacity for building alliances crossing conventional party, social and geographical barriers. These cross-cutting alliances, expressed in BOND and the County Councils Coalition, had exposed the vulnerability of the Government to carefully orchestrated political pressure. The need for public acceptability of radioactive waste management policies had been established.

A second element of the political transformation brought about by the conflict was at the institutional level. The government had been forced to make a number of significant concessions. Local opposition brought the abandonment of Billingham, the naming of alternative sites and the withdrawal of ILW from shallow disposal. Despite the denial of a local planning inquiry through the use of the SDO, the Goverment had ultimately abandoned its shallow burial proposals. Local government and local communities had successfully asserted their role in decision making.

Thirdly, in terms of policy, the need for a more rational strategy had been acknowledged. The attempt to secure sites at Billingham and Elstow by a method of Decide, Announce, Defend (DAD) had been undermined. The government belatedly tried to retrieve the situation by introducing a comparative evaluation of four shallow burial sites. But this approach assumed the case for shallow burial without an analysis of alternative methods of waste management. Both the announcement of specific sites and, subsequently, the insistence on a specific disposal method were examples of premature legitimation. When this failed the government was forced to consider a more rational approach based on careful evaluation of waste management options before examining potential sites.

The scale of the political transformation in radioactive waste management policy was made manifest with the publication by NIREX of the discussion document, *The Way Forward*, five months after the abandonment of shallow burial. It signalled a new approach aiming 'to promote public understanding of the issues involved and to stimulate comment which will assist NIREX in developing acceptable proposals'.[153] Public acceptability would be achieved through 'open discussion and feedback from a wide audience on the various aspects of site selection, investigation and development, environmental protection, monitoring and control of repository options'.[154] The open, participative and rational strategy advocated during the conflict was now being endorsed by NIREX.

Three concepts for deep burial were under consideration (Figure 4.2); under land, accessed from a land base; under the seabed, accessed from the coast; and under the seabed, accessed from a platform as in the oil industry or from an artificial island. The most likely hydrogeological environments for finding suitable sites were hard rocks in low relief terrain mainly in Scotland; small islands mainly around the western Scottish coasts; and seaward-dipping and offshore sediments stretching along eastern England from Cleveland to Norfolk and along the coast of north west England. Thus a debate about options would be encouraged before specific sites were identified.

A year after the publication of *The Way Forward*, a fifty page report on the consultations was released.[155] Altogether 2526 written replies had been received though the majority were letters (1714) and petitions (71). It was impossible to weight the responses which 'cannot be taken as anything more than an expression of the nature of the concerns which people held' (p. 6) (Figure 4.3). Not surprisingly, there was no overall unanimity of view. While local authorities gave some support to deep disposal, environmental groups argued the case for on-site storage which was not presented as an option by NIREX. It later emerged NIREX had secretly studied the on-site storage option, but rejected it on the grounds that 'This could result in increased opposition to the development of new nuclear facilities', as well as perceived problems with planning permission.[156] There were criticisms of the consultation process, concern about the method of decision-making and expressions of the NIMBY syndrome.[157] There

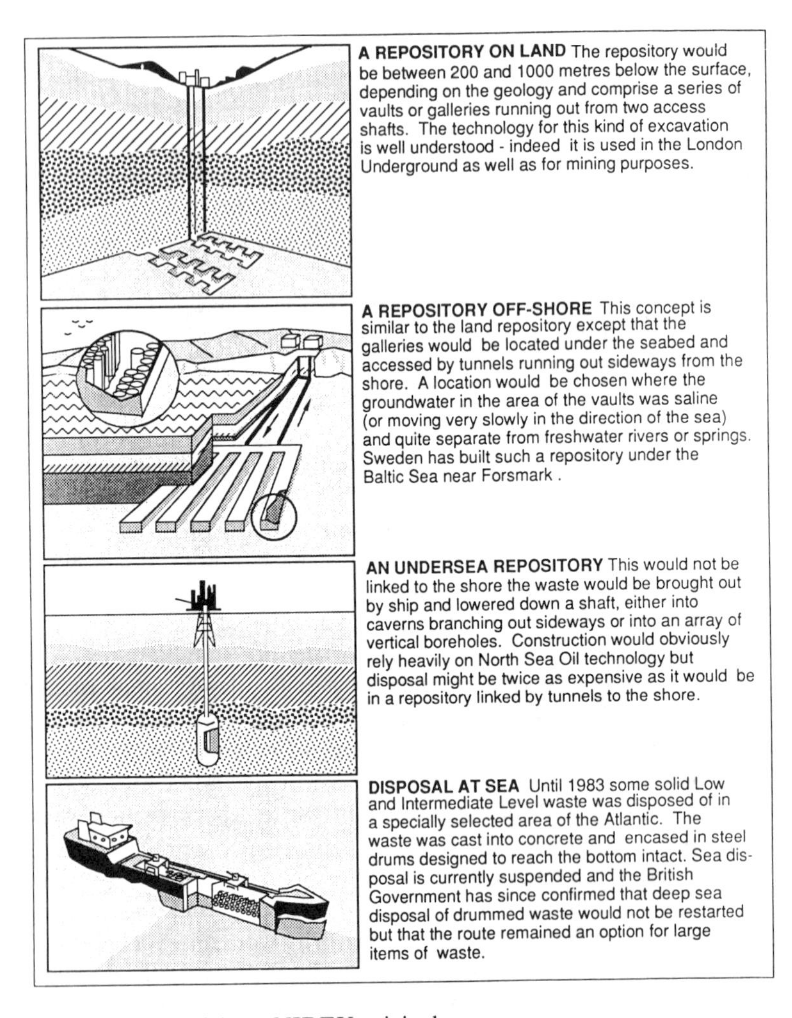

SOURCE Adapted from NIREX original.
FIGURE 4.2 *Concepts for deep burial of nuclear waste*

was, however, some comfort to NIREX in the responses from the Highland Regional Council which recognised the potential of Dounreay as a regional site for handling wastes; from Caithness which indicated that NIREX should be invited to test for a repository; and from Copeland (the council which includes Sellafield) which welcomed 'the possibility of developing a new approach towards the storage and disposal of LLW and ILW' (p. 11).

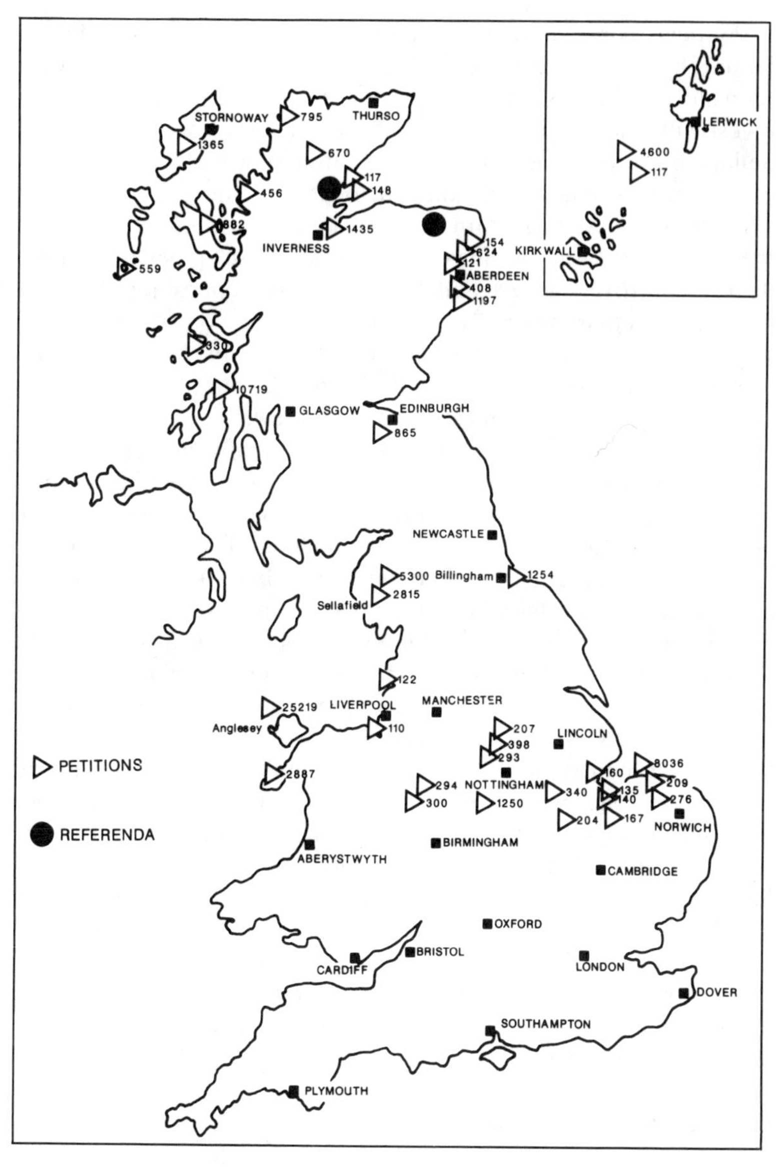

SOURCE NIREX (1988).

FIGURE 4.3 *Regional and local responses to NIREX's disposal options*

Political considerations pointed inexorably to Sellafield and Dounreay as the areas most likely to be chosen for nuclear waste repositories. In 1987, British Nuclear Fuels had indicated plans to investigate the possibility of a sub-seabed repository close to their Sellafield reprocessing works.[158] In Caithness speculation centred on the Dounreay plant itself and on an inland site offered by Lord Thurso on his estate at Altnabreac[159] (see Figure 3.1). But, even in these nuclear oases the idea of a nuclear waste repository proved deeply controversial. At Sellafield there was opposition especially from local environmentalists who argued for on-site retrievable storage in above-surface repositories.[160] In Caithness the local district council reversed its original decision in favour of exploration for a site in the area, as political opposition grew.[161]

When Sellafield and Dounreay were eventually announced as the sites for further investigation on 21 March 1989 no attempt was made to camouflage the political nature of the choice.[162] The then Secretary of State for the Environment, Nicholas Ridley, in a written statement accepted the NIREX conclusion that, 'it would be best to explore first those sites where there is some measure of local support for civil nuclear activities'.[163] In making the statement this way, Mr Ridley overlooked Sellafield's significant military role, and, in replying in written form, avoided any immediate parliamentary debate on the issue. NIREX emphasised the 'dependence and support for the nuclear industry in these areas. Availability of land, coupled with that nuclear background, infrastructure and minimising of transport were plus points'.[164] Applications to drill boreholes were refused by the local authorities in both areas but allowed on appeal by the Secretary of State. In both communities opposition to the proposals was mobilised and another long conflict was in prospect.[165] NIREX's technical appraisals[166] were challenged by expert consultants working for national environmental groups,[167] and wider opposition even included the Nordic nations concerned about radioactive pollution of their fishing grounds.[168]

Lessons from the Conflict

During the late 1980s in the UK the search for a solution to the problem of nuclear waste had become more urgent. About 6000 tonnes of ILW was accumulating each year at the power stations and other sites and there was ten times as much LLW arising each year.[169] Once Drigg was full there would be a gap until a new site could be opened around the year 2005. The legacy of the past was creating further demands for disposal capacity. In 1989 the CEGB began decommissioning the first generation of Magnox power stations, starting with Berkeley, thereby creating further waste that would eventually have to be managed.[170] The high costs of decommissioning led the government to retain the nuclear stations in the public sector for fear of undermining the privatisation of the supply industry. Reprocessing also imposed burdens of waste and cost on the nuclear industry.[171] The House of Commons Energy Committee criticised the early (pre-1976) reprocessing contracts which meant that Britain was saddled with 1500 tonnes of foreign wastes which it would have to manage rather than return to the country of origin.[172] High costs also threatened the reprocessing industry as its major customer, the CEGB, announced plans to construct a dry store for spent fuel at Heysham.[173] This potential loss of trade was offset by a contract to reprocess spent fuel from the Federal Republic of Germany (see chapter 6). But this would involve storage and transportation of highly radioactive fuels and the creation of large volumes of LLW and ILW, thus raising once more the spectre of Britain becoming the 'nuclear dustbin' of the world.[174] A detailed re-evaluation of the costs of nuclear waste management and decommissioning undertaken by the Department of Energy as part of the privatisation process of the electricity supply industry, led to increased costings being made for these 'back-end' activities.[175] According to one estimate the costings had jumped from £4.5bn to £15bn.[176]

The government's failure to find an acceptable solution for these mounting problems during the 1980s was ascribed to its vacillation and indecisiveness by a House of Lords Committee reporting in 1988.[177] The conflict over nuclear waste had baulked the government's plans but had also established the political criteria for decision making. The possibility of developing a land-based repository on a greenfield site had been virtually eliminated. Geologically suitable conditions for exploration for nuclear waste repositories are more extensive in the UK than politically acceptable locations. Communities already familiar with the nuclear industry, or heavily dependent upon it, are most likely to accept a nuclear waste facility. By ensuring a more rational and politically sensitive strategy, the opposition had strengthened the government's prospects of achieving a publicly acceptable solution. By open discussion, careful research and adequate compensation the government could hope to divide its potential opposition. After all, if it is accepted that a solution is necessary, then the majority will be grateful that it is found somewhere else. Sellafield and Dounreay apparently satisfied both criteria.

Opponents of these sites could present some familiar arguments. They could argue that there was a need for a public inquiry to consider possible alternatives but the options had already been narrowed down to deep disposal. They could demand that the list of ten other sites regarded as acceptable should be published. If they were, and one or more of the other sites were better suited geologically, opponents of Sellafield and Caithness would be given suitable ammunition. On the other hand publication might encourage the other threatened communities to support the proposed sites at Sellafield and Dounreay. The battle of the dumps had not only defeated shallow burial and thereby removed the prospect of a nuclear waste dump from the four communities; it had also provided the government with a politically acceptable alternative that helped to legitimate the choice of Sellafield and Caithness.

During the battle of the dumps the secretive, centralised and incremental character of decision making had been tilted towards a more open, participative and rational process. The government and NIREX had been opposed by a united front able to mobilise support across party, class and community. The county councils amassed formidable technical expertise and were served by the detailed work of their consultants. Planning officials coordinated the case effectively and politicians provided the leadership and cooperation essential to focus the work of the campaign. The local pressure groups, and later BOND, applied publicity, pressure and tactical sophistication over a long period. The campaign exposed a government strategy that was premature and inadequate compared to other countries, and activated alternative disposal routes or management strategies. In this way the self-interest in preventing an undesirable activity in any one of the communities became submerged in an appeal to general principles which struck a chord with popular sentiment against the secretive decision making of the nuclear industry.

In the short-run the protests over shallow burial were a conspicuous success. By demonstrating the effectiveness of a coordinated and focused campaign the protesters had provided an example that could be followed by other communities. But it was unlikely the government would be so unprepared in the future or that circumstances would so conspire to favour the opposition. The government and NIREX, too, had drawn lessons from defeat. In the long-run the victory of the dumps may prove to have facilitated the development of policies for nuclear waste disposal that command a broad consensus. Paradoxically, the very success of the protestors may, ultimately, have enabled the nuclear industry to overcome a potential obstacle to its future expansion. But this remains open to question. The shift in political power which both creates political conflict but also opens the way for consensus can also be perceived in the United States to which we now turn.

NOTES

1. John Bunyan, *The Pilgrim's Progress*.
2. *Bedfordshire Times*, 24 November 1983.
3. *Bedfordshire Times*, 24 November 1983.
4. *Bedfordshire Times*, 27 October 1983.
5. *Bedfordshire Times*, 10 December 1983.
6. *Bedfordshire on Sunday*, 6 November 1983.
7. Bedfordshire County Council, Agenda, 15 December 1983.
8. NIREX advertisement, 'Why Nuclear Waste?'
9. *Bedfordshire Times*, 1 December 1983.
10. *Bedfordshire on Sunday*, 20 November 1983.
11. *Bedfordshire Times*, 3 November 1983.
12. *Bedfordshire Journal*, 30 November 1983.
13. *Bedfordshire Times*, 3, November 1983.
14. *Bedfordshire on Sunday*, 6 November 1983.
15. *Bedfordshire Times*, 10 November 1983.
16. *Bedfordshire on Sunday*, 27 November 1983.
17. Letter to *Bedfordshire Times*, 22 December 1983.
18. *Bedfordshire on Sunday*, 18 March 1984.
19. *Bedfordshire Times*, 13 December 1984.
20. *Bedfordshire Times*, 3 January 1985.
21. *Bedfordshire Times*, 9 February 1984.
22. *Milton Keynes Mirror*, 9 February 1984.
23. *Bedfordshire on Sunday*, 26 February 1984.
24. Bedfordshire County Council, Notes of meeting with William Waldegrave, 15 February 1984.
25. Bedfordshire County Council (1984), County Structure Plan, Proposed Alterations to Policies, Policy 97, Nuclear Waste.
26. Judgement of Mr Piers Ashworth QC, in the High Court of Justice, Royal Courts of Justice, 20 June 1984, p. 15.
27. Appeal from the High Court of Justice, Revised Judgement, Royal Courts of Justice, 26 June 1984.
28. Statement by Dr Jerry Fitch in *Bedfordshire Times*, 9 February 1984.
29. *Bedfordshire Times*, 9 August 1984.

30. *Band Wagon*, Number 2, 2 February 1984.
31. *Bedfordshire Times*, 24 May 1984.
32. *Bedfordshire Times*, 26 July 1984.
33. *Bedfordshire Times*, 25 October 1984.
34. *Bedfordshire on Sunday*, 28 October 1984.
35. Bedfordshire County Council, Oxford Clay sub-Committee, Review of the NIREX proposal, 21 December 1984.
36. *Hansard*, 24 January 1985, col. 1146.
37. Pearce, F., 'Britain plans new nuclear dumps', *New Scientist*, 24 February 1983.
38. Cleveland County Council, Industrial Promotions and Land Committee, 31 March 1983.
39. Stockton Borough Council, Extraordinary Council Meeting, 31 October 1983.
40. *Northern Echo*, 16 April 1984.
41. NIREX, 'The Disposal of Low and Intermediate-Level Radioactive Wastes: the Billingham Anhydrite Mine', A Preliminary Project Statement, Harwell, October 1983.
42. *Hansard*, 8 March 1985, col. 1298.
43. *Northern Echo*, 16 April 1984.
44. *Hartlepool Mail*, 27 October 1983.
45. Interview with Peter Curd, NIREX Information Director, 1 November 1988.
46. *Hansard*, 24 January 1985, col. 1154.
47. House of Commons, Environment Committee, Session 1985–6, *Radioactive Waste*, London, HMSO, 1986, para. 251.
48. HMSO, 'Investigation of possible increased incidence of cancer in West Cumbria', Report of the Independent Advisory Group chaired by Sir Douglas Black, London, 1984.
49. *Hansard*, 24 January 1984, col. 1146.
50. *Bedfordshire on Sunday*, 27 January 1985.
51. *Bedfordshire Journal*, 31 January 1985.
52. *Bedfordshire Times*, 31 January 1985.
53. *Bedfordshire Times*, 31 January 1985. The other papers were *Bedfordshire on Sunday, Bedfordshire Journal* and *Bedford Record*.
54. Marplan-Anglia Reports Poll, conducted 2 March 1985.
55. *Hansard*, 24 January, 1985, col. 1150.
56. *Hansard*, 8 March 1985, col. 1292.
57. Ibid., col. 1332
58. Interview in *Bedfordshire Times*, 28 March 1985.
59. *Hansard*, 8 March 1985, col. 1340.
60. Bedfordshire County Council, Briefing Notes for Delegation to the Minister, 13 June 1985.
61. Bedfordshire County Council, Oxford Clay sub-Committee, Report of meeting with Minister, 1 August 1985.
62. House of Commons, Environment Committee, Session 1984–5, *Radioactive Waste*, Minutes of Evidence, Bedfordshire County Council, p. 472.
63. Ibid., p. 482.

64. Ibid., p. 467.
65. Ibid., p. 472.
66. Bedfordshire County Council, 'A review of research and development in the field of intermediate and low-level radioactive waste management', Additional Written Evidence to the House of Commons Environment Committee, July 1985, p. 3.29.
67. *Bedfordshire Times*, 24 October 1985.
68. Friends of the Earth (1984) *The Gravedigger's Dilemma*, pp. 21–2.
69. Radioactive Waste Management Advisory Committee, *Sixth Annual Report*, August 1985.
70. HMSO (1984) '*Report of the Independent Review of Disposal of Radioactive Waste in the Northeast Atlantic*', Chairman, Professor F. G. T. Holliday, pp. 47, 55.
71. Radioactive Waste (Professional) Division of the Department of the Environment, '*Assessment of Best Practicable Environmental Options (BPEOs) for Management of Low- and Intermediate-Level Solid Radioactive Wastes*', Draft Final Report, November 1985; Report 1986.
72. Bedfordshire County Council, Oxford Clay sub-Committee, BPEO Exercise, Environmental Resources Ltd., 12 November 1985, p. 7.12.
73. Environmental Resources Ltd. '*Review of the BPEO Report*', prepared for Bedfordshire County Council, March 1986, p. 13.
74. Ibid., p. 2.
75. Environmental Resources Ltd (1986), op. cit., p. 2.
76. TUC Response to the Draft Report on Best Practicable Environmental Options for Management of Radioactive Wastes, Produced by the Department of the Environment, 6 January 1986.
77. Bedfordshire Against Nuclear Dumping (BAND) 'Evidence against a shallow land burial nuclear waste repository at Elstow in Bedfordshire, or any similar Oxford Clay based repository', 20 January 1986.
78. *Bedfordshire Times*, 23 and 30 January 1986.
79. *Bedfordshire Times*, 5 September 1985.
80. *Hansard*, 25 February 1985, col. 813.
81. NIREX, 'Disposal of Low and Intermediate Level Radioactive Wastes, A Preliminary Project Statement'. (Separate statements for each site), February 1986, p. 2.
82. *Lincolnshire Standard*, 14 March 1986.
83. Aspinwall and Company, 'Possible Sites for Shallow Burial of Radioactive Waste; Report on Programme of Research', Consulting Hydrogeologists Hydrologists, Shrewsbury, February 1981.
84. *Lincolnshire Standard*, 12 April 1985.
85. *Lincolnshire Standard*, 28 February 1986.
86. *Hansard*, 13 March 1986, col. 1094.
87. *Lincolnshire Standard*, 14 March 1986.
88. *Lincolnshire Standard*, 14 March 1986.
89. *Grimsby Evening Telegraph*, 19 September 1985.
90. Humberside County Council (1986), Humberside Structure Plan, Proposed New Policy, 1985.

91. Humberside County Council, 'Focus on the Humber Estuary', March 1986.
92. *Hansard*, 25 February 1986, col. 816.
93. Ibid., 13 March 1986, col. 1110.
94. Ibid., 25 February 1986, col. 818.
95. *Essex County Standard*, 19 August 1955.
96. Ibid., 2 February 1956.
97. Ibid., 25 October 1955.
98. Ibid., 11 May 1956.
99. Ibid., 18 November 1955.
100. *Maldon and Burnham Standard*, 27 February 1986.
101. Essex County Council, 'The Defence of Bradwell', 1986.
102. *Bedfordshire Times*, 17 April 1986.
103. County Councils Coalition, 'Radioactive Waste Disposal', Policy Statement, April 1986.
104. *Maldon and Burnham Standard*, 13 May 1986.
105. *Bedfordshire Times*, 10 April 1986.
106. Britain Opposed to Nuclear Dumping (BOND), Aims, 1986.
107. House of Commons Environment Committee (1986), *Radioactive Waste*, Session 1985–6 First Report, Vol. 1, para. 2.
108. Ibid.
109. Bedfordshire County Council, Press Release, 14 March 1986.
110. 'The Government's first stage response to the Environment Committee's Report on Radioactive Waste', 2 May 1986.
111. HMSO (1986), *Radioactive Waste: The Government's Response to the Environment Committee's Report*, Cmnd 9852.
112. Bedfordshire County Council, Press Release, 8 May 1986.
113. County Councils Coalition, 'County Councils Response to Government's White Paper on Radioactive Waste', 29 July 1986.
114. *Hansard*, 13 May 1986, col. 624.
115. *Hansard*, 21 May 1986, col. 476.
116. Ibid., col. 499.
117. *Bedfordshire Times*, 29 May 1986.
118. Consultants were appointed as follows: Elstow – Mott, Hay and Anderson (contractors, NorWest Holst £2.2M); Fulbeck – Sir Alexander Gibb and Partners (contractors, Environmental Investigation Group £3.2M); Bradwell – Sir William Halcrow and Partners (contractors, Environmental Investigation Group £2.3M.); Killingholme – Howard Humphreys and Partners (contractors, Foundation and Exploration Services £1.9M).
119. Mott, Hay and Anderson, consulting engineers, 'NIREX site investigation CEGB Storage Depot – Elstow', 27 May 1986.
120. UK NIREX Ltd, 'Proposed investigations of short-listed sites showing potential for development of a near surface disposal repository', May 1986, para. 21.
121. *Bedfordshire Times*, 21 August 1986.
122. *Guardian*, 19 August 1986.
123. *Lincolnshire Standard*, 22 August 1986.
124. *Guardian*, 19 August 1986.

125. *Lincolnshire Standard*, 22 August 1986.
126. *Guardian*, 29 August 1986.
127. *Evening Gazette*, Colchester, 5 September 1986.
128. *Lincolnshire Standard*, 15 August 1986.
129. *Maldon and Burnham Standard*, 3 September 1986.
130. *Bedfordshire Times*, 4 September 1986.
131. *Bedfordshire Times*, 14 August 1986.
132. *Bedfordshire Times*, 11 September 1986.
133. *Bedfordshire Times*, 21 August 1986.
134. *Grantham Journal*, 26 September 1986.
135. *Lincolnshire Standard*, 26 September 1986.
136. *Bedford Express*, 5 November 1986.
137. *Grantham Journal*, 7 November 1986.
138. *Hansard*, 9 February 1987, col. 44.
139. *Bedfordshire Times*, 13 November 1986.
140. ERL, *The Disposal of Radioactive Waste in Sweden, West Germany and France*, prepared for the County Councils Coalition, January 1987.
141. *Hansard*, 9 February 1987, col. 34.
142. Ibid., col. 57.
143. Ibid., col. 63.
144. Ibid., col. 64.
145. UK NIREX Ltd, 'Response to the County Councils Coalition Report on the Disposal of Radioactive Waste in Sweden, West Germany and France', February 1987.
146. Department of Energy, Sizewell B public inquiry. Report by Sir Frank Layfield, London, HMSO, January 1987.
147. *Grantham Journal*, 17 April 1987; *Bedfordshire Times*, 9 April 1987.
148. Personal conversation with A. Blowers.
149. *Bedfordshire on Sunday*, 19 April 1987.
150. *The Times*, 2 May 1987.
151. Letter to County Councils Coalition from Parliamentary Monitoring Services, consultants.
152. Letter from John Baker, Chairman of NIREX to Nicholas Ridley, Secretary of State for the Environment, 30 April 1987.
153. NIREX (1987) *The Way Forward: a Discussion Document*, Harwell, p. 4.
154. Ibid., p. 29.
155. University of East Anglia (1988), *Responses to The Way Forward*, Environmental Risk Assessment Unit, Norwich, November (the UEA report was prepared under consultancy to NIREX).
156. Edwards, R. 'Cover-up allegation on nuclear waste', *Guardian*, 10 July 1989.
157. They were expressed both at the release of 'The Way Forward' (at a House of Commons press conference) and in various meetings and seminars subsequently. Fishlock, D., Offshore N-waste dumps 'win little public support', *Financial Times*, 29 November 1988. 'NIREX accused of rigging dump report', the *Orcadian*, 17 November 1988. Kemp, R., 'The politics of siting radioactive wastes', Paper to 1989 Conference on Radioactive Waste Management, 22/23 February 1989, London.

158. Hooper, J., 'Seabed N-Waste burial divides critics', *Guardian*, 15 September 1987; 'Dumping Shock for Sellafield', *The People*, 10 May 1987; Highfield, R., 'Sellafield plans nuclear waste dump on seabed', *Daily Telegraph*, 15 September 1987.
159. Hetherington, P., 'Nuclear Peer takes role of local villain', *Guardian*, 30 June 1988.
160. Cumbrians Opposed to a Radioactive Environment, Greenpeace and Friends of the Earth, 'Radioactive Waste Management, the Environmental Approach', Briefing paper, November 1987.
'Well, do you want more nuclear waste?' *Whitehaven News*, 8 December 1988. 'Chairman talks on waste storage', 'No one else would tolerate this', *Whitehaven News*, 15 December 1988.
161. Nuclear Waste Agency well aware of 'social stigma', *John O'Groats Journal*, 2 December 1988; 'Anger as NIREX survey team heads for Caithness', *John O'Groats Journal*, 9 December 1988.
162. Clouston, E. and Edwards, R. 'Jobs key to nuclear dustbin', *Guardian*, 22 March 1989.
163. *Hansard*, 21 March 1989, cols. 505–506.
164. BNFL News, April 1989, No. 1778;
Monckton, N., 'UK Nirex: going forward', *ATOM*, No. 392, June 1989, pp. 14–17.
165. NENIG Briefing 26; NENIG Action Briefings 9 + 10 1988/9, Northern European Nuclear Information Group, Shetland Islands Townsley, M. 'Why the sound of silence echoes around Dounreay', *Observer* (Scotland), 25 June 1989.
166. UKAEA, Safety Studies, NIREX radioactive waste disposal. Presentation of the NIREX disposal safety research programme. NSS/G108, November 1988; UK NIREX Ltd, Deep Repository Project. Preliminary Environmental and Radiological Assessment and Preliminary Safety Report, No 71, March 1989.
167. Richardson P. J., Exposing the faults; the Geological Case Against the Plans by UK NIREX to Dispose of Radioactive Waste, Greenpeace/Friends of the Earch, March 1989.
168. *Hansard*, 26 July 1989, col. 714. The Prime Minister stated she had received approximately 2400 objections from Norway to the deep disposal plans.
169. Radioactive Waste Management Advisory Committee (1988, 1989), Ninth and Tenth Annual Reports.
170. Donovan, P., 'Start of the Spider's 100 year death', *Guardian*, 1 April 1989. *Hansard*, Statement on electricity privatisation and nuclear power (decommissioning and waste) 24 July 1989, cols. 744–757.
171. Donovan, P. 'Nuclear waste disposal doubts cast shadow over privatisation', *Guardian*, 22 June 1989; Gribben, R., 'State to keep nine nuclear plants', *Daily Telegraph*, 25 July 1989; *Hansard* (Lords) 25 July 1989, Electricity Bill (amendments) Final Reading, cols. 1310–1353.

172. House of Commons (1989), Select Committee on Energy, *British Nuclear Fuels Plc*, 5 April; *Hansard*, Atomic Energy Bill, 13 April, 1989, cols. 1105–1136.
173. Samuelsen, M., 'Lancashire site studies for nuclear waste dump', *Financial Times*, 16 September 1988; *Lancashire Evening Post*, Nuclear dump plan is attacked, 18 April 1989.
174. Ryan, S., 'Nuclear Waste Scandal', *Daily Mail*, 13 April 1989. Harrison, M., 'MPs say Britain must not be world's N-waste dump', *Independent*, 13 April 1989. Lowry, D., 'Buried reports on nuclear waste', *Guardian*, 9 June 1989.
175. Electricity Bill, Standing Committee E, sitting for 7 March 1989. House of Lords, Select Committee on the European Communities, *Radioactive Waste Management*, Report and Evidence, Session 1987–88.
176. Brown, P., 'Magnox bill could be £15bn', *Guardian*, 29 July 1989.
177. House of Lords (1988) Select Committee on the European Communities, *Radioactive Waste Management*, Report and Evidence, Session 1987–88.

5 The United States: In Search of the Nuclear Oasis

BACKGROUND AND GEOPOLITICAL/INSTITUTIONAL CHARACTERISTICS

The American approach to radioactive waste management can be explained by four geopolitical and institutional characteristics. The first and most obvious characteristic is the geographical scale of the country: the US is 38 times the size of Britain, and nine American states are larger than the UK in area. Moreover, the UK is far more densely populated than the US. California, the most populous state, has but half the UK's population and the seven states with the smallest population can barely muster together the population of Greater London. Geography thus confers upon the US much greater opportunity for finding remote sites for radioactive waste disposal. It also imposes greater distances between the various stages in the nuclear fuel cycle, making transportation a significant issue.[1] To take an extreme example, the Palo Verde nuclear power plant 50 miles west of Phoenix, Arizona, obtains uranium from New Mexico, which is converted to uranium hexafluoride in Illinois, sent to be enriched in Ohio, Kentucky or Tennessee, and thence to be fabricated into fuel rods in Connecticut before being shipped back to Arizona.[2] This uranium would thereby travel over 4000 miles before the fuel rods are loaded in the nuclear reactor. Power from the plant is distributed to four states and spent fuel rods are stored pending transport to a deep repository whose location has yet to be determined. Whereas Palo Verde is in the West, most of the country's 110 commercial nuclear power plant units are in the East, more accessible to front-end processes of the nuclear fuel cycle but farther from uranium sources and the potential high-level waste repository in the West. This regional imbalance underlines the geopolitical problem, namely, that *the East–West conflict is a key element in the politics of nuclear waste in the USA*.

The second characteristic, which arises from the nature of the US nuclear industry, is the political focus on spent fuel and high-level waste (HLW). The USA has almost three times as many operating reactors as the UK, with a capacity of 97.9 GW in 1990, rising to 106.5 GW by the end of the century. But the USA has no operating civil reprocessing works, which produce large volumes of HLW through chemical separation and low-level wastes through material contamination. Reprocessing was undertaken at West Valley, New York from 1966–72 until the plant was closed for commercial and environmental reasons.[3] In 1977 President Jimmy Carter declared civil reprocessing to be commercially unnecessary and likely to promote nuclear weapons proliferation; two civil reprocessing works, at Morris, Illinois, and Barnwell, South Carolina were never completed. Though former President Ronald Reagan favoured civil reprocessing it has not been able to attract sufficient private investment capital. Therefore, HLW from the civil nuclear programme is primarily in the form of spent fuel currently accumulating at power plants across the country. In 1987 an estimated 15 700 metric tons of spent fuel were stored at nuclear plants, with an estimated rise to 41 000 tons by the year 2000.[4]

Reprocessing is undertaken in the military sector within the reservations managed by the US Department of Energy (DOE). Approximately 369 100 cubic metres of reprocessing HLW is stored in liquid and solid form at Hanford in Washington State, at the Savannah River Plant (SRP) in South Carolina, and at the Idaho National Engineering Laboratory (INEL)[5] with a small amount awaiting treatment at the former civil reprocessing works at West Valley.[6] There have been problems of leakage of the military HLW, and growing public fears near the sites.[7] Furthermore, in 1987 there were 191 837 cubic metres of transuranic wastes (broadly similar to the UK's longer-lived, intermediate-level wastes) from military spent fuel reprocessing that were buried at Hanford, INEL, SRP, Los Alamos and Sandia National Laboratories, New Mexico, and Oak Ridge National Laboratory, Tennessee (Figure 5.1).[8] Some of these wastes are destined for the Waste Isolation Pilot Plant (WIPP), scheduled to be opened in late 1990 or 1991 in salt caverns near Carlsbad, New Mexico. WIPP may provide permanent disposal for the military transuranic wastes. The DOE's present estimate of 369 100 cubic metres of HLW at the military reservations is about seven times the volume of commercial spent fuel. But the volume will remain stable due to concentration, and the total radioactivity is less than 10 per cent of that from the civil programme. The military HLW will be vitrified and eventually disposed of in a commercial

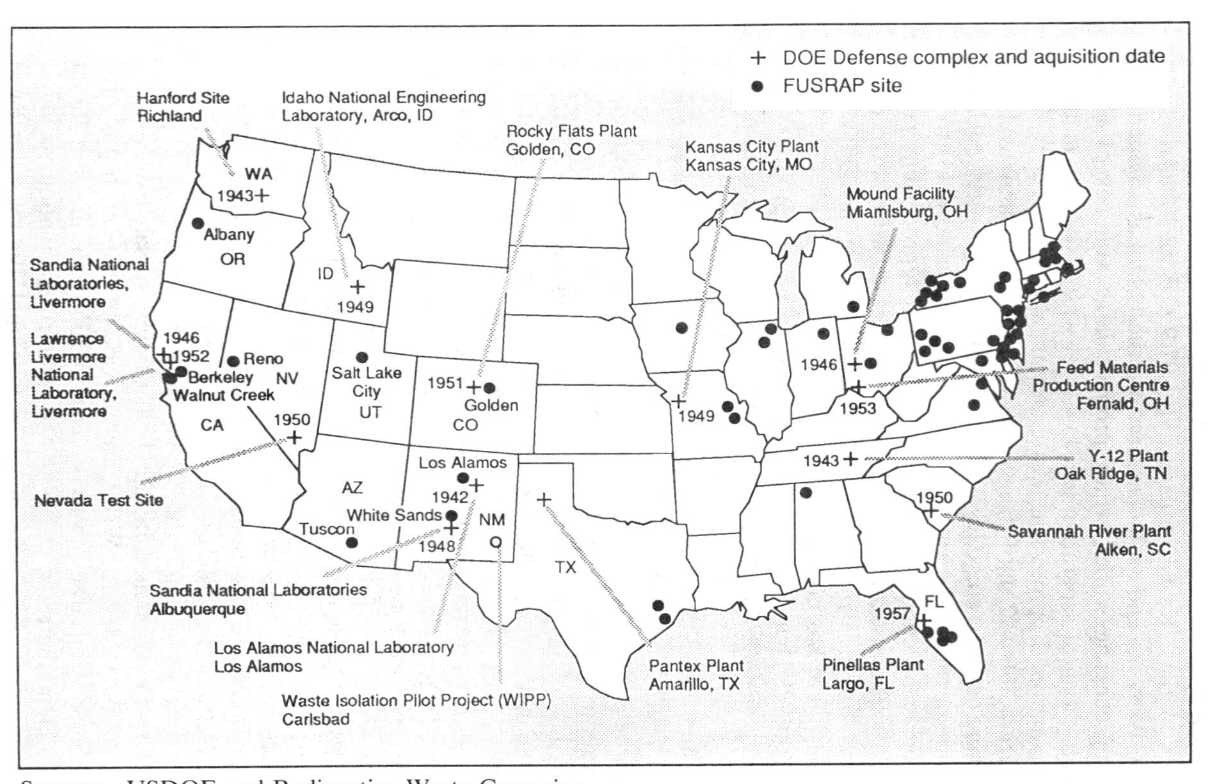

SOURCE USDOE and Radioactive Waste Campaign.
FIGURE 5.1 *Existing military nuclear waste sites in the United States*

repository. Hence, the name of the DOE's programme office – Office of Civilian Radioactive Waste Management – is somewhat deceptive.

In the USA low-level wastes (LLW) are defined as 'anything not classified as HLW' among commercial radioactive wastes,[9] and are therefore defined by what they are not. Volumes of these wastes are much greater than the HLW and it was estimated that around 1.2 million cubic metres were already in the six commercial shallow burial sites in 1986,[10] with about 2.3 million cubic metres of LLW in the military reservations. The civil wastes come mainly from nuclear power plants but also from industry, hospitals and the universities. Increased compaction of industrial radioactive wastes has held the volume steady in recent years. As a result, whereas 50 per cent of the civil LLW volume came from power plants in 1979, the proportion had risen to 80 per cent by 1982.[11]

Whereas LLW is buried in commercial and military landfills, HLW disposal constitutes a more immediate problem as existing storage areas at power plants are limited and there are fears about leakage at the military sites. Consequently: *the most urgent radioactive waste problem requiring a political solution in the USA is the disposal of high-level wastes*.

The third characteristic is the institutional separation of the civil and the military radioactive waste streams. Under the Atomic Energy Act of 1946, the US Atomic Energy Commission (AEC) was created with wide-ranging regulatory powers over the nuclear weapons programme. While the AEC and congressional Joint Committee on Atomic Energy held a monopoly over nuclear affairs for three decades, the creation of the civil nuclear power programme gradually brought the decentralised political institutions of the US more into play, as we will show later. The AEC was succeeded first by the Energy Research and Development Administration (ERDA) and Nuclear Regulatory Commission (NRC) in 1974, and then ERDA by the DOE in 1977, in an attempt by President Carter to better coordinate the energy functions of the government. During the 1960s and 1970s there was considerable regulatory reform, partly in response to growing environmentalist pressures.[12] Among these reforms was the National Environmental Policy Act of 1969 and President Richard M. Nixon's creation of the Environmental Protection Agency (EPA) in 1970, responsible *inter alia* for promulgating environmental standards, guidelines, and criteria to protect the public and the environment from radiation hazards. The EPA has issued environmental standards for spent fuel, HLW and transuranic wastes under Federal Code 40 CFR 191,[13] and was preparing a similar rule for civil LLW. While the EPA sets the standards beyond the fence, the regulation of nuclear plants and civil radioactive waste disposal facilities are the responsibility of the NRC. These standards must accommodate those set by the EPA. Furthermore, a problem of regulatory oversight has arisen between the NRC and the EPA over hazardous wastes mixed with LLW. The NRC has published its standards for the disposal of HLW in deep geological repositories (10 CFR Part 60) and its licensing requirements for shallow land burial of LLW (10 CFR Part 61).

The picture is further complicated by the role of the DOE, which has jurisdiction over military nuclear facilities: the EPA and the NRC thus have no authority over the military reservations. The EPA and the NRC are open to political participation in the debates about their standards while the DOE's nuclear weapons programme operates in the interest of national security. Thus: *political debate over radioactive waste policy has, until recently, been largely confined to the civil nuclear programme where the political process is open to greatest challenge.*

The separation of powers is not complete. Indeed, the tendency for overlapping responsiblities is a fourth characteristic of nuclear waste management policy in the USA. Within a federalist system (as in the FRG and Switzerland), the individual US states have often sought to increase their powers over nuclear matters, despite a history of federal pre-emption. Nonetheless, the deep disposal of HLW is a federal responsibility with the DOE recommending and building the repositories. Both civil and military HLW (following vitrification) will be disposed of together and the repositories will be licensed by the NRC. By contrast LLW disposal from the commercial programme is devolved to the states and either directly controlled by the NRC or indirectly by agreement with the NRC (there are 27 so-called agreement states).

The potential conflict between federal and state interests first manifested itself in the case of the WIPP. The role of the WIPP has changed numerous times (see below). Intended for disposal of military transuranic wastes in the mid 1970s, the WIPP would be suitable for spent fuel from the civil programme too, according to the DOE in 1978. 'By deciding to combine the two types of wastes at the WIPP. DOE inadvertently overlapped the legitimate activities of the federal and state governments and gave itself the hopeless task of negotiating a new boundary between them.'[14] New Mexico was 'nuclear-friendly' territory and the WIPP was viewed as both a complement to existing nuclear facilities (Los Alamos and Sandia National Laboratories, uranium mines, and nuclear-capable air force bases), and as a means of rescuing Carlsbad from economic depression caused by the run-down of its potash mining and other industries. But the prospect of civil wastes and possibly of permanent disposal was the camel's nose under the tent. New Mexico insisted on a role in decision making. The federal authorities were equally insistent that the WIPP was a military project in the national interest not subject to state concurrence or veto. Eventually the State achieved limited involvement with the DOE based on an ill-defined consultation and co-operation agreement,[15] and the project was initially to be an experiment. This dispute was merely the prelude to the Nuclear Waste Policy Act of 1982, which secured state rights over decisions on HLW disposal.

Within the institutional structure addressing radioactive wastes in the USA there is ample scope for overlapping powers and consequent conflicts between agencies or between federal and state governments. Add to this the country's predilection for detailed standard setting[16] with built-in opportunities for participation and legal challenge at every stage and the policy-making process is vulnerable to prevarication, deferral and reversal. Therefore, a fourth characteristic can be identified, namely that *the plurality of interests that can legitimately participate in policy making has resulted in policy drift.*

Given these four characteristics, it is clear that the politics of nuclear waste in the USA has a strong regional East–West dimension; is focused on the problem of high-level wastes; has been largely confined to the civil nuclear programme; and is subject to complex constitutional checks and balances that respect the plurality of interests that have to be considered. The outcome has been political *impasse* and an uncertain future, as consensus has been replaced by conflict.

FROM CONSENSUS TO CONFLICT

In the USA opposition to nuclear power and more recently to radioactive waste disposal policies has intensified, and has disrupted the policy initiatives of government and its nuclear agencies.[17] The near disaster at Three Mile Island in 1979 proved a traumatic turning point,[18] though even before this 'utilities were cancelling plants and new orders were not forthcoming as costs soared, and revenues remained depressed in the wake of the energy price shocks of the 1970s' and the resulting emphasis on energy conservation.[19] While there has been no official abandonment of the nuclear power option, for commercial reasons there have been no new reactor orders in the United States since 1978 and several plants remain unfinished or if completed, have not been commissioned. Expansion is unlikely for the foreseeable future and decommissioning is the more immediate problem.[20]

As in the UK, nuclear power contributes about 20 per cent of the country's electricity, but the scale and momentum of the US pro-gamme produces a more rapidly increasing volume of wastes. Early methods of radioactive waste disposal were rudimentary (see chapter 2). For a time some wastes were jettisoned into the oceans but the practice was outlawed by the 1983 amendment to the Marine Protection Research and Sanctuaries Act of 1972, though ocean dumping had already ceased by 1970. Much of the early LLW was buried at federal sites in Idaho and Tennessee, but in 1962 civil wastes began to be disposed of in commercial shallow-land facilities. There were originally six of these facilities but three were closed as a result of poor siting, leakage, and commercial problems – West Valley, New York (open from 1963 to 1975), Maxey Flats, Kentucky (1963–77), and Sheffield, Illinois (1967–78).[21] Significantly, the three remaining facilities at Beatty, Nevada (opened 1962), Richland, Washington (1966) and Barnwell, South Carolina (begun 1969, licensed 1971) are geographically displaced from the major regions of LLW generation in the North and East. With the closure of the three eastern sites, the LLW disposal burden falls on two sites in remote western locations and one in the Southeast. Hence, the location of operating LLW disposal facilities is inequitable and provides a geographical context for contemporary political conflict over disposal policy.

Aside from the military reservations there have been no disposal routes for HLW. West Valley contains a small volume of liquid HLW and spent fuel from its short era of experimental civil reprocessing; there is also spent fuel at the cancelled reprocessing plant in Morris, Illinois. In the early years of commercial development of nuclear power the AEC favoured long-term disposal in salt formations and its Project Salt Vault (1963–67) was to have demonstrated the safety of handling and storing solidified radioactive wastes at Lyons, Kansas. This site was selected in June 1970 for a national HLW repository, but almost immediately revelations about problems of plugging the salt in the former mine and prevention of groundwater intrusion resulted in mounting public opposition, and ultimately forced abandonment of the project.[22] One of the outcomes was the designation of the WIPP project near Carlsbad in 1972.

In 1976 President Gerald Ford announced that experts had concluded that deep geological storage in stable formations underground was the best method for disposal of HLW. By the early 1980s, the DOE was studying nine sites in six states for a potential repository. The sites were in different rock types (Figure 5.2); three in salt domes (Cypress Creek Dome and Richton Dome in Mississippi, and Vacherie Dome in Louisiana), four in bedded salts (Deaf Smith and Swisher counties in Texas, and Davis Canyon and Lavender Canyon in Utah), one in volcanic tuff (Yucca Mountain, Nevada), and one in basalt (Hanford, Washington). All of these sites were in the South or West, a long distance from the major sources of HLW in the North and East. As with LLW the stage was set for political conflict over the geography of radioactive waste disposal.

By 1980, with wastes accumulating at power stations and other nuclear sites threatening their shutdown, with growing environmental and safety concerns, and with Governors of States containing existing LLW, HLW or potential HLW repositories flexing their political muscles, the problem of radioactive waste management had reached an impasse. The Congress stepped in to break the deadlock and produced two laws that were intended to provide the basis for a national policy for radioactive waste management. High-level wastes were the subject of the Nuclear Waste Policy Act (NWPA) of 1982 and LLW was addressed by the Low Level Radioactive Waste Policy Act of 1980.[23] Although the NWPA was enacted later, HLW has been the major focus of political attention and so will be considered first.

HIGH-LEVEL WASTE POLICY – THE PROBLEM OF POLITICAL CREDIBILITY AND THE NEVADA SOLUTION

The NWPA is a subtle and sophisticated piece of legislation attempting a rational solution to a basically political and geographical problem.[24] Under the Act the DOE is responsible, *inter alia*, for developing deep geological repositories and for undertaking research and development on the disposal of HLW and spent fuel. At

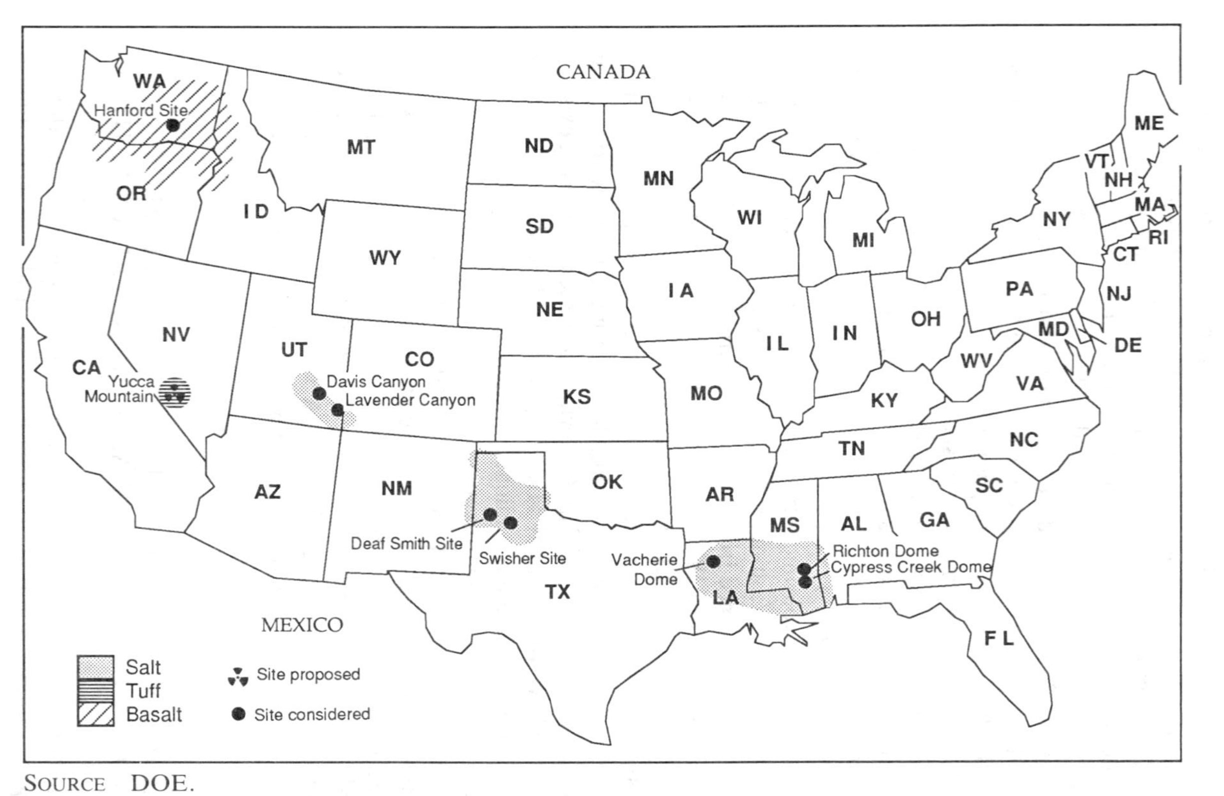

SOURCE DOE.

FIGURE 5.2 *Geological areas deemed potentially suitable for HLW disposal by the US DOE in the early 1980s*

$27 billion for the total life-cycle costs, the geological repository programme was to be one of the largest civil construction projects ever undertaken in the USA. The cost was being met from a Nuclear Waste Fund levying one tenth of a cent per kwh on all commercial nuclear electricity generators.

The Act is implemented by the EPA and NRC regulations. The EPA standard limits the release of radioactivity to the environment for 10 000 years after disposal, a time frame without precedent (40 CFR Pt. 191). The health risks should be no higher than 1000 premature deaths from cancer in the first 10 000 years of disposal of 100 000 tons of waste at an average of one death per ten years. The level of risk is comparable to the risks that future generations would face from the amount of uranium ore needed to make the fuel had it never been used. The NWPA requires assurance on marking the repository and multiple barriers, and requires waste retrievability so that future generations shall have options to correct any mistakes this generation might unintentionally make. Active institutional control is assumed for a hundred years.

The authority with which such uncertain predictions are made is also echoed in the NRC regulations for the repository in 10 CFR Pt. 60. These regulations have siting criteria emphasising the need for geological stability, depth, low population, and hydrologic isolation (the fastest path of travel time from the repository to the accessible environment 'shall be at least 1000 years'). Crucially, sites shall be located on lands that are either acquired lands under the jurisdiction and control of the DOE, or lands permanently withdrawn and reserved for its use. With such easy assurance and heroic assumptions the two agencies were able to plan for a time horizon longer than recorded human history.

This legislative judgment and technical certitude quickly fuelled public apprehension, which fed political conflict. The carefully orchestrated HLW disposal programme began to miss its deadlines. Under the Act three sites had to be recommended to the President by January 1985. He would choose one site by 31 March 1987 to begin receiving wastes in 1998. The capacity of the first repository was limited to 70 000 tons. Before construction could begin at the nominated first repository in the West, five potential sites were to be nominated from the eastern states by 1 July 1989 for a second national repository. The vital decisions on this second repository would lag seven years behind the first but it would be built well before the first repository reached capacity. The provision of a second repository was intended to achieve the essential geopolitical compromise between East and West.

There was a set of deadlines for the site selection process for the first repository. When the NWPA was passed there were nine western sites under review, which the DOE had been considering for HLW disposal for many years.[25] At the next stage in late 1984 the short-list was reduced to five, which were, in rank order, Yucca Mountain, Nevada; Richton Dome, Mississippi; Deaf Smith County, Texas; Davis Canyon, Utah; and Hanford, Washington. From these five, three were chosen on 28 May 1986 for the next stage of site characterisation before the final choice was to be made. The three sites named were those in Nevada, Washington, and Texas.

To many observers this decision had more to do with politics than technical criteria, in part because the DOE's Hanford Reservation, which originally ranked a distant fifth, appeared in second place on the short-list of three. The official reason for this decision was that the NWPA required the DOE to consider sites in diverse geological media: the selection of Hanford and Yucca Mountain were virtually guaranteed, since the three other short-listed sites were in salt. The law allowed that when the preferred site was eventually chosen the Governor or the legislature of the host state could veto the decision. Such a veto could only be overriden by a majority vote of both houses of Congress. The intention of this provision was to have the DOE consult with the States throughout the process while maintaining the primacy of the federal government acting in the national interest.

In its original form the NWPA had some prospect of achieving a first HLW repository, although delays were likely given the strength of opposition from the earmarked states. In Texas local feelings were especially strong since Deaf Smith County is in the rich farmlands of the Panhandle, and a leading agricultural producer in the State and the nation. Both the State of Texas and the local pressure group STAND (Serious Texans Against Nuclear Dumping) had campaigned to prevent Deaf Smith County from being selected. A survey revealed that over 80 per cent of a sample of residents in the area opposed the dump, 44 per cent had considered moving away and 80 per cent expected land values to drop if the HLW repository was constructed locally. Comments such as 'I don't think it is good to put it in an area where people are trying to make a living from the land, or in any populated area' or 'This is the best agricultural land in the world. The world will go hungry one day and look to us for food, but they won't take contaminated food' underline the farming preoccupations of the locality.[26] Since this is Texas, opposition groups intended to employ the three Ls – Lawyers, Legislation and Lead – to ram home their case. A lawsuit against the DOE's site selection process was filed before the short-list was announced.

The two other sites are in remote desert areas. Hanford is a nuclear oasis with a heavy local commitment to the military reservation.[27] The Yucca Mountain site in the barren Nevada desert is in a State where most land is federally owned, and where underground testing of nuclear weapons occurs nearby at Yucca Flats, the Ground Zero. Yet there has been opposition to the HLW repository in both states, especially by state officials and state residents from outside the local areas. A state veto over the final site selection was inevitable in all three short-listed states. Given the alliances that the three states could build with other western states and those politicians sympathetic to their cause, especially in the US Senate, there is no guarantee that the two-house override would occur. Furthermore, given the perceived need for an HLW repository, western interests could easily be sacrificed in favour of the national interest.

If selection of the first site was proving more difficult than imagined, the choice of a second site in the East was impossible. To comply with the 1989 deadline for the nominations for the second site, the DOE had begun studying crystalline rocks[28] in 17 states in the north-central, northeastern, and southeastern regions. In January 1986 the DOE announced 12 potentially acceptable sites in seven eastern states. There were three in Minnesota, one in Wisconsin, two in Maine, one in New Hampshire, two in Virginia, two in North Carolina, and one in Georgia. During the ensuing consultations the DOE met with its most entrenched opposition to date, at meetings, through the media, and through various forms of innovative protests – vigils, truck blockades[29] and so on. At a stroke the DOE succeeded in uniting liberal and conservative, urban and rural, and traditional and mobile communities in these areas. Canada protested against one of the proposed sites in Maine. Everywhere citizen action groups developed broad coalitions and mounted pressure on state and federal politicians. It was one thing to select federally-owned sites in the comparatively unpopulated West, quite another to violate long-settled privately-owned land in the East. The DOE had created a political earthquake in some of the most powerful states in the Union, which rippled toward the White House in a year of mid-term elections (1986), when the Republicans were on the defensive.

The opposition to the second repository could not be ignored and was felt to be a key factor in several crucial Senate and House election contests. Accordingly, on 28 May 1986, five months before the Congressional elections when the DOE announced the three western sites they also proclaimed that work to find a site for a second HLW repository was suspended indefinitely. It was unclear, however, what 'indefinite suspension' meant. While the DOE maintained that work would resume in the 1990s, if there was still a need for a second repository, critics alleged that indefinite suspension was merely a euphemism for cancellation, since the DOE had indicated that a second facility might not ultimately be needed because of lower-than-expected spent fuel projections.[30] What did emerge was that the HLW repository had been postponed for political rather than technical reasons, as would also happen in the UK in 1987 (chapter 4). Documents released by the DOE to Democratic Representatives Edward Markey of Massachusetts and Jim Weaver of Oregon, then Chairmen of the House Subcommittee on Energy Conservation and Power, and Oversight Subcommittee of the Interior and Insular Affairs Committee, respectively, revealed the thinking behind the decision. In favour of the decision was the 'political relief from second-repository states'. 'It would appear

SOURCE *Dallas Times-Herald.*

that immediately terminating the second repository would give a great deal of political benefit to DOE via the second-repository states'.[31]

The DOE recognised the potential dangers in such an approach. It might be construed as an 'obvious political ploy', would demonstrate the success of resistance and could lead to a 'loss of utility confidence in the program'. Nevertheless, a second repository could still be pursued, if needed, with first round sites. Although the DOE anticipated a 'severe political backlash from first-repository states', this was thought to be short-term. This assumption proved maladroit as the decision immediately stirred the latent East–West conflict that the NWPA had been carefully engineered to avoid. The decision unleashed a political firestorm according to Rob MacDougall of the NRC, and opponents of the western sites nominated for the first repository 'went bananas' in the words of Carol Peabody of the DOE,[32] and assaulted the programme with lawsuits, and efforts to delay the site characterisation process.

The decision to postpone the second repository succeeded in intensifying resistance to the first. A National Nuclear Waste Task Force of citizen groups was formed in the summer of 1986 to oppose and reform the DOE's siting programmes, and successfully lobbied for a cut of almost half the Congressional funding. In the light of these developments, the DOE postponed the opening date for the first HLW repository to 2010. The outcome of subsequent debates was highly uncertain though four possibilities seemed to emerge. One possibility was identification and construction of the first repository regardless of whether, when and where a second repository was reintroduced into the programme. Such a strategy could render a decay of political opposition, through a prolonged and determined site characterisation process. Moreover, given the essential nature of a HLW repository, it was highly conceivable that the three western states could be isolated by a majority of states seeking to avoid being earmarked themselves. Such a strategy could also lead each of the three states resisting the proposed repository to sacrifice their common interest for self-interest. The combination of the 'not in my backyard' syndrome on the part of individual states and the strategy of divide and rule by the federal authorities might yet secure a hapless victim from one of the three selected states. If such was the case, Hanford with its support for the nuclear industry or, more likely Nevada, the first-choice site with its remote location, would ultimately be chosen. The prospects for this outcome could be improved (some analysts felt) if some steps were taken to re-introduce a second HLW repository into the process.

The second possibility was that the western states would secure sufficient support to undermine the whole site selection process. The abandonment of the second repository had aroused hostility in the West on grounds of interregional equity, possible geological and technical problems and the long distances involved in transport, which would affect many states on the routes to the repository. The site selection process was commonly perceived as a foregone conclusion despite the elaborate procedures involved. There was every prospect that the two-House override would fail, particularly in the Senate, whose membership is drawn equally from the constituent states in the union. If such an impasse were to be reached the provisions of the NWPA could have been effectively vitiated.

There was a third possibility implicit in developments that could perhaps provide a pragmatic solution, at least in the medium term. This would be to use the WIPP near Carlsbad for transuranic wastes and to accept prolonged storage of spent fuel and HLW in a Monitored Retrievable Storage (MRS) site. The MRS concept was built into the NWPA as a facility for receiving HLW prior to disposal at the repository, although disposal could proceed regardless of any construction of a monitored retrievable storage facility. The MRS would prepare and package wastes and was to be sited in the East, Central, or Southeast part of the country near the major sources of spent fuel in order to reduce transportation. The DOE had identified three sites, all in Tennessee, with preference for the Clinch River site of the cancelled breeder reactor (near Oak Ridge), and with alternatives at Oak Ridge and Hartsville. Tennessee was suitably located, being a state with considerable existing nuclear commitments. Indeed, some business interests in economically depressed Morgan County (also near Oak Ridge) lobbied for the facility, as did a businessman in southern West Virginia.

Opposition to an MRS in Tennessee came from anti-nuclear groups who feared the prospect of spent fuel reprocessing in the State, and from the Governor, who regarded the facility as unnecessary. Another opponent was the local Democratic Congresswoman Marilyn Lloyd, otherwise a strong supporter of the 'atomic city' that dominates her Oak Ridge constituency. Although the Governor's opposition may have been a means of ensuring financial concessions from the DOE, delays in finding a HLW repository have raised fears about the real role of the MRS. When then-DOE Secretary, John S.

Herrington, announced his 28 May 1986 decision to suspend the eastern siting analysis, the whole HLW programme was in stalemate for a year and a half and the MRS took centre stage. The DOE sent its MRS proposal up to Congress in March 1987 which, with consideration of the stalled HLW siting programme, was the only game in town. The MRS, although proclaimed to be an interim site, could easily become a *de facto* permanent waste dump if no final repository were established, providing a false illusion of a solution to the waste problem.

A fourth possibility was continued storage at nuclear power plants achieved through waste compaction, reracking of spent fuel and increasing the storage space. This solution was favoured by most environmentalists. Doing nothing is a reasonable interim solution; it requires little if any transportation of HLW and allows monitoring and retrieval of the spent fuel. The matter of ultimate disposal could be reconsidered in a decade in light of new technology and changed economic conditions, with the spent fuel cooler and much less radioactive.[33]

Toward the end of 1986 it was clear that the HLW disposal policy, so carefully orchestrated through the NWPA, was in disarray. The politics of East–West conflict, the tension between federal and state authorities, and the development of alliances against the policy wherever sites have been identified had overturned consensus and promoted uncertainty.[34]

Faced with the prospect of the HLW disposal programme dead in the water, Congress acted in late 1987 to amend the NWPA. Two alternative legislative courses of action emerged. The first was a proposal by Representative Morris Udall, a Democrat from Arizona and often the champion of environmentalist interests, to place a moratorium on the siting programme and to have a study commission consider anew a broad range of HLW siting options and technologies. This alternative was supported by the National Nuclear Waste Task Force. The other option was a proposal by Senator J. Bennett Johnston Jr., a Democrat from Louisiana (who had fought in the past to keep an HLW repository out of the salt domes in his own state), to restrict site characterisation to Yucca Mountain and to add financial incentives for Nevada ultimately to accept the HLW repository. This second alternative was popularly known as the 'Johnston bill' or the 'Screw Nevada bill', depending on one's political persuasion.[35]

The Johnston bill was railroaded through the Congress with surprising ease, as part of a required appropriations bill, and was approved as the Nuclear Waste Policy Amendments Act (NWPAA) in December 1987. Johnston's victory was attributed to his growing political savvy and power, but also to the illness of Congressman Udall. In addition, no small role was played by science journalist Luther Carter, a strong advocate of the Yucca Mountain site, whose work was cited by Johnston in the Senate hearings on the NWPAA. While there were some differences between Carter's proposals and the NWPAA, the release of Carter's book a few months before the law's passage was obviously timely, though coincidental.[36]

The NWPAA was also seen as a way to save billions of scarce federal dollars by dropping the Hanford and Deaf Smith County, back-up western sites, and eastern sites from consideration for at least a decade (indeed, thousands of angry consultants were fired as their firms lost lucrative DOE contracts in the process). A small benefits package is included in the NWPAA for Nevada, to the tune of $10–20 million a year, but to receive the funds the State must waive its right to disapprove of the siting of the repository. The new stark reality, however, is that the DOE would be stuck without any immediately viable backup sites if the Yucca Mountain site ultimately proved unlicensable. To many observers, the Yucca Mountain site looked attractive on the surface (Figure 5.3): it was in a desert, far from heavily populated areas, and adjacent to the DOE's test site for nuclear weapons. Predictably, US Senator (and former Governor) from Nevada, Democrat Richard H. Bryan, among many others, took a different view. The Senator was quick to point out the potential at Yucca Mountain for earthquakes, fault movement, volcanism, rapid groundwater movement and degradation, structural

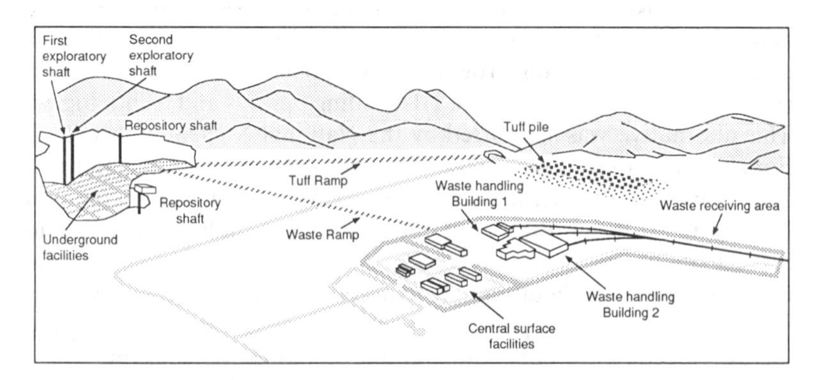

SOURCE Adapted from US DOE original.
FIGURE 5.3 *The Yucca Mountain facility in Nevada*

problems with tuff rock at high temperatures, the potential for gold and silver recovery at the site, and the repository's incompatibility with the underground testing of nuclear weapons.[37]

While there has been vocal opposition to the Yucca Mountain proposal from the Governor's Agency for Nuclear Projects and anti-nuclear groups such as the Reno-based Citizen Alert (among others), the Nevada State Legislature took an unprecedented action in May 1987: they created a new county with *no people* to surround Yucca Mountain. Bullfrog County, as it was called, was 144 square miles in area, and lasted just nine months. The idea was to get a large windfall for the State from the DOE, just in case the 'unwanted' repository was ever developed. The NWPA required the US government to pay 'Grants Equal To Taxes' (or GETT) that the project, if it was privately owned, would pay in real estate taxes.[38] Bullfrog County had the State's highest permissible tax rate and since there was no local population any unused funds could revert to the State Treasury. But in February 1988, neighbouring Nye County, which stood to lose the potential GETT payments, won a crucial judicial victory when Bullfrog County was found by a Nevada District Court to violate the State Constitution. The short life of Bullfrog County is a clear indication that when it comes to HLW disposal, even the State of Nevada has its price for acquiescence to the DOE.

The immediate effect of the NWPAA was to eliminate much of the political furore that had beset the repository siting programme from its inception. With only one site to focus on (which has no local community) most opponents of HLW disposal in the West (and elsewhere) were forced to concentrate on the risks of accidents in spent fuel shipments.[39] The DOE also received legal relief from the NWPAA, as most of the 40 plus lawsuits challenging the repository siting programme and its procedures were nullified. Most troubling to those in the scientific community about the new Act was that it contradicted one of the technical bases of the NWPA: the purpose of characterising multiple sites was to develop the knowledge base for selecting the most suitable hydrogeological regime for HLW disposal. The new focus on Yucca Mountain in effect makes that decision prematurely, which could ultimately prove to be a critical mistake.

THE MRS COMMISSION: BACK TO TENNESSEE?

The NWPA required the DOE to consider an MRS as an 'option' in the HLW management scheme. Often referred to as an away-from-reactor (AFR) storage facility, an MRS has been proposed as a sort of 'relief valve' for the nuclear power industry, whose power plants are quickly running out of spent fuel storage space. Officially, an MRS would be a centralised packaging and processing plant for spent fuel shipped from the 110 nuclear reactor units throughout the country. It probably would be located at the surface, though some analysts such as Luther Carter have advocated a below-ground vault. The DOE's search for an MRS site in Tennessee was delayed by court challenges to the siting procedures, until a Federal Appeals Court ruled against the State of Tennessee in 1986 and cleared the way for the MRS.

The opponents of a Clinch River MRS received a reprieve when the Congress passed the NWPAA in December 1987. The DOE's existing proposal was 'annulled and revoked', though the Department was authorised to build and operate an MRS following a new round of siting analyses. A three-member MRS Commission was established by the new Act for purposes of reporting to Congress by 1 November 1989 on the need for an MRS. After the inevitable political manoeuvring, the Commission members were appointed in May 1988. The Commission's prime considerations were to be the economic trade-offs of an MRS as against at-reactor storage of spent fuel, and the inherent advantages of an MRS in the national waste management system for HLW.

After the MRS Commission reported to Congress, the DOE could begin a new site survey for the MRS. It was possible that the site, presuming of course that the MRS Commission gave the expected green light, could end up back in Tennessee, and would almost certainly be proposed in the East near the majority of nuclear plants. An environmental impact statement on an MRS would not be required until a construction licence was granted by the NRC; thus, no rigorous site comparisons were required. Once the site selection is made, a benefits package of $5–10 million a year would be available to a state or Indian Tribe hosting an MRS, but as in the case of the HLW repository, only if they surrendered their right to oppose the project. The financial and institutional terms of the arrangement would be worked out with a Nuclear Waste Negotiator, established within the Executive Office of the President by the NWPAA.

A crucial aspect of the NWPAA and a concession to Congressman Udall was the linkage between the MRS and the HLW repository. Specifically, construction of the MRS could not begin until a licence for building the HLW repository was issued by the NRC. The capacity of the MRS was also limited to 15 000 metric tons of spent fuel, well below the 1989 inventory and only half the capacity the DOE had previously sought. Finally, construction of the MRS or its acceptance of nuclear waste would be prohibited in the event that there was a permanent end to repository construction. Thus, the NWPAA lowered the chance that the MRS, if built, could turn into a permanent HLW repository. According to Caroline Petti, then Coordinator of the National Nuclear Waste Task Force,[40] the nuclear industry saw these restrictions as much too severe, and were not sure if they wanted an MRS any more. Consequently, time will tell if the MRS Commission paves the way for the MRS to return to Tennessee.

THE WIPP: A NUCLEAR OASIS GETS THE JITTERS

Located in the far southeastern corner of New Mexico, the USA's first deep geological repository has been constructed to receive long-lived radioactive wastes. Once the Land Withdrawal bill is approved by the Congress in 1990 there may be no further obstacles to commissioning the Waste Isolation Pilot Plant (WIPP), 2150 feet below the featureless rangelands of the Chihuahua Desert (Figure 5.4), 26 miles east of the town of Carlsbad (population 29 000). In New Mexico there have been hopes and fears that the WIPP may simply be the prelude to a far larger project, the country's first HLW repository.[41]

This prospect may look improbable in light of the NWPAA, which focuses HLW disposal analysis on Yucca Mountain. But the Nevada solution could be overturned as a result of technical problems at the site and the political opposition mobilised by State leaders. As K. David Pijawka of Arizona State University (who has studied socioeconomic impacts of the Yucca Mountain proposal) put it,[42] 'If Nevada fails you'll have no waste management system – it's the last throw'. In this situation locating the deep repository at the New Mexico site may well become the only politically-feasible solution to the problem of disposing of the country's HLW. Critics were quick to point to heat experiments at the WIPP, which are only relevant for HLW disposal.

While efforts to promote nuclear waste management programmes elsewhere in the USA were met by intense political conflict, the WIPP arrived in New Mexico by stealth. The National Academy of Sciences had in 1957 proclaimed salt to be the most promising geological medium for radioactive waste disposal because of its dryness

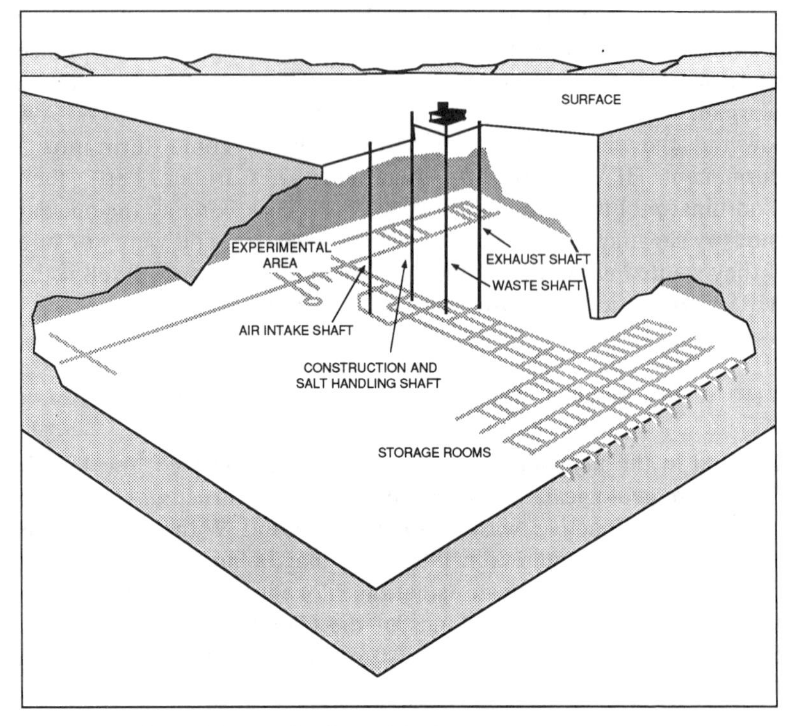

SOURCE Adapted from USDOE original.
FIGURE 5.4 *The WIPP facility in New Mexico*

and plasticity, which enables natural encapsulation of nuclear wastes. The failing potash industry of Carlsbad was casting around for alternative employment in the area, and, with the full support of local politicians, invited the AEC (the DOE's predecessor) to come to New Mexico. Initial investigations in the Delaware Basin (the bedded salt formation underlying the area) revealed drilling problems but eventually the present site was selected and construction of the $700 million project began in 1981.

Opposition to the WIPP has been minimised by federal control and limited concessions to state interests. The WIPP was designated as a research and development facility, for defence transuranic (non-heat producing) wastes only. This excludes it from the provisions of the HLW Acts and also eliminates the participation of the NRC, whose licensing powers only extend over civilian wastes. Furthermore, during the first five years of disposal the wastes must be retrievable, thus initially eliminating the need to comply with the EPA standards for the permanent emplacement of nuclear wastes. The standards were challenged by environmentalists as inadequate, but the EPA requirement that a repository should not pose a significant health risk for 10 000 years is the most stringent to date.

The State of New Mexico has been kept at arm's length by the DOE and its predecessors, though this has not been easy.[43] In 1972 the WIPP was to be a pilot plant for commercial HLW disposal, with later disposal of military transuranic wastes from nuclear facilities. In 1975 ERDA redefined the WIPP as an unlicensed facility for just transuranic wastes. Next, when the DOE was created in late 1977, it was rumoured that the Department wanted to change the WIPP into a repository for the more plentiful military HLW. Then, in 1978, the DOE's Deutch Report recommended demonstrating commercial spent fuel disposal at the WIPP, along with the transuranic wastes *without* retrievability.[44] Finally in 1979, Congress passed the WIPP Authorisation Act, which again restricted the WIPP to military transuranic wastes: the retrievability requirement, too, was eventually restored. Using the WIPP for spent fuel disposal would have brought the State into direct conflict with federal authorities. When President Reagan gave the project the go-ahead in 1981 (President Carter had tried to cancel it) the deadlock was broken.

The New Mexico Environmental Evaluation Group (EEG) was established in 1978 with DOE funding to provide an independent scientific appraisal of the project. While the EEG has been generally supportive of the WIPP, its reports have urged a more cautious approach while problems of brine seepage, waste processing and containment are further investigated.[45] The EEG's independence was undermined in late 1987 when the state's pro-WIPP Governor, Garrey E. Carruthers, tried to impose a pay freeze, a move of the EEG from Santa Fe to Carlsbad, and a condition that all EEG reports be approved by the State's administration before publication. The situation was saved in March 1988 by a formula agreed upon by the Governor and the State's two US Senators. This agreement removed the EEG from State administrative control and placed them under the New Mexico Institute of Mining & Technology in Socorro and moved them to Albuquerque. The move was a face saver for the Governor but, according to EEG geologist Lokesh Chaturvedi,[46] it was a 'silly solution' since Albuquerque is neither the State capital nor the site of the WIPP and would bring the EEG 'right under the noses of the major environmental opponents of the projects'.

Opponents of the WIPP have emphasised the technical flaws in its design. Roger Y. Anderson, formerly with the Sandia National Laboratories and now a professor of geology at the University of New Mexico in Albuquerque, points to evidence of salt dissolution above the site that would speed up the movement of water, as well as evidence that further penetration of the oil-bearing formations beneath may release pressures that could carry plutonium-contaminated wastes up into the overlying aquifer. More vexing is the problem of brine seepage, which might eventually saturate the wastes and create a pathway to the biosphere,[47] a possibility also acknowledged by the EEG and the National Academy of Sciences. Anderson and his colleagues argue that a slurry of liquid wastes under pressure or released by drilling operations might be carried into the biosphere within a few hundred years, rather than beyond the 10 000 years required by the EPA standards. The Albuquerque-based Sandia Laboratories have published figures refuting these claims and have attacked the credibility of Anderson and other scientists who question the suitability of the project. As Anderson says,[48] the WIPP's proponents insist that 'if there are problems we'll fix it' and 'nothing cannot make it work'.

Although the EEG believes that the site would be safe from the effects of salt dissolution, Chaturvedi acknowledges that any possible problems from brine seepage could be averted by engineered barriers and by ensuring that all wastes are compacted and solidified. The cost of injecting a million drums of liquid wastes, however, at \$5000 per drum would prove prohibitive. Moreover, the need to ensure the integrity of the wastes is likely to be overwhelmed by the political imperative of commissioning the country's first deep repository for radioactive wastes. As Chaturvedi says, 'The entire future of nuclear waste in this country is riding on WIPP'.

The nuclear industry in New Mexico has a political momentum that may transcend technical uncertainties. While the enthusiasm of Carlsbad's previous Mayor, Walter T. Gerrells, who was once quoted as saying that 'It'll be good for the economy if it doesn't kill us all', might not have been shared by everyone, there is little doubt that the project was widely welcomed in the local community. In the view of Donald Hancock of the Southwest Research and Information Center,[49] an Albuquerque-based citizens group, and a longstanding opponent of the WIPP, there was no good reason for the Carlsbad site to be chosen other than the political one – they want to get rid of the waste and Carlsbad wants the jobs. In a relatively poor state with only

1.5 million people the contemporary nuclear commitment was powerful and aided by the pro-WIPP stance of Governor Carruthers.

Opposition to the project was largely confined to groups located in Albuquerque and Santa Fe 250 to 300 miles away. Groups such as the Committee to Make WIPP Safe (a group of local scientists and health professionals that was formed in 1987) and the Southwest Research and Information Center accepted that the site would open and have focused on environmental and transportation hazards. Transportation hazards are the source of opposition from corridor states through which the waste shipments would have to travel, notably from neighbouring Colorado, astride the route carrying about four-fifths of the waste consignments. Anxieties could eventually be defused by adequate emergency planning, highway investments providing routes around the major cities, and NRC certification of transport packaging.

New Mexico's support for the WIPP has been bought relatively cheaply. The WIPP represents a $700 million capital investment, and an annual income of $60 million for a total over the expected lifetime of the project of $2 billion. Although the WIPP provided 671 peak construction jobs, much of the investment has leaked away in contracts to contractors based as far away as Pittsburgh and California. Chaturvedi estimated that only $16 million of the capital investment found its way to New Mexico. He observed, 'High-level waste is where the real money is'. Under the NWPAA the host state for the nation's first HLW repository is eligible for a $10–20 million 'bribe' per year in addition to a $30 million supercomputer discussed during Senate hearings (which may end up at the University of Nevada-Las Vegas). While Yucca Mountain is currently the favoured site the rapidly changing politics of nuclear waste disposal could eventually lead to another impasse. If that happens, Chaturvedi reckoned, 'the impetus for looking at New Mexico will be great indeed'.

Yucca Mountain has stiff political obstacles to surmount: the WIPP already exists and could pave the way for a HLW repository in the salt formation at or near the site. The financial incentives make it an attractive proposition in a state where the nuclear industry is welcomed. Until the WIPP's opening became imminent, technical rather than political hurdles needed to be overcome in New Mexico. Indeed, a last minute obstacle to the commissioning of the WIPP was provided by a 1988 report of the National Academy of Sciences, which urged great caution because of the uncertainties regarding brine accumulation.[50] But many of the citizen groups opposed to the WIPP seized upon this delay and have intensified their campaigns. As

Myla Reson of the Santa Fe-based Concerned Citizens for Nuclear Safety said, 'it has taken years of work and the efforts of thousands of people working against WIPP to get us where we are now. We are in this fight to stay'.[51] But, as Hancock ruefully observed, 'In this country when you have insurmountable problems you do one of two things – you declare it a non-problem, or you shove it down somebody's throat'.

LOW-LEVEL CIVIL WASTE POLICY: THE PARADOX OF TOO MANY SITES?

With the three eastern LLW burial sites closed in the wake of public anxiety over the Three Mile Island accident in Pennsylvania in March 1979, the Governors of South Carolina, Washington, and Nevada were concerned over the inequities of providing LLW disposal facilities for the remaining 47 states. In 1978, Governor Richard Riley of South Carolina had announced a huge reduction in the volume of wastes allowed into the Barnwell site, while the Richland and Beatty sites were closed for short periods in 1979. These actions followed a series of transportation mishaps at the sites. The three Governors made it clear that if there was no reform of federal policy they would limit or prohibit LLW at their sites. This precipitated the passing of the Low Level Radioactive Waste Policy Act (LLRWPA) at the end of the lame-duck Congress of 1980. The provisions of the Act were very simple – states would be responsible for the disposal of LLW through regional compacts to be formed by the states and approved by the Congress. Existing LLW dumps were to be closed by 1986 and each compact would provide a site to take its wastes and would be able to exclude wastes from other compacts or 'unsited' states. This action-forcing legislation set specific deadlines for planning and building LLW disposal facilities but left states with the freedom to form compacts as they saw fit or to go it alone.

The LLW policy relies on states to handle technical issues (which are often beyond their expertise) through co-operation with compact member states, few of which have any experience with membership in interstate bodies addressing toxic substances.[52] Although compacts have been invoked where neither the federal government nor individual states claim jurisdiction, they have hitherto addressed matters of mutual benefit. The proposal for LLW compacts was the first attempt to establish a national system of regional compacts to solve a problem that few states wished to confront. Moreover, it was to operate under federal legislation that delegated responsibility to the states. In effect both the federal and state governments had delegated the function to quasi-governmental regional authorities, which were not subject to the Administrative Procedures Act and that were not, therefore, obliged to hold hearings or engage in public participation. Nobody quite knew the legal or political implications of this proposal. All that was clear was that a compact required two or more states, which did not even need to be geographically contiguous.

Furthermore, little was known about the sources, volumes and radioactivity of LLW, or about the acceptable techniques of disposal. Estimates of LLW quantities and burial capacities varied depending on how amounts were measured and on the assumptions made regarding volume reduction through compaction, incineration, and so on. Prior to passage of the LLRWPA, in 1979 the states providing the largest volumes of LLW were, in order, New York, South Carolina, Pennsylvania, Illinois, North Carolina and Massachusetts, all in the eastern half of the country with only one state operating a disposal site (Barnwell). In fact, in 1979 Barnwell received nearly four-fifths of the total LLW.[53] By 1986 the leading LLW-generating states were, in rank order, Illinois, Pennsylvania, New York, South Carolina, California and North Carolina;[54] among these states again only South Carolina had an operating disposal site.

The NRC regulations under 10 CFR Part 61 were intended to promote those practices that experience has shown to be acceptable and to prohibit those that have not.[55] The preferred disposal technique has been land burial within 30 metres of the surface. 'A cornerstone of the system is stability – stability of the waste so that once emplaced and covered, the access of water to the waste can be minimised.' Institutional control is intended to last for 100 years and containment for the most radioactive wastes (classified as 'C' wastes) must be secure for 500 years. In 1988 the NRC also determined that a small quantity of reactor wastes referred to as 'greater than Class "C" ' LLW would be disposed of in the HLW repository.[56] Standards being prepared by the EPA are based on exposure limits with the 'As Low As Reasonably Achievable' (ALARA) concept.[57] The EPA is also considering criteria for identifying LLW 'Below Regulatory Concern' (BRC), which can be disposed of at municipal sanitary landfills. These BRC wastes account for about 35 per cent of all commercial LLW by volume. The NRC established its criteria for demonstrating which wastes qualify as BRC in 1986.

The process of compact formation was dilatory within the general framework of the LLRWPA and NRC regulations. The freedom for states to choose partners was a recipe for promoting self-interest. Those states that neighboured the existing three LLW sites were most anxious to form compacts, knowing that they would have access to a disposal site and could exclude wastes from unsited compacts. In the Northwest, Oregon, Idaho, Montana, Utah, Hawaii and Alaska linked with Washington where the Richland site was operating (Figure 5.5). The Rocky Mountain compact consisting of New Mexico, Colorado and Wyoming formed around non-contiguous Nevada with its site at Beatty. In the Southeast a compact based on the Barnwell site in South Carolina includes Virginia, North Carolina, Tennessee, Mississippi, Alabama, Georgia and Florida. At the time of their formation, these three sited compacts accounted for 41 per cent of the volume of civil LLW and 25 per cent of the radioactivity. The other compact that formed early was the Central States, composed of Nebraska, Kansas, Oklahoma, Arkansas and Louisiana, which together generated only 3 per cent of the nation's civil LLW and could thereby avoid receiving the waste of the large producers.

The large LLW generators have tried desperately to avoid responsibility for high volumes of wastes from other states. Illinois, for example, pursued its own best interests and hooked up with Kentucky, a small waste generator with the shutdown site at Maxey Flats. Initially, Illinois was part of a putative Midwest compact covering ten states. When it became clear that Illinois would be the target for a disposal site the State found a single willing partner in Kentucky. In this marriage of convenience Illinois accepted responsiblity for finding a site for its own waste and the small amounts from Kentucky. The Illinois Department of Nuclear Safety began site characterisation in late 1988 near Martinsville, in the eastern part of the State near Terre Haute, Indiana. Three backup sites were also identified. Farther East, Pennsylvania was involved in the possible formation of a large Northeast compact involving 11 states, including two of the top three waste producers. Like Illinois, Pennsylvania broke away and agreed to host a site in a compact with small LLW generators, in this case the Appalachian compact, with West Virginia, Maryland and Delaware. Two other major waste producers, New York and Massachusetts, are unaffiliated though New York may join the smaller New England states, and Massachusetts might harbour tiny Rhode Island. In the interim Rhode Island and

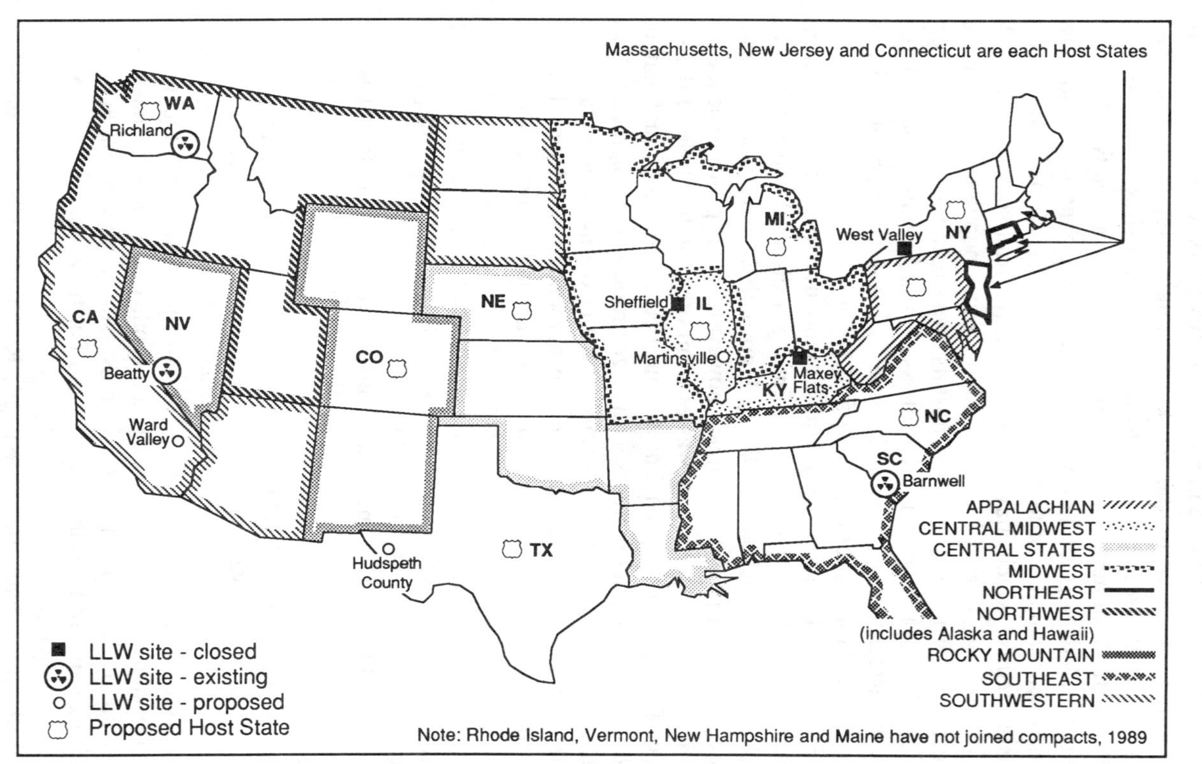

FIGURE 5.5 *The inter-state compact arrangements for LLW management*

Washington, DC have contracted for two years to send their small quantities of LLW to the Rocky Mountain site. New York and Massachusetts were prepared to find their own sites, though in the latter state voter approval is required to develop a facility. The rump of the failed Northeast compact consists of Connecticut and New Jersey, two smaller producers of wastes. Yet *each* state will host a waste facility, which may require extensive transshipment of LLW through New York City, the nation's most populous city. In the West, California had agreed to host its own site in the Mojave Desert (Ward Valley) despite local protests,[58] and eventually signed the Southwestern compact with the small LLW-generators of Arizona and the distant Dakotas (Figure 5.5).

Elsewhere there was considerable manoeuvre but little real action. Texas decided to go it alone by finding a disposal site for its own LLW, which it picked in 1987 in Hudspeth County near El Paso. Since it has not joined a compact, however, it may prove to be legally difficult for Texas to refuse wastes from other states. Five of the six New England states (all but Connecticut) remained unaffiliated, though they sought sanctuary with large producers that agreed to host a LLW site, or tried to join compacts with existing disposal facilities. For example, Maine tried to get an agreement for disposal in the Northwest, dallied with the idea of joining a large Northeast compact or a small New England compact with New Hampshire and Vermont, but ended up staying unaffiliated. Ultimately the outcome of a 1985 referendum required statewide approval for storage and disposal of any radioactive wastes in the State.

By 1984 it had become clear that the 1986 deadline for LLW site closures would certainly be missed. Procrastination was a political inevitability for, 'not only are the issues difficult but, for many states, the strategy of delay has made sense'.[59] The three sited compacts had been content to wait while the remaining states formed up so that they could then invoke the exclusionary clause. The majority could have sabotaged the LLRWPA but the possiblity that the federal government would take over responsibility for LLW management precluded this option. Instead most states played an elaborate waiting game of bluff and counter-bluff as they jockeyed for position in the compacts.

With these inadequacies evident, Republican US Senator Strom Thurmond and Democratic Governor Richard Riley of South Carolina, with Congressman Udall successfully pushed for a series of Amendments to the LLRWPA in 1985. In return for keeping the existing disposal sites open until the end of 1992, the amendments instituted a new set of deadlines (called milestones), incentives and penalties. A cap of 11.9 million cubic feet was placed on the total LLW disposal during the seven years 1985–92, with annual limits at each of the three existing sites (that is, Barnwell 1.2 million ft^3, Richland 1.4 million ft^3, and Beatty 200 000 ft^3). Monthly allocations were given to the utilities with less generous and declining allowances for those in unsited regions in order to encourage volume reduction, site selection, and facility construction. For similar reasons escalating surcharges (1986–87 $10/$ft^3$, 1988–89 $20/$ft^3$, and 1990–92 $40/$ft^3$) would be levied on utilities in unsited regions, 75 per cent going to the three host states and 25 per cent being refunded either to states that built a disposal facility or to utilities that continued waste storage. By 1996 states would be required to take possession of their LLW. In a letter to all utilities the Governors of South Carolina, Washington and Nevada warned,

> All nuclear utilities must recognize that further efforts to make substantial changes in this delicately balanced compromise package are likely to destroy the opportunity to resolve the low level waste issue (16 September 1985).

They insisted that the existing situation was 'totally unfair' and that their 'patience was running out'. Attempts to alter the amendments would be resisted,

> If such changes are made, we will have no choice but to withdraw our support promptly from this legislation and to employ all our available resources to prevent its enactment and to take other actions to remove the unfair disposal burden for our states.

By 1989 all nine compacts had been ratified by the US Congress, while seven states remained unaffiliated (Figure 5.5). But the pattern of compacts obviously owes more to politics than to any notion of geographical balance. At one extreme is the Rocky Mountain compact with less than 1 per cent of the nation's LLW volume, and at the other extreme the Southeast compact accounts for 33 per cent. Moreover, sites in those compacts with low waste volumes may not prove to be commercially viable.

While compact formation had proved difficult, finding states to host sites and then selecting sites within those states is proving to be a political minefield, even with the more realistic milestones of the amended Act. Few states have volunteered to host a site. In the

Northwest, with only 5 per cent of the nation's LLW, Washington is the host and the Richland site will be kept open. In the Rocky Mountain compact Colorado is the host, although the Beatty site will remain open until 1993. Illinois with 90 per cent of the Central Midwest LLW is considering sites and Pennsylvania will provide the first site in Appalachia. Three other large LLW generators, California, New York and Massachusetts, have accepted the need to find sites in their own states; California has already proposed a site. The Midwest compact had called for volunteers, but ultimately Michigan was selected as its first host state. The Central States compact invited developers to find a site since no volunteer came forward; a very reluctant Nebraska eventually was chosen as the host state.[60]

It was in the Southeast, the largest and one of the first compacts, where the problems of host state selection first became manifest. North Carolina was chosen from a short-list of three states to replace the Barnwell site after 1992, but threatened to pull out of the compact on the grounds that the judgement was unfair. The North Carolina Governor was prepared to 'take our turn', but wanted to make sure that 'states that are going to take advantage of our hospitality won't withdraw after 20 years'. However, if the State withdrew they would still be responsible for their own waste and could not stop LLW coming in from elsewhere. This prompted North Carolina to remain in the Southeast compact, the rules of which were amended in 1987 to limit the operating period of their LLW site to 20 years, followed by a new site in another state.

In the East, the technique of shallow land burial, preferred by the NRC, has been resisted by many states. Experience with leaks in the former sites in Kentucky, Illinois, and New York underlined the problems of waste burial in high rainfall regions. Shallow burial has been ruled an unacceptable technology by the Central Midwest, Appalachian, and Midwest compacts, and the states of New York, Massachusetts, and Texas. In each of these regions alternative technologies are being explored, such as the use of mines (New York) and above or below-ground vaults (Texas) (Figure 5.6). Shallow burial is the preferred technique only in the drier West.[61]

It seems highly unlikely that all problems concerning LLW site selection and disposal technology can be resolved by the new 1993 deadline. The LLRWPA represents, according to one of its architects, Holmes Brown (formerly with the National Governors' Association),[62] 'a brave experiment in a new type of federalism'. But by adopting a system heavily dependent on interstate co-operation,

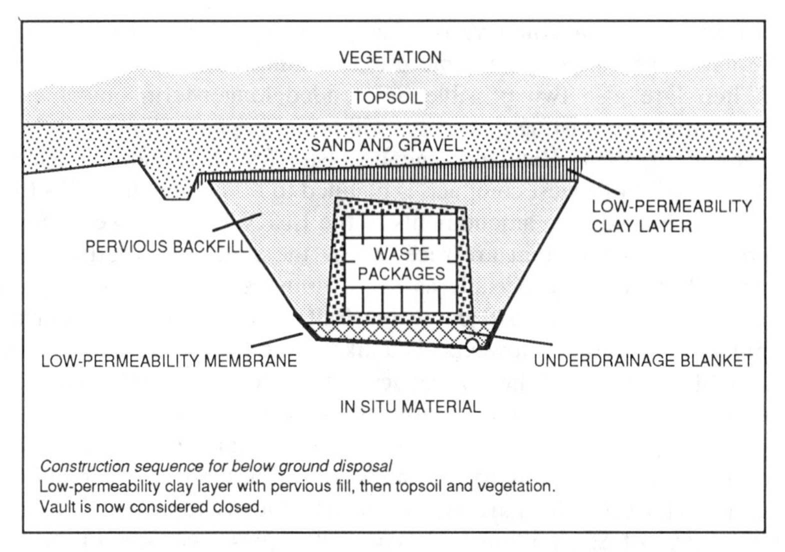

SOURCE Adapted from US Nuclear Regulatory Commission original.
FIGURE 5.6 *Conceptual representation of near-surface LLW repository*

the Act has transferred the acrimonious politics of radioactive waste from an impasse in the Congress to unstable coalitions where self interest overrides common purpose. The pattern of conflict is complex, involving compacts with existing sites and those without, 50 state legislatures and governors, a host of utilities, and pressure groups. Responsibility for ensuring implementation of the LLRWPA is formally with the compacts but, in practice, depends on the willing compliance of the states. Ultimately, if the Act fails, the problem of LLW disposal will resort to the federal government. As Holmes Brown indicates the Act leaves each state with a Hobson's choice: 'We'll shoot all six or you decide which one'.

While the LLRWPA may eventually lead to new disposal facilities in regions where high waste volumes make such a proposition commercially viable, a balanced regional system of LLW disposal seems, at the time of writing, unlikely. Instead, continued storage at nuclear reactor sites and other waste-generating establishments is a likely outcome, at least in the short term. The implication is that the nation's 110 nuclear power units, already likely to become *de facto* repositories for used fuel rods because of the failure of the HLW policy, could inherit the role as above-ground storage centres for LLW as well.[63] Rather than the 'six or seven sites strategically located around the country, as initially envisaged by the LLRWPA, the nation could easily have more than a hundred'.[64]

There are also two possible unintended longer-term outcomes. One is that the existing three sites will remain open well beyond the deadline of 1993. Richland, the host site for the small (at least LLW-wise) Northwest compact, is planned to remain open for 10–15 years. In Nevada the amendments to the LLRWPA were welcomed and the surcharges that are received by the state are an attractive financial incentive, especially as LLW volumes are increasing. The surcharge will rise to $40 per cubic foot, just as closure becomes imminent. Colorado, however, is unlikely to have an alternative site available in time and there is capacity at Beatty for at least 20 more years. In South Carolina, Barnwell is regarded as a successful operation providing a fifth of the revenue of Barnwell County and contributing $4.8 million to the state's education fund, and $6 million from surcharges, which are increasing. The facility represents a total investment of $52 million and Barnwell enjoys local support in a distressed agricultural region where the nuclear industry at the nearby Savannah River Plant provides 11 000 jobs. An interesting aspect of the surcharge system is that, while the charges are still relatively small on a per-reactor basis, they will be quite a windfall for the three sited states, which may at long last be tempted to stop discouraging disposal of LLW within their borders.[65]

The other alternative could be reversion to federal control of LLW. If the LLRWPA fails to produce a 'solution', or if the existing sites are closed, the federal government will be forced to step in. It is unlikely that *long-term* storage of LLW at nuclear power plants, supported by some environmentalist groups, will be condoned at the federal regulatory level. A federal solution to the LLW problem could be to use sites on federal (military) reservations controlled by the DOE. Two obvious areas would be the Savannah River Plant (SRP) and Hanford Reservation. It would be ironic indeed if the efforts of the three states to remove LLW succeeded only in moving the problem to nearby federal sites outside state control.

Despite innovative legislation and intensive negotiations, the prospect for a truly regional solution to LLW disposal, as of 1990, was remote. The LLRWPA has resulted in small compacts that include some non-contiguous states, potentially resulting in over a dozen LLW disposal sites, twice as many as had been anticipated. This Act has demonstrated the power of politics to produce irrational and sub-optimal solutions to the nuclear waste problem. As with HLW, the government's attempt to secure a comprehensive policy has festered in political conflict.

THE LOCATION AND STATUS OF MILITARY LLW[66]

The fact that the massive US nuclear armaments production complex also makes substantial quantities of radioactive wastes of all levels is often overlooked. But with nuclear weapons production beginning in the mid 1940s, military wastes were the first nuclear waste problem in the USA (chapter 2). Indeed, the military produces twice as much LLW as the civil sector, with three times the radioactivity.[67] This contrasts with the military HLW which, though much more voluminous, is considerably less radioactive than its civilian counterpart. The DOE, however, has long delayed waste clean-up at the military sites, while they have focused on the alleged need for new defence reactors to produce plutonium and tritium. The DOE has argued that such new facilities are needed in the wake of the mothballing of the Hanford N-Reactor in February 1988, and problems with the Savannah River Plant (SRP). The three reactors at the latter plant came under heavy fire and were closed for safety reasons in 1988, when a long history of secret accidents at the SRP was uncovered by the Congress.[68] But according to the former DOE Secretary, John Herrington,[69] the nation is 'awash in plutonium', and tritium requirements (last supplied by the SRP) for nuclear weapons are debatable, and in any event could be met without nuclear fission. Similarly troubling is the concern that spending perhaps $7 billion to build more military production reactors could reduce funding for radioactive waste cleanup at the DOE's *existing* weapons sites. The cost of such cleanup was conservatively estimated in 1988 to exceed $100 billion, with over 40 per cent of that for Hanford alone.[70]

The Hanford and SRP Reservations, though they account for over half of the existing military LLW volume, are not the only DOE facilities that produce LLW. Facilities involved in some aspect of nuclear weapons production are located in 13 states: Washington, California, Nevada, Colorado, Idaho, New Mexico, Texas, Missouri, Ohio, Kentucky, Tennessee, South Carolina and Florida. Once immune from the politics surrounding nuclear power and the nuclear arms race, many of these plants came under very intense scrutiny in

1988. This is partly attributable to the publicity and notoriety generated by four publications: the National Academy of Sciences' report, following the Chernobyl disaster, which led to the mothballing of the superficially similar Hanford N-Reactor;[71] a report by a major environmental group, the Natural Resources Defense Council, which described the US nuclear weapons facilities;[72] the Radioactive Waste Campaign's report on military nuclear wastes;[73] and the DOE's own assessment of clean-up problems at its military nuclear sites.[74]

Apart from the SRP, the major military LLW sites in the USA are concentrated in the West. Over 99 per cent of the military LLW that has been accounted for since generation began in World War II is buried at six locations: Hanford, the SRP, the Nevada Test Site, Oak Ridge National Laboratory, Los Alamos National Laboratory, and the INEL (Figure 5.1). Much smaller LLW volumes are buried at the Lawrence Livermore National Laboratory (California), Sandia National Laboratories (New Mexico), and the Pantex Plant (Texas). Other LLW is either poorly accounted for or released to the environment at seven other facilities: the Y-12 Plant at Oak Ridge, Rocky Flats (Colorado), the three uranium enrichment plants (Oak Ridge, Paducah, Kentucky and Portsmouth, Ohio), and Ohio's Mound Laboratory and Fernald Feed Materials Center.

Anti-nuclear groups such as the Radioactive Waste Campaign (based in New York City) have advocated closing the facilities that comprise the nuclear weapons complex. Some plants are already being phased out, such as the Hanford N-Reactor (first shut down for repairs in January 1987). The Oak Ridge Gaseous Diffusion plant was similarly placed on standby in 1985 because of lowered demand for enriched uranium. The DOE also recommended closing Rocky Flats and Fernald in a classified report to the National Security Council in December 1988. Commenting on the waste storage problem at Rocky Flats, 16 miles north-west of the state capital at Denver, Colorado Governor Roy Romer said 'This stuff is like a time bomb with the clock ticking'.[75] Yet these are very old facilities that are expected to be replaced by other plants. Whether other military nuclear facilities continue to operate, the task of cleaning up the existing sites remains monumental. It should be noted that the weapons facilities began producing radioactive wastes a decade or two before civil LLW burial grounds opened, and until very recently had not been subject to the scrutiny now commonly given to civil nuclear power plants and waste disposal sites.

PRO-NUCLEAR COMMUNITIES

During the period that government has attempted to implement policies for radioactive waste management in the United States several communities have stepped forward and put out the welcome mat to nuclear waste dumps. While surprising to most observers, these areas typically have several characteristics in common that stimulate their interest in a waste disposal facility: a past that has included uranium mining and milling, as in New Mexico; Edgemont, South Dakota; and Naturita, Colorado; or an existing LLW facility, as in Richland, Washington; Beatty, Nevada; and Barnwell, South Carolina. Geographical isolation, economic dependence on the nuclear industry, and high unemployment, combined with this familiarity with the industry, are other common threads that have made some towns into nuclear oases.[76]

Edgemont, South Dakota is an economically-depressed small town where a uranium mill closed in 1972 and an abandoned US Army ammunitions depot has been proposed as a potential LLW facility. On 5 June 1984 voters in Fall River County (where Edgemont is located) passed a non-binding referendum by a nearly two-to-one margin to permit construction of a LLW disposal site at the munitions depot. Opponents of nuclear waste disposal who lived elsewhere in South Dakota organised a statewide campaign to reverse the Fall River County initiative, culminating in a November vote. The November 1984 initiative, which required that any proposal to site a nuclear waste facility anywhere in the State must be approved by a majority of South Dakota's voters in another referendum, passed with more than 60 per cent of the statewide vote.[77] Unable to secure a LLW disposal facility, Edgemont opted for a less controversial sewage-ash processing plant. Perhaps the most interesting point lost in all of this is that the State does not even need a LLW site. South Dakota, with just 708 000 residents in 1987, has no nuclear power plants and as one of nation's lowest LLW generators only belatedly joined a LLW compact.

Three other nuclear oases in the United States are the communities that have the existing LLW disposal facilities: Richland, Beatty, and Barnwell. The Richland region is especially comfortable with nuclear materials, since it is adjacent to the DOE's Hanford Reservation, which was established in 1943 as part of the Manhattan Project.[78] Despite persistent leaks of the liquid HLW stored in tanks at Hanford, the local 'tri-cities' area of Richland, Pasco and Kennewick was heavily in favour of a permanent HLW repository there, in the face of massive opposition elsewhere in Washington State. The DOE officials had dubbed the area a 'halo community' for its willingness to host the HLW repository – the only such community among the five finalists. But the hydrogeological problems at Hanford are serious, the basalt is dangerous to work in, it was the most expensive of the potential HLW sites, and the DOE is now focusing on Yucca Mountain. Richland, however, will keep its LLW facility open into the next century. Thus the area has received a consolation prize, though they would have preferred the much larger financial recompense that would have come with a HLW site.

Beatty and Barnwell have much in common. Both host LLW facilities that have operated since the 1960s, and are scheduled to close in 1993. In both cases federal-state politics explains why the waste sites will close against the desires of the local residents.[79] Nevada is, of course, wrapped up in the fracas with the DOE over Yucca Mountain. If State officials came out in favour of retaining the Beatty site, it could be used to help the DOE's argument to licence Yucca Mountain. Ostensibly, closing Beatty and siting a new LLW facility in Colorado is more equitable, but Beatty has performed relatively well and could be operated for 15–20 more years. Similarly, Barnwell is the 'victim' of the ex-Governor of South Carolina, Richard Riley, who pressured Congress to pass the LLRWPA. The State did not want to be the nation's LLW dumping ground, so North Carolina will now reluctantly try to find a LLW site. Although Barnwell also has had a reasonably good performance record, especially more recently, South Carolina has its hands full with the environmental problems that beset the SRP.[80]

ANTI-NUCLEAR COMMUNITIES

A more familiar community on the US political landscape is the one that flatly rejects the prospect of living with a nuclear waste facility, as residents involved in BOND have done in Britain. These communities, too, have a common thread, which is the enormous fear of the stigma that a radioactive waste repository may bring and the resulting disruption of local livelihoods. Five areas that have been considered for a HLW repository illustrate this stigma. Predictably, none of these areas has had any previous *favourable* experience with radioactive materials.

Deaf Smith County, Texas and Richton, Mississippi share many things: rural life-style, southern location, close proximity to medium-sized cities, abundance of salt deposits, and high ranking on the original short-list of five for the western repository. Both sites have favourable geology, in many respects, for the HLW repository, but the residents did not see it that way at all. Deaf Smith County, with its productive farmland, was concerned about the proximity of the proposed repository site to the prolific but steadily declining Ogallala aquifer, which supports the area's and much of the Great Plains region's irrigated agriculture. Whether the aquifer would ever flood the repository shaft is irrelevant to most citizens, since the prospect of worthless crops (that is to say, no buyers) following HLW disposal is real enough. The pressure groups STAND (actually started by a farmer in nearby Swisher County, also once a site for the proposed HLW repository) and People Opposed to Wasted Energy Repositories, or 'POWER', were formed to fight the DOE's HLW siting proposals in Texas.

The problem with the Richton site is that it is next to a town of 1100 people. The *Jackson Clarion-Ledger*, a local newspaper, suggested as early as 1981 that a HLW site there could force the entire town to relocate. The Mississippi field office of the DOE eventually became a no-man's land, as distrust became deeply seated. Local opposition to repository siting was apparently successful, as the Richton Dome proposal somehow slipped out of the top three in May 1986.

The DOE has investigated the suitability of several sites in the aptly-named Paradox Basin of Utah for a HLW repository. While the region's salt beds and domes presented rather complex geology to investigators, the DOE had a more immediate problem: the proximity of the sites to the Canyonlands National Park and the Colorado River raised the ire of many Utahans and conservationists.[81] Though residents of Moab (population 5400), the largest city in the basin, were sharply divided on the acceptability of a repository, several of the national environmental groups put up a united front to oppose siting in Utah. If the Canyonlands is 'sacrificed', they wondered, what would come next?

Two other areas that vehemently opposed a HLW repository have done so while accepting activities that generate radioactive wastes, thus underscoring the difference between nuclear facilities and waste dumps in the public mind. Portland, Maine and surrounding areas were frantic over the possibility of a repository at the nearby Sebago Lake batholith, since Sebago Lake is the source of Portland's drinking water. Before the second-round HLW repository siting was stopped, a public hearing on the proposal in February 1986 drew 3000 angry people,[82] a huge crowd for Maine. Maine also has a law preventing the construction of additional nuclear power plants until a permanent HLW disposal site is available (a similar law in California was upheld as constitutional in 1983). Yet in three referendums the State's voters have voted against the closure of their largest producer of radioactive waste, the Maine Yankee nuclear reactor (the last vote was in 1987).

A similar attitude is found in Las Vegas, Nevada and, by extension, with the politicians in the State capital at Carson City over 400 miles to the north west. The proposed Yucca Mountain site is 95 miles north west of the Las Vegas metropolitan area, where about 60 per cent of the State's one million people live. Many residents and business interests fear the stigma of being near the nation's dumping ground for HLW and predicted that tourism would suffer greatly. 'People don't want to see lethal garbage pass by their schools, homes, farms and businesses', said Bob Fulkerson of Citizen Alert. Transportation is the weakest link in the whole nuclear waste debate.[83] Of course, there have been many other reasons to oppose the Yucca Mountain project, as clearly articulated by the ex-Governor Bryan. Yet, ironically he also has reiterated the State's firm support for the testing of nuclear weapons at nearby Yucca Flats,[84] which also generates radioactive wastes. Apparently, some nuclear wastes are more equal than others!

The lesson here is clear. While there are certainly legitimate grounds to oppose radioactive waste disposal in technically unsuitable places, these reasons are often obfuscated by negative perceptions of a 'dump' and disruptions in livelihood. Just as prevention is the best cure for disease, the best solution to conflict over nuclear waste disposal is to avoid selecting controversial sites in the first place. Needless to say, this is not always possible, as concern for geological suitability and interregional equity (however defined) may well require site selection in controversial areas.

THE POLITICS OF RADIOACTIVE WASTE DISPOSAL IN THE USA

This survey of the contemporary history of radioactive waste disposal policy in the United States leads to several conclusions. These

conclusions buttress the four geopolitical and institutional characteristics we outlined at the outset of this chapter.

The East–West dimension

The political geographical conflict between the eastern and western regions has been most pronounced in the debate over HLW repositories. While remote western sites offer a solution acceptable to the East, the West has been unwilling to accept sole responsiblity for wastes that are mainly generated in the East. In the case of LLW, states in the Southeast and West with existing waste sites insisted that major waste-generating areas in the North and East should accept responsibility for their own LLW. In both cases it has become clear that *an acceptable solution must be based on the principle of interregional equity, whereby areas that benefit from nuclear power must bear a fair share of the costs or burdens incurred.*

The politicisation of conflict

The problem of HLW disposal has created the most controversy. As sites were identified in the West, opposition intensified, particularly in Texas. The identification of potential sites for a second repository in the East met with opposition sufficient to halt the project. Low-level waste disposal policy has had a lower political profile as states have manoeuvred for favourable positions in compacts. As the process has moved toward selection of host states and specific sites the political temperature has risen. The experience in the USA confirms that *political opposition to nuclear waste disposal develops when waste management policy moves from general principles to site-specific proposals in greenfield locations.*

The civil-military connection

Local politicisation has concentrated, until recently, almost entirely on the problem of radioactive wastes generated by the civil nuclear programme. In the USA, civil and military wastes have been kept separate though the HLW repository is now intended for both streams. Military nuclear reservations are under federal control through the DOE which, until 1988, had resisted populist public opposition. Civil wastes are primarily regulated by the EPA and NRC; characteristic of the US's decentralised political institutions, the states have the right to participate in the HLW decision-making process and are largely responsible for LLW management. There is ample provision for political participation through public hearings and an elaborate review procedure. This indicates that *political conflict arises where there is opportunity for local interests to participate and is generally quelled when federal control is exercised in the interest of national security.*

Plurality and fragmentation

Most opposition to radioactive wastes in the US tends to be fragmented. But, wherever sites are proposed, spontaneous perceptual and emotional reaction has provided a basis for the development of broad anti-nuclear coalitions. This was especially clear in the conflict over the Eastern HLW repository, where coalitions cut across all political ideologies, income levels, and lifestyles. Therefore calling the termination of the Eastern siting programme anything but political is ridiculous,[85] since this emotion was genuine and spontaneous. United opposition coheres at the state level, and this ensures support for those communities opposed to waste disposal sites such as in Deaf Smith County. But it also ensures that the support from nuclear-friendly communities such as Edgemont or Hanford is overwhelmed by opposition in the state as a whole. This feature suggests a fourth conclusion, that *broad coalitions operating at the state level have developed sufficient political power to frustrate national waste management policy and to submerge localised support for the nuclear industry*.

WHITHER NUCLEAR WASTE?

US radioactive waste disposal policy is a classic instance where the goals are generally agreed to, but the means of reaching them are disputed. Consequently policy-making has reached stalemate. Given the powers of the states and the principle of consultation and cooperation, it is unlikely that a federal takeover of decision-making would be politically viable. Therefore a solution has to be found that embraces full public participation. This will require a rational (and national) site selection process that minimises transport and secures geological and technical security. It implies that areas that benefit from nuclear power accept some of its burdens.[86] And it assumes that appropriate incentives and compensation will be found for those areas eventually designated (whether willing or not) sufficient to justify site selection. Elsewhere, a process of 'constitutional choice' has been suggested whereby these criteria are used to resolve the siting problem on a regional basis.[87] But thus far the DOE has been unable to gain sufficient public trust to make such a process feasible for HLW disposal. And the jury is still out on LLW disposal, since by 1989 only three new sites had been selected and no new facilities had been built. Failure to achieve a rational and fair solution points to the probability of a pragmatic solution whereby existing LLW sites continue to accept much of the waste, and the WIPP and possibly a Tennessee MRS provide a medium-term solution for transuranic and HLW management.

NOTES

1. Resnikoff, M., *The Next Nuclear Gamble: Transportation and Disposal of Nuclear Waste* (New York, Council on Economic Priorities, 1983); and Kirby, A., and Jacob, G., 'The politics of transportation and disposal: hazardous and nuclear waste issues in Colorado, US', *Policy and Politics*, Vol. 14, 1986, pp. 27–42.
2. Pasqualetti, M. J., 'Nuclear power impacts: a convergence/divergence schema', *The Professional Geographer*, Vol. 35, No. 4, November 1983, pp. 427–36.
3. Resnikoff, M., *Living Without Landfills* (New York, Radioactive Waste Campaign, 1987).
4. Leigh, I. W., *International Nuclear Fuel Cycle Fact Book*, PNL-3594, Rev. 8. Prepared by the Pacific Northwest Laboratories for the US Department of Energy (Richland, Washington, 1988).
5. The distribution of the military HLW at the three DOE sites in 1986 was about 63 per cent at Hanford, 34 per cent at the Savannah River Plant and 3 per cent at the INEL. The Savannah River Plant, however, contains 56 per cent of the radioactivity. The civil spent fuel inventory has about eleven times the radioactivity contained in the military HLW, though the latter is much more voluminous because of its largely liquid form. Source: US DOE (1987); see note 10.
6. The West Valley plant, which was operated from 1966–1972 by Nuclear Fuel Services, Inc., is now known as the West Valley Demonstration Project and is run by the Westinghouse Corp. The purpose of the project is to solidify 600 000 gallons of liquid HLW now in storage tanks, and to decommission and decontaminate the plant. The DOE estimates that 2320 cubic metres of liquid HLW are at West Valley. The facility, unfortunately, has experienced leaks of radionuclides from the burial grounds.
7. Marshall, E., 'The buried cost of the Savannah River plant, *Science*, Vol. 233, 8 August 1986a, pp. 613–5.
8. Transuranic wastes are defined by the DOE as 'material having no significant economic value which, at the end of institutional control periods, is contaminated with alpha-emitting radionuclides with atomic numbers greater than 92 and half-lives greater than 20 years, in concentrations greater than 100 Ci/g (DOE Radioactive Waste Management, DOE Order 5820.2,6 February 1984). The main storage sites for transuranic wastes, pre-WIPP, are Hanford with 8328 cubic metres (513 000 curies of alpha-emitting radioactivity), Savannah River Plant with 2590 (2 051 000), the INEL with 30 432 (118 000), Los Alamos with 7133 (227 000), Oak Ridge with 1830 (550 000), and the Nevada Test Site with 602 (2000). Source: US Department of Energy (1987).
9. Stewart, J. C., and Prichard, W. C., 'Institutional aspects of siting nuclear waste disposal facilities in the United States', in Blowers, A., and Pepper, D., eds., *Nuclear Power in Crisis* (London, Croom Helm, 1987) pp. 164–77.
10. US Department of Energy, *Integrated Data Base for 1987: Spent Fuel and Radioactive Waste Inventories, Projections, and Characteristics*

DOE/RW-006, Rev. 3, Springfield, VA, National Technical Information Service, October 1987.

11. White, I. L., and Spath, J. P., 'How are states setting their sites?', *Environment*, Vol. 26, October 1984, pp. 16–20, 36–42.
12. Fernie, J., and Openshaw, S., 'A comparative analysis of nuclear plant regulation in the US and UK', in Blowers, A., and Pepper, D., eds., *Nuclear Power in Crisis* (London, Croom Helm), 1987, pp. 98–128.
13. Federal Register, 'Environmental standards for the management and disposal of spent nuclear fuels, high-level and transuranic radioactive wastes', Final Rule, 40 CFR Pt. 191, Vol. 50, No. 182, 19 September 1985.
14. Downey, G. L., 'Politics and technology in repository siting: military versus commercial nuclear wastes at WIPP 1972–1985', *Technology in Society*, Vol. 7, 1985, pp. 47–75.
15. The 'consultation and cooperation' process is described in the NWPA of 1982. It follows from a recommendation for an even vaguer and more controversial 'consultation and concurrence' process made by President Jimmy Carter's Interagency Review Group on Nuclear Waste Management (1979) in their report to the President. For details, see Carter (1987, fn. 36) or Solomon and Cameron (1985, fn. 87).
16. Vogel, D., *National Styles of Regulation: Environmental Policy in Great Britain and the United States* (Ithaca, New York, Cornell University Press, 1986).
17. Kasperson, R. E., Berk, G., Pijawka, K. D., Sharaf, A. B., and Wood, J., 'Public opposition to nuclear energy: retrospect and prospect', *Science, Technology, and Human Values*, Vol. 5, Spring 1980, pp. 11–23.
18. Reader, M., ed., *Atom's Eve: Ending the Nuclear Age, An Anthology* (New York, McGraw-Hill, 1980); and Cutter, S. L., 'Emergency preparedness and planning for nuclear power plant accidents', *Applied Geography*, Vol. 4, 1984, pp. 235–45; Zeigler, D. J., and Johnson, J. H., 'Evacuation decision-making at Three Mile Island', in Blowers, A., and Pepper, D., eds., *Nuclear Power in Crisis* (London, Croom Helm 1987) pp. 272–94.
19. Fernie and Openshaw, 1987, op. cit., pp. 109–10.
20. Pollock, C., *Decommissioning: Nuclear Power's Missing Link*, Washington, DC, Worldwatch Institute Paper Number 69, 1986.
21. Resnikoff, 1987, op. cit.
22. Barlett, D. L., and Steele, J. B., *Forevermore-Nuclear Waste in America*, (New York, Norton, 1985).
23. HLW disposal was initially taken up by Congress with the bill for LLW but, being the more contentious problem, was deferred for later consideration.
24. Solomon, B. D., Shelley, F. M., Pasqualetti, M. J., and Murauskas, G. T., 'Radioactive waste management policies in seven industrialized democracies', *Geoforum*, Vol. 18, No. 4, 1987, pp. 415–31.
25. Jacob, G., 'Conflict, politics and location: siting a nuclear waste repository', unpublished PhD dissertation, University of Colorado, 1988.
26. Texas Department of Agriculture, *Panhandle Residents' Views of High-level Nuclear Waste Storage*, (Austin, Texas, May 1985).

27. Loeb, P. R., *Nuclear Culture: Living and Working in the World's Largest Atomic Complex*, (New York, Coward, McCann & Geoghegan, 1982).
28. Crystalline rock formations have been favoured in HLW repository siting studies in many of the world's nuclear states. These include Sweden, Finland, Switzerland, Spain, South Africa, India, Canada and Argentina. France is also investigating a granite site for possible HLW disposal.
29. Maynard, J., 'The story of a town', *The New York Times Magazine*, 11 May 1986.
30. SWUCO, 'Radwaste report', No. 2, 15 August 1986.
31. US House of Representatives, Staff Memorandum on the Second high-level radioactive waste repository, prepared for the Oversight subcommittee of the House Interior and Insular Affairs Committee, Washington DC, 30 October 1986.
32. R. MacDougall and C. Peabody were interviewed by A. Blowers in Washington in September 1986.
33. Singer, S. F., 'High-level nuclear waste disposal', letter in *Science*, Vol. 234, 10 October 1986, pp. 127–128.
34. Marshall, E., 'Nuclear waste program faces political burial', *Science*, Vol. 233, 22 August 1986, pp. 835–6.
35. Koenig, D., 'Hecht guest of nuclear group which supports Screw Nevada bill', *Las Vegas Review-Journal*, 3 December 1987, pp. 1B.
36. Carter, L. J., *Nuclear Imperatives and Public Trust: Dealing with Radioactive Waste*, (Washington, D.C., Resources for the Future, 1987).
37. Bryan, R. H., 'The politics and promise of nuclear waste disposal: the view from Nevada', *Environment*, Vol. 29, October 1987, p. 34.
38. Shapiro, F. C., 'Yucca Mountain', *The New Yorker*, 23 May 1988, pp. 61–67.
39. Resnikoff, 1983, op. cit.
40. C. Petti was interviewed by B. D. Solomon in Washington, DC in October 1988.
41. Blowers, A., 'Radioactive waste in the United States – will New Mexico draw the short straw?', *Environment Now*, Vol. 9, October 1988, pp. 26–7.
42. K. D. Pijawka was interviewed by A. Blowers in Tempe, AZ in March 1988.
43. Downey, op. cit.
44. US Department of Energy, *Report of the Task Force for Review of Nuclear Waste Management*, Washington, DC, Government Printing Office, February 1978.
45. Neill, R. H., Channell, J. K., Chaturvedi, L., Little, M. S., Rehfeldt, K., and Spiegler, P., *Evaluation of the Suitability of the WIPP Site,* EEG-23 (Santa Fe, New Mexico, Environmental Evaluation Group of the State of New Mexico, 1983).
46. L. Chaturvedi was interviewed by A. Blowers in Santa Fe, NM in April 1988.
47. Anderson, R. Y., 'Open letter to the U.S. Congress on radioactive waste disposal at WIPP' (University of New Mexico, Department of Geology, 1988)

48. R. Y. Anderson was interviewed by A. Blowers in Albuquerque, NM in April 1988.
49. D. Hancock was interviewed by A. Blowers in Albuquerque, NM in April 1988.
50. WIPP Panel, 'Report on brine accumulation in the WIPP facility', National Academy of Sciences, Board on Radioactive Waste Management, Washington, DC, unpublished report, 1988.
51. Tichenor, J., 'WIPP site delayed', *RWC Waste Paper*, Vol. 10, No. 3, Fall 1988, pp. 7, 12.
52. The compact concept has been used since US independence in 1783 to cope with transboundary problems between states, covering such issues as boundaries, tunnels, bridges, river catchments and basins, and air pollution. Some compacts such as the Colorado River Compact covering seven states were established to resolve regional problems, while others concerned problems that affected just two neighbouring states. Occasionally there have been compacts covering the whole nation, the first of which was the Interstate Compact for Supervision of Parolees and Probationers in the 1930s. Kearney, R. C., and Stucker, J. J., 'Interstate compacts and the management of low level radioactive wastes', *Public Administration Review*, Vol. 45, January/February 1985, pp. 210–20.
53. NUS, 'The 1979 state-by-state assessment of low-level radioactive waste shipped to commercial burial grounds'. Prepared by the NUS Corp. for the US Department of Energy, Washington, DC, November 1980.
54. SWUCO, op. cit.
55. White and Spath, op. cit., p. 38.
56. NIRS, *Alert*, Washington, DC, Nuclear Information and Resource Service, 28 June 1988.
57. Galpin, F. C., and Meyer, G. C., 'Overview of EPAs low-level radioactive waste standards development program'. Paper presented at the DOE Low-level Radioactive Waste Participants Meeting, Denver, Colorado, 22–26 September 1986.
58. DiMento, J. F., Lambert, W., Suarez-Villa, L., and Tripodes, J., 'Siting low-level radioactive waste facilities', *Journal of Environmental Systems*, Vol. 15, No. 1, 1985–6, pp. 19–43.
59. White and Spath, op. cit., p. 40.
60. An initiative that if passed would have required Nebraska to secede from the Central States compact was turned down by the State's voters in November 1988.
61. Blake, E. M., 'Alternatives to shallow-land burial', *Nuclear News*, Vol. 30, March 1987, pp. 61–4.
62. H. Brown was interviewed by A. Blowers in Washington, DC in September 1986.
63. Resnikoff, 1987, op. cit.
64. Barlett and Steele, op. cit., p. 249.
65. Raudenbush, M. H., 'US struggles to site disposal facilities', *Nuclear Engineering International*, Vol. 31, December 1986, pp. 49–50.
66. Material for this section is derived from Coyle, *et. al.*, *Deadly Defense:*

Military Radioactive Landfills (New York, Radioactive Waste Campaign, 1988).
67. US Department of Energy, 1987, op. cit.
68. Schneider, K., 'Accidents at a U.S. nuclear plant were kept secret up to 31 years', *The New York Times*, 1 October 1988, pp. 1, 7.
69. Herrington, J. S., Testimony before the Subcommittee on Interior and Related Agencies of the Committee on Appropriations for FY 1989, Part VII, US House of Representatives, Washington, DC, 23 February 1988.
70. US Department of Energy, *Environment, Safety, and Health Report for the Department of Energy Defense Complex*, Washington, DC, Government Printing Office, 1 July 1988.
71. National Research Council, *Safety Issues at the Defense Production Reactors* (Washington, DC, National Academy Press, 1987).
72. Cochran, T., *et. al.*, *U.S. Nuclear Warhead Facility Profiles: Nuclear Weapons Databook, Vol. 3* (Cambridge, MA, Ballinger, 1987).
73. Coyle, D., *et. al.*, op. cit., 1988.
74. US Department of Energy, 1988, op. cit.
75. Romano, M., 'Romer inspects Rocky Flats plant', *Rocky Mountain News*, 5 December 1988, p. 6.
76. Stewart and Pritchard, 1987, op. cit.
77. Murauskas, G. T., and Shelley, F. M., 'Local political responses to nuclear waste disposal', *Cities*, Vol. 3, No. 2, May 1986, pp. 157–62.
78. Loeb, 1982, op. cit.
79. Blowers, A., 'Way out West and down in the dumps', *New Scientist*, Vol. 112, 4 December 1986, pp. 71–2.
80. Marshall, E., 1986a, op. cit.; Schneider, 1988, op. cit.
81. Jacob, 1988, op. cit., pp. 219–21.
82. Carter, 1987, op. cit., p. 410.
83. Wingard, L., 'Opponents of dump angry', *Las Vegas Review-Journal*, 19 December 1987, p. 1B.
84. Bryan, 1987, op. cit., pp. 14–17, 32–8.
85. See US House of Representatives, 1988, op. cit.
86. Kasperson, R. E., Derr, P., and Kates, R. W., 'Confronting equity in radioactive waste management: modest proposals for a socially just and acceptable program', in Kasperson, R. E., ed., *Equity Issues in Radioactive Waste Management* (Cambridge, MA, Oelgeschlager, Gunn & Hain, 1983) pp. 331–68.
87. Solomon, B. D., and Cameron, D. M., 'Nuclear waste repository siting: an alternative approach', *Energy Policy*, Vol. 13, No. 6, December 1985, pp. 564–80.

6 Nuclear Waste in Western Europe

A COMMON PROBLEM, DIFFERENT APPROACHES

In continental western Europe, as in the UK and the USA, the political problem of managing radioactive wastes arrived by stealth. In the early years of development of nuclear energy as a *soi-disant*, cheap, safe and clean source of power the drawbacks, particularly the gradual accumulation of long-lived dangerous radioactive wastes, were largely discounted. As the nuclear programmes of various countries expanded so wastes accumulated and the technical problem of their management had to be tackled. At the same time throughout Western Europe (and increasingly in Eastern Europe too[1]) there has been growing concern about the health and environmental risks from radioactive wastes. Failure to secure public support for waste management policies would be a major factor in limiting the development even continuance of the nuclear energy option. Thus the political stakes in the radioactive waste issue are high. Radioactive waste became an increasing source of political conflict in several Western European countries during the 1980s, especially in the aftermath of the Chernobyl accident in 1986.[2]

This conflict is experienced at three different institutional levels. There is, first, the general political conflict at *national* level over the nuclear issue which embraces the whole nuclear cycle from uranium mining through fuel fabrication, energy production, reprocessing and waste management. Public opinion and political parties are divided on the issue of nuclear power. These divisions increase when the military dimension is included. The degree of conflict varies. In some countries, notably Sweden, anti-nuclear opinion has achieved a degree of coherence across the range of nuclear issues leading to a non-nuclear strategy being implemented by government. In some, and France is perhaps the best example, pro-nuclear interests have secured an ascendancy over a weak opposition. Elsewhere, and the Federal Republic of Germany (FRG) is a prime example, fragmentation of issues and divisions among interests have instilled uncertainty into the future of nuclear energy.

The second level of conflict is site-specific and occurs between the central state and *regional* and *local communities*. Such conflict is aroused by proposals to develop nuclear facilities. In areas with existing nuclear plants there may be local support for the nuclear industry based on familiarity and the jobs provided, as we saw in the UK's choice of nuclear oases. This is also a major reason why new nuclear power stations are constructed alongside existing ones. But, where greenfield sites are chosen, resistance by local communities has developed in recent years, with perhaps the exception of France in the 1980s. This is especially the case with radioactive waste which involves unfamiliar technology, unknown risks and provides few local benefits. The ability to resist proposals for nuclear facilities at specific sites will, in part, reflect the degree of participation and power available to local authorities. The nuclear industry is most likely to achieve its objectives where a strongly centralised state is able to ensure that local opposition to projects are subordinate to presumed national interests.

The third level of conflict is brought about by the cross-frontier transport of nuclear wastes, which may create political conflict between states or raise concern at the *international* level. The international dimension of conflict was highlighted by the political and economic scandal surrounding the Federal German Transnuklear company during 1987–8. Transnuklear (TN) was part of a considerable complex of internationally interlocked companies (see Figure 6.1). The Transnuklear affair, as it became known, involved systematic bribery, embezzlement and deception and also revealed the significance of the various levels of corporate and operational integration of the radioactive waste and nuclear fuel cycle industry in Western Europe.[3]

The Transnuklear affair had specific implications for nuclear waste politics in the FRG as will be explained later in this chapter. But, it also had international repercussions in the European Commission[4] and the European Parliament.[5] With plans to deregulate border controls within the European Economic Community (EEC) in 1992, the issue of an expanded trade in radioactive materials, including wastes, became a matter of priority following the Transnuklear revelations.[6] There were political and legal ramifications for countries outside the EEC, notably Sweden, which had contracts to treat LLW and to take spent fuel for long term store from countries within the EEC.[7]

The interaction of local, national and international levels is a feature of political conflict over nuclear waste in western Europe. In

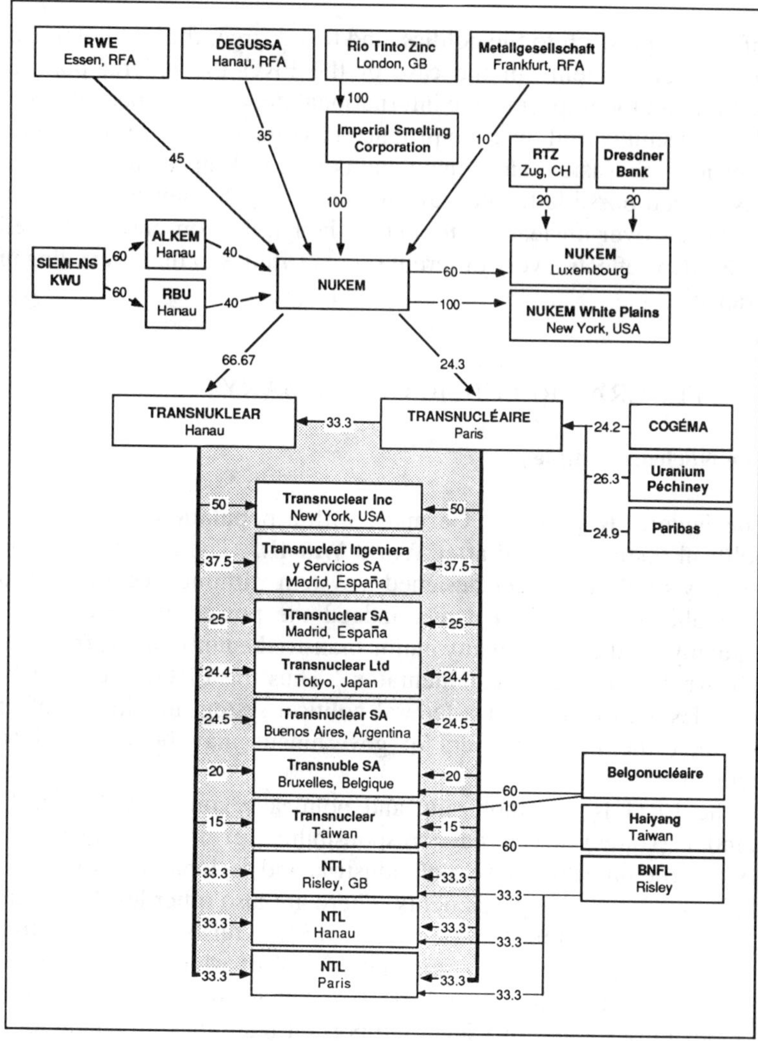

NOTE The numbers indicate percentage of share holdings.
SOURCE WISE, Paris (1988)
FIGURE 6.1 *The Transnuklear corporate 'interlock'.*

this chapter we will concentrate on the implications of these different levels of conflict for nuclear waste management policy and prospects in three Western European states: the Federal Republic of Germany (FRG), Sweden and France. They are (with the UK) the largest producers of nuclear energy in Western Europe, and each exemplifies a different approach to the politics, and to a degree, the technology, of waste management. In the case of the FRG the federal political system and the impact of the international dimension especially from the Transnuklear affair have produced a complex pattern of conflicts over nuclear waste management and policy. In France and Sweden it has proved possible to subordinate conflict by the development of a consensus over nuclear waste policy, though the consensus achieved in Sweden reflects a very different set of attitudes to that prevailing in France.

THE FEDERAL REPUBLIC OF GERMANY

The Nuclear Complex

The Federal Republic of Germany with a population of 61m has a political system devised after World War II.[8] 'The architects of the country's political order designed a system without a centre. It is a Republic in more than name, indeed the power of the centre is arguably weaker than in any major industrial country apart from the US' wrote one expert commentator.[9] Thus the FRG, like the US, provides a case study of a federal political system in which conflict between the different tiers of government may be expected to flourish.

The FRG is geographically and politically constructed from 11 Länder (states) which have responsibility for most services and control the important levers of industrial and regional policy through their own legislative parliaments. There are also other levels of local government, in the Landkreis (county), Kreisfreie Städte (free towns) and Gemeinden (communities), all of which have different powers, responsibilities and relative budgetary autonomy. The Länder in particular provide an important restraint on Federal level decisions, both through their own frequent elections, and because the Länder and Gemeinde together take up more than half the national tax revenues, and also pay out more in regional subsidies. The Länder exercise considerable control over nuclear decision-making. Federal authority is relatively weak.

The role of nuclear energy in the Federal Republic is controversial and complex. This partly relates to the federal political system but also to the historic concerns that civil/peaceful nuclear technology might be diverted covertly to military use.[10] In 1954 the Federal

Government made a solemn renunciation of nuclear weapons production or ownership. Many politicians and scientists in the FRG believe that their atomic energy industry has been severely handicapped by the greater R&D freedom allowed in competitor countries such as France, the UK or US.[11] Nuclear energy accounts for between 33–36 per cent of electricity in the FRG which makes it sixth in the world for nuclear capacity.

Nuclear generation is spread unevenly across the FRG: mining areas such as the northern Ruhr obtain less than 10 per cent of electricity from nuclear plants, whilst utilities in the rural south-west distant from coal mines, have a nuclear share of 50–70 per cent.[12] Until 1960, the FRG relied almost entirely on hard or brown coal for fuelling power plants. The first commercial reactor (at Gundremmingen) began operation in 1967. Between 1970 and 1986 there was a tenfold increase in the percentage nuclear contribution, growing from 3.7 to 36 per cent.[13] By the end of the 1980s the 21 commercial nuclear plants in the FRG contributed virtually the same to electricity capacity as coal.[14] Three further PWR plants were under construction. As in the UK, post-1973 oil crisis, government plans were made for massive nuclear capacity increase. The 1974 plan for 45 000 MW by 1985 turned out to be hopelessly over ambitious, only 16 000 MW being achieved.

The commitment to nuclear energy was confirmed by the Federal Chancellor, Helmut Kohl, early in 1987, stating 'If we were to move out of nuclear energy immediately, it would mean the collapse of the (West) German economy'.[15] Industrialists in energy intensive sectors such as chemicals and metallurgy agreed, arguing that electricity charges were 30–40 per cent higher than in competing countries such as nuclear dominated France. They also pointed out that the costly retrofitting of desulphurising technology designed to reduce acid rain caused by coal-fired plants were being passed on to industry's electricity bills. But, as the Transnuklear affair unravelled Kohl admitted he had 'huge doubts about the overall safety (in the FRG) in the light of the scandal'.[16] The whole matter was a deep shock to the political system and the nuclear industry where huge amounts of investment capital were tied up serving the FRG utilities and conducting widespread business abroad.

By the late 1980s the FRG had developed a very complex nuclear fuel supply, reactor construction and nuclear operations industry (Figure 6.2). Reflecting the federal distribution of power, the various parts of the industry are located in various Länder. All of the operations produce radioactive wastes. Uranium processing companies are based in Bonn, Frankfurt and Saarbrucken respectively; uranium enrichment is undertaken in Gronau, and Essen; fuel fabrication is primarily centred on Hanau, whilst nuclear fuel cycle services have been concentrated in Hanau, Frankfurt, and Dusseldorf.[17] Apart from its commercial nuclear power plants, the FRG has small scale experimental breeder reactors, KNK II, started in 1978, and the SNR-300, at Kalkar, started in 1973 and costing DM6.5 billion, as yet unopened for economic reasons, although completed in 1986. There is also a demonstration scale thorium high temperature reactor (THTR) at Hamm-Uentrop which was opened in 1985 on which it is hoped to base the next generation of reactors.[18] A pilot reprocessing plant, WAK (Wiederaufarbeitungsanlage) at Karlsruhe, was opened in 1967. A commercial scale reprocessing plant at Wackersdorf (WAW) had been planned[19] with an annual capacity of 350 metric tons (mt) heavy metal, but the project was cancelled in spring of 1989.[20]

Wackersdorf had proved controversial with its financial backers through escalating costs.[21] The FRG anti-nuclear movement, along with campaigners from nearby Austria, made the site a centre for demonstration[22] and the cancellation of the plant threatened to have widespread ramifications for the FRG nuclear industry.[23] Political controversy has also focused on fuel fabrication at Hanau where in July 1987, temporary closure was brought about as a result of corruption charges involving the circumvention of the licence procedure for plant modification.[24]

The Radioactive Waste Burden: Legislation and Management

The management of radioactive waste, including the construction and operation of federal installations for long term storage and disposal of radioactive wastes and the safe transportation of nuclear materials, is the responsibility of a national institute for science and technology, Physikalisch Technische Bundesanstalt (PTB), under the federal government. Federal concern for safety and environmental protection is exercised by the office for radiation protection, Bundesamt fuer Strahlenschutz (BAS).[25] Under the Atomic Energy Act of 1959 the federal government may fix the methods of intermediate storage and final disposal of radioactive waste. The 1976 Amendment divided responsibility for waste management between waste producers who have to take care of intermediate storage and the federal government

NOTE Not all existing reactors are marked.
FIGURE 6.2 *Nuclear facilities and disposal sites in the Federal Republic of Germany*

which has to ensure safe final disposal.[26] It also stipulates that for an electrical utility to get its operating licence renewed year by year it must have proof that for the subsequent six years the spent fuel discharged is covered by a management contract with a reprocessing company, an intermediate storage facility, or a final disposal repository.[27]

The radioactive waste inventory in the FRG in the year 2000 is expected to be between 2 000 000–3 000 000m^3 of 'conditioned waste', of which 43 per cent will be from nuclear power plants; 30 per cent arising from reprocessing (including returned waste from France and the UK); 22 per cent from research centres; and 5 per cent from industrial research, medicine, fuel fabrication, and decommissioning. About 3 per cent of this total volume will be HLW.[28] At the end of 1987 there were 37 000m^3 of LLW and 424m^3 of HLW in the FRG.[29] A stockpile of over 20 000 barrels of low activity waste had been building up at nuclear plants since the closure, in 1978, of the temporary disposal facility at Asse in Lower Saxony. The closure was brought about by political pressure preventing an extension of the operating licence.[30]

Radioactive Waste Disposal Sites

Since the early 1970s, the development of radioactive waste management and disposal policy has been fraught with difficulties, primarily political and legal. Out of a series of political and planning battles which will be explained shortly, the following projects have emerged (see Figure 6.2).

Gorleben: In Lower Saxony, near the village of Gorleben, is the location of three nuclear waste projects. One is an away-from-reactor (AFR) interim store (similar in concept to the USA's MRS 'monitored retrievable store') for spent fuel. This is an air ventilated, dry store for 420 fuel casks containing 1500 tonnes, constructed above ground. This capacity will be taken up by 300 metric tons of spent fuel and 1200 metric tons of vitrified waste. The second project is a low activity waste facility with 35 000 drum storage capacity.[31] In addition, PTB have been test drilling the Gorleben salt dome for a HLW store, below ground.[32] The below ground repository is being assessed for all types of radioactive waste disposal. Test drilling started at Gorleben in 1984, and is planned to continue until 1992–93. If approval were given, construction of a repository at Gorleben could begin in 1995. The repository is planned to open in the year 2000.[33] The size of the test bore programme strongly suggests that the site could easily be converted directly into a full repository if the tests prove suitability.

There are more than 200 salt domes in northern FRG. Salt domes are attractive owing to their durable impermeability to groundwater, the significant transport medium in geological systems. Some, such as that at Gorleben cross the border with the German Democratic Republic (GDR). In 1987 the city authorities in Helmstedt expressed concern that their water supplies were being radioactively contaminated by water flowing through the former Bartensleben salt mine, just across the border, where the GDR has been disposing of LLW and ILW at Morsleben, along with toxic wastes, since 1972. The political conflict between the two countries has been headed off following an agreement covering 27 issues, including radioactive waste, concluded in 1987 after 14 years of negotiation.[34]

Ahaus: In North-Rhine-Westphalia, has another AFR interim spent fuel store under construction. Although it was begun in October 1983, construction was temporarily suspended in May 1985 when a Munster (state) court issued a temporary injunction on construction.

Until Gorleben and Ahaus are opened the FRG is relying substantially upon the managing and processing of its commercial spent fuel abroad. FRG utilities have contracts with Cogéma in France to cover up to 3160 tonnes at La Hague, a large reprocessing plant on France's north western coast; and with BNFL in the UK to process 760 tonnes at Sellafield. Apart from the plutonium due for return to the FRG, both French and British contracts stipulate the return of the radioactive waste arisings.[35] So it is a problem postponed, not solved. The FRG strategy that was to have involved the so-called 'closing' of the nuclear fuel cycle, with the construction of an indigenous reprocessing plant, was thrown into considerable chaos with Wackersdorf's cancellation (and conversion into a solar energy cell production and recycling plant).[36] The decision to halt the plant came after Federal Environment Minister, Dr Klaus Töpfer, consulted with the utilities over their future plans to place reprocessing contracts with Wackersdorf. The collective response, but in particular the decision of the major utility VEBA AG to extend contracts with Cogéma in France, made the project financially unviable.[37] Altogether, the Wackersdorf plant cost the FRG industry and government over $2100 million (DM4 bn).[38] The FRG also signed an intergovernmental co-operation agreement with the UK two months after the Wackersdorf cancellation, which allowed for a further 4000 tonnes of spent fuel – about half Wackersdorf's planned total capacity – to be reprocessed at Sellafield from the end of the 1990s.[39] In agreeing this, the FRG increased its dependence on the UK, superseding its earlier reliance on France.[40]

The Wackersdorf developments followed earlier attempts by utilities in 1983 and 1985 to try to complete export agreements to send spent fuel to China and Sudan respectively.[41] Sweden and Austria had also made preliminary agreements with Sudan. Each of these proposals constituted desperate attempts by the nuclear industry to fulfil legal requirements under the 1976 amendments to the Atomic Law.

In addition to its construction of repositories and interim export of wastes, the FRG has three major research projects for all categories of waste disposal.

Konrad: This is a deep, very old dry iron ore mine in the Salzgitter region of Lower Saxony whose workings have been backfilled and retained. The research project, begun in 1975, is intended to assess the suitability of the mine for all non-heat generating wastes, about 95 per cent of the total volume,[42] especially the large volumes created in decommissioning, as space is available for 500 000m^3.[43] PTB has spent over £80m on Konrad since its iron ore operational period (1965–76) halted, compared to around three times that amount at Gorleben.[44] Although PTB had planned to open Konrad in 1991, both the Lower Saxony environmental ministry and the Federal institution for geological science have criticised its suitability.[45] Moreover critics have pointed out that a Gruppe Okologie (Hanover) report of November 1983 had already claimed the unsuitability of Konrad and criticised the bias in official reports.[46]

Asse: The Asse II Salt mine in Lower Saxony was purchased in the early 1960s and research was started in 1964. For a decade more than 100 000 special drums were routinely deposited in Asse and one of the difficulties has been the failure of proper labelling of the contents of drums. Since the stack-storage of LLW was halted at the end of 1978, it has been used as an *in situ* R&D facility. Over 141 000 barrels of LLW are already in place.[47] The plans included a field test for the borehole emplacement of vitrified high-level waste. Plans are also underway for ILW retrievable emplacement and direct disposal of spent fuel.[48]

Karlsruhe: At Karlsruhe in Baden-Wurttemberg an experimental project to vitrify HLW from reprocessing began in 1986[49] in conjunction with the PAMELA facility at Mol (Belgium). A final direct disposal cask, 'Pollux', is being developed following a joint resolution by Federal government and Länder premiers in 1979 and a study on direct disposal began in January 1985.[50]

The Multi-faceted Nuclear Politics of the FRG

As in the US and elsewhere, in the early 1970s public concern and action transformed nuclear energy from being an internal technical and economic debate[51] among the nuclear elite into a significant political issue involving citizen participation. Arising from the so-called APO movement (extra-Parliamentary) in the late 1960s which 'modernised' politics in a stagnant state,[52] came the Burgerinitiativen (BIs) – the citizen initiatives groups which embraced the civic awareness philosophy promulgated in the earlier BIs from the 1950s and revitalised it into one of citizen empowerment and action. The BIs took up ecological, transport, peace and nuclear issues. During the 1970s they challenged the civil nuclear programme and, through widespread demonstrations, effectively halted the NATO plans to deploy the neutron bomb in the FRG. Along with these 'actions' of the BIs were the extremely violent challenges to the State led by the Red Army Faction (RAF), which, for a time in the mid-1970s, seemed to pose a potential challenge to the stability and authority of the Federal State itself.[53]

This stability was fostered by the German tradition for an ordered civic society in which there is both formal respect for expertise and authority, and a politico-legal system permitting public involvement in local decisions. When atomic energy programmes were first developed, this consensual system worked without undue problem. There was consensus on the benefits of nuclear power for a rapidly growing economy between the main political parties, industry and the trades unions. There was, in consequence, no political division in the Federal Parliament in Bonn over the planned growth of nuclear energy in the 1973 programme. Indeed in the Bundestag, even in 1986 a report on nuclear energy by the Conservative Christian Democratic Union (CDU)-led federal government gave strong commitment to continued development of the substantial programme, pushed ahead by its predecessor the Social Democratic Party (SPD) government between 1969 and 1982.[54]

Despite the apparent political consensus in Bonn throughout the 1970s and early 1980s, the nuclear energy establishment encountered political difficulties. Plans to sell eight PWR plants and fuel cycle facilities and services to Brazil, originally agreed in June 1975, brought to public attention the FRG's complex nuclear links abroad, including those with South Africa.[55] It succeeded, too, in adding to the dissent already developed against the FRG's indigenous nuclear reactor construction programme. The protest spread partly as a result of dissatisfaction with the formal planning and administrative system.

The administrative system has traditionally sought to preserve its neutrality through the promotion of a concept characterised as 'a social partnership in which the state integrates representative employers' and employees' oranisations in making social and economic policy'.[56] But in the sphere of environmental concerns, such well defined groups with identifiable interests did not emerge and smoothly fit into the planning process.

The Politics of Dissent

The planning of various nuclear facilities in the FRG has been disrupted by mass protests (Figure 6.2). From the early 1970s the primary focus for extra-parliamentary and grass roots protest and dissent was the construction of new nuclear plants, at Wyhl (1976), Brokdorf (1976), Grohnde and Kalkar (1977). In what some participants proudly called 'eco-sabotage', the largest reactor site demonstrations gathered at Wyhl in Baden-Wurttemberg in the south-west in February 1976. Over 20 000 people, including French and Swiss, and German farmers occupied the proposed construction site after two years of struggle in the Länder courts. At Brokdorf, in the north east near Hamburg, in 1976 around 45 000 demonstrators tried to occupy the site and ended up in a major violent confrontation with police.[57] The following February (1977) demonstrations were held at Grohnde in Lower Saxony numbering around 20 000 and again at the Brokdorf site, but to avoid direct (and violent) confrontation, the 60 000 demonstrators split themselves between Brokdorf itself and the nearby hamlet of Itzehoe. In September 1977 over 50 000 gathered at the site of the Kalkar fast reactor, near the Rhine and Dutch border. 10 000 or so Dutch people crossed the border to join the protest. That was the last sizeable nuclear reactor demonstration of the 1970s. However, in June 1986, a month and a half after the Chernobyl accident, 40 000 gathered once more at the Brokdorf reactor, by then virtually complete. Several smaller demonstrations at reactor sites have been organised since.

An important feature of the mass 'manifestations' against nuclear plants in the 1970s was the heterogeneous mix of the opposition: students, farmers and housewives joined together in a common cause. There was also the international aspect brought about by the tendency to build nuclear plants on major rivers near the national borders. From 1980 onwards, the mobilisation of anti-nuclear protest incorporated the anti-nuclear weapons movement which had waned after the victory against the neutron bomb in 1978, with the campaign against the intermediate range Cruise and Pershing missile deployment. The re-emergence of the peace movement in the 1980s was the most important anti-nuclear development. Both the peace movement and the protests against nuclear energy faced similar problems of organisation and confrontation with the state security authorities.[58]

By the late 1980s anti-nuclear opposition had increasingly turned towards spent fuel and nuclear waste management facilities, led strongly by the BIs and the Green Party (founded in 1978/79),[59] which has taken the complex issue up at all levels of government and in the European Parliament too. In the 1980s, major demonstrations have taken place at the Wackersdorf site. There were 30 000 protestors when site clearance began in December 1985, 20 000 in June 1986, 30 000 in March 1987. More demonstrations, coinciding with the now traditional mass anti-nuclear mobilization over the Easter weekend, have since been held, culminating in a successful outcome for the protestors.

It is the transport and disposal of nuclear waste that in FRG, as elsewhere, has captured the public concern most strongly, very often because it means the introduction of the nuclear industry to remote and conservative rural regions. One opinion poll taken in the FRG in July 1986, showed 75 per cent of the population perceived nuclear waste as a 'serious drawback' to nuclear energy. From 44 per cent worried in 1975, the 75 per cent concern has been constant since 1981. Despite Chernobyl, in contrast only 50 per cent regard the possibility of a major accident as a serious drawback.[60]

Running in parallel with demonstrations have been the use of the courts system to halt or delay the licensing procedure and construction. From the nuclear industry's viewpoint the process is protracted and expensive and 'has caused repeated deferrals of power plant projects under construction and of planned fuel cycle facilities'.[61] From the objector's perspective the problem lies in the dual licensing procedure whereby the first partial construction licence indicating that the proposal is in order is issued by the State government but the plant operating licence allowing full completion although also issued by the State government, may be over-ruled by the Federal government. Once a licence is granted 'it can usually be executed immediately, notwithstanding some other party filing a law suit against it. Ordinarily therefore, there are two law suits; first against the immediate execution of the licence; and then against the facility itself'.[62] A lawyer who has acted on behalf of BI objectors considers that the court system ultimately works against the intervenor because of its complexity and cost.[63] Notwithstanding the problems, in September 1988 a local farmer won an historic first case to obtain a withdrawal of a completed plant's licence, for the Muelheim-Kaerlich power plant in Rhineland Palatinate. This decision had major ramifications for various projects planned for the back end of the fuel cycle, because it showed intervenors could achieve victory in the court systems.[64]

Gorleben: the Politics of Persistence

The Gorleben project illustrates the complexities of the FRG planning system for atomic projects and the opportunities for protest to which it gives rise.[65] The Gorleben project was begun in April 1974 with the announcement of the 'Integrierte Entsorgungskonzept' (the Integrated Waste Management Concept) by the Federal Ministry of the Interior.[66] The concept included spent fuel management, a reprocessing plant, and disposal facilities for nuclear wastes centred on one site. Two years later, in spring 1976, this Ministry announced new legal measures in order to force the electrical utilities to tackle the nuclear waste problem more determinedly. These measures came into force in 1977.[67] The formal announcement of the choice of Gorleben was made on 22 February 1977 and Gorleben became an issue of potential conflict between federal and Länder governments.[68]

Following a strong hint at a press conference by the Christian Democrat (CDU) Premier of Lower Saxony, Ernst Albrecht in the autumn of 1976 that Gorleben would be chosen as the 'nuclear park', protest by environmentalists began. The site was on the edge of a nature reserve, Luneburg Heath: soon the local protest slogan read 'Nature reserve or nuclear park?'. Local farmers with 40 tractors immediately blocked traffic in the centre of Hanover, the capital of Lower Saxony, in protest.[69] The reaction of the sparsely populated area of eastern Wendland, chosen to host the 'Entsorgungzentrum' was one of shocked disbelief.[70] For some, such as Mrs Lieselotte (Lilo) Wollny, a housewife in Gorleben, it was the beginning of a mid-life politicisation that took her to the Bundestag as one of a number of Green Party representatives chosen from the region for the Länder, Federal and European Parliaments. 'Once you start thinking, once you realise what it means, for this thing to take place in front of your door, you take notice. And you begin to make links ... a nuclear plant here, a military camp there ...'.[71] The driving force behind the local resistance was Mrs Marianne Fritzen, whose original political involvement was in the peace movement and community involvement with the Protestant Church. After a decade at the forefront of political confrontation she was elected as a Green Party local councillor; 'I used to have to pay to protest, now they pay me to do it'.[72] Mrs Fritzen became a *cause célèbre* across the FRG and kept the Gorleben struggles in the public mind, when a photograph of her surrounded on all sides by visored and armed police in a demonstration in March 1979, was made into an election poster.

By the end of July 1977 both Deutsche Gesellschaft fur Wiederaufarbeitung von Kernbrennstoffen (DWK) (the company responsible for spent fuel storage and reprocessing) and the federal agency PTB (with responsiblity for the final waste repository) had applied for a partial construction licence. DWK's report was cleared in October by the FRG Federal advisory bodies on nuclear safety, Reaktorsicherheits Kommission (RSK) and the Strahlenschutz Kommission (SSK). In November, the Federal Government, which had agreed to the Gorleben site selection in July, told the Bundestag it had accepted the reports and the problem of spent reactor fuel had been 'solved'.[73]

The village of Gorleben is geographically isolated, just 3km from the FRG-GDR border, formed by the River Elbe (Figure 6.2). The people of the area are traditional conservative monarchists (kaisertreu). In the late 1970s, 68 per cent of the vote went to the conservative CDU. But a decade later, CDU support had dropped to only 40 per cent. The region of Wendland is dominated by agriculture. The plans advocated by Premier Albrecht to introduce a high technology industrial park challenged the peace and traditions of the people. It created a strong cultural contradiction between the farming people and outsiders.[74] The contradiction also surfaced within the regional resistance, for radical students from Hamburg and Berlin were attracted to the struggle, as much to confront State authorities as to defend the traditions of Wendland.[75]

Within a year of the initial announcement, which had brought a demonstration of 20 000 people,[76] Premier Albrecht, in the period before the Land elections in Lower Saxony in June 1978, ordered a technical re-examination of the DWK proposal. Moreover, following the demands of the regional opposition, he agreed to the establishment in early 1978 of the Gorleben International Review (GIR), an expert committee supported by Land funds to appraise the entire project.[77] The GIR started its examination in July 1978, one month after Albrecht had won his election and in February 1979 submitted its report. The primary attention at the GIR hearing (as with the Windscale inquiry in the UK two years before) was on the reprocessing plans, the technical problems involved and the implications of the 'plutonium economy', with its sensitive overtones of possible future military use.[78]

Radioactive waste received more cursory attention. Albrecht insisted that he would wait for the evidence of his special test drilling programme.[79] Existing evidence that the salt dome was unsuitable was ignored. Albrecht did, however, request a special session on alternative spent fuel management strategies. The GIR 2200 page draft report was discussed with experts (mostly from the nuclear community) over six days of hearings in Hanover. The symposium undertook its scientific deliberations against an extraordinary contemporary background. It was opened at 9am on 28 March, at exactly the same moment when the accident at Three Mile Island (TMI) in Harrisburg, USA occurred. The coincidence was poignant. The TMI accident soon dominated discussion of technological nuclear risk. Outside a huge demonstration of 140 000 people gathered, many of whom had begun to trek the 200km from Gorleben two weeks earlier. The placards and posters from Gorleben had a common leit-motif 'Harrisburg ist uberall!'[80] The GIR itself expressed doubt over the suitability of the Gorleben saltdome for a repository. In early May the press began to print stories suggesting that the Federal Government would be forced to modify its Ensorgungszentrum, especially following TMI.[81]

Albrecht, in his official reply on 16 May 1979, said that the reprocessing plant was being rejected for political reasons, despite his belief in its technical feasibility and despite his own and the Lower Saxony Land governments' strongly pro-nuclear leanings.[82] He made a commitment that there would be no reprocessing plant at Gorleben, though later between 1981 and 1983 attempts were made to resurrect the reprocessing plant at a site other than Gorleben in Lower Saxony, in what was[83] called a political game of 'site poker' with Wangerhausen and Volkmarschen (Northern Hessen), Schwandorf (Bavaria), Dragahn (Lower Saxony) and Kaisersesch (Rhineland Palatinate) all potentially competing. Eventually Wackersdorf in Bavaria was chosen, in 1984, partly because of the powerful support given by Franz Josef Strauss the then (CSU) Prime Minister of Bavaria.

Albrecht gave the go ahead for deep drilling at Gorleben. PTB's application to conduct the work received a State (Land) licence on 27 June 1979. Actual test drilling began in September 1979. The Gorleben site was occupied by protestors, who built a makeshift village called 'the Free Republic of Wendland'. The protest was broken up in violent fashion by the police, supported by border patrol helicopters, in June 1980.[84] There were two important reasons for the direct action initiative of setting up the Free Village encampment. Firstly, Germans traditionally believe the forest to have a sacred natural quality, and that forests should be open access to citizens. The nuclear plans would violate a tranquil forest area. Secondly, the local councillors and mayors of the area, although initially opposed to the Gorleben plans, began to show increased acceptance as they were fêted by the Bonn government and the nuclear industry and became accustomed to featuring in the national media, whereas before they had only parochial matters to consider.[85] The local people, assisted by activist outsiders, felt they had to take matters into their own hands.[86]

The drilling programme at Gorleben faced various problems, not all of which were revealed by PTB or the Land administration in Lower Saxony.[87] For a year after drilling began, the Federal Government also kept the results secret from local Luchow-Dannenberg citizens and independent scientists. However, two years later, in June 1982, a study by Kiel University's Professor Duphorn, conducted on behalf of the Federal government, explicitly drew the conclusion from his 5300 drillings that the site should be abandoned.[88] In June 1984, a few months after another 12 000 people had demonstrated against Gorleben, the Bundestag Commission for Internal Affairs held hearings but the site was not abandoned.[89] Following the collapse of the test drilling shaft as a result of problems in freezing the unstable ground around the bore shaft in May 1987,[90] angry local protestors reminded PTB and the Lower Saxony and Federal Governments of the Duphorn report.[91] The Gorleben shaft remained shut for over two years.

The Gorleben project illustrates the capacity of protesters to mobilise effective opposition to nuclear facilities in rural, conservative areas, a feature that we have already encountered in the UK. The political pressure they brought to bear succeeded in persuading a pro-nuclear Land government to reject the grandiose plans of the federal government for the integrated reprocessing and waste disposal facility. The conflict had wider ramifications. Initially the focus was in Bavaria, at Wackersdorf, unleashing mass protests in another deeply rural and conservative area. The rejected spent fuel pool storage plans led to a modification of the planned facility at Ahaus leading to objections at a six day hearing (21–27 June 1983) including many from the Netherlands. The Ahaus hearing highlighted a weakness in the partial licensing procedure whereby permission for construction of the building could be sought without direct reference to its intended end use. As one radical newspaper, TAZ, commented, 'A nuclear temporary storage depot where tons of highly radioactive fuel elements will be stored, can be licensed like a cowbarn'.[92] Finally, even if drilling confirms the technical suitability of the Gorleben site the political hurdles, already clearly demonstrated, will have to be overcome before a repository could be constructed.

The Transport Issue and the Transnuklear Affair

Apart from confronting planning problems in securing an array of regionally distributed, but centralised, facilities to process, store and dispose of radioactive waste, the nuclear planners were challenged by the anti-nuclear opposition over transport plans. The regional BIs organised effective blockades of road routes to facilities such as the Gorleben AFR store, to highlight transport risks.[93] These actions successfully halted traffic, but also caused friction within the BIs, arising from the level of violent action shown by the 3000 police brought out to disperse demonstrators. There was also recourse to the courts in attempts to block or delay nuclear materials transported by rail. In one such intervention the municipal authorities of the city of Nuremburg, 90km from the proposed Wackersdorf plant, objected to the rail transport plans, arguing that any accident along the rail line through the city would make the land adjoining the track unuseable for years. In a lawsuit Nuremburg city authorities claimed that if nuclear materials were transported through the city by rail, this would deny them their plans to construct a new hospital close to the tracks. This would create a 'zone of building denial', effectively sterilising the prime trackside building land.[94] The Nuremburg intervention caused interest in the Konrad area, where worry 'spread like ripples',[95] for by the year 2000 about 600 railroad cars filled with nuclear waste would travel to Konrad. Such statistics more than ever focused attention on nuclear transport in the mid 1980s. The Transnuklear affair in 1987–88 made matters far worse. The Transnuklear

(TN) company was, at the time, responsible for four-fifths of radioactive waste transit in the FRG, and had over twenty years of experience in designing, and operating, systems for nuclear materials transport, which Transnuklear described in a contract tender to the US Department of Energy as always being 'in compliance with international regulations'.[96] This claim was put into serious doubt by the TN affair.

The scandal began in April 1987 with investigations into evidence emerging in the press[97] of systematic bribery ('Schmiergeld') and embezzlement concerning the management of contracts to return LLW from Belgium to disposal sites in the FRG. The affair began when a TN container lorry crashed in Belgium, on 21 October 1986, 20km from the Belgian LLW treatment and research centre at Mol SCK/CEN, where the FRG firm Transnuklear (TN) had located two mobile cementation plants to 'fix' the treated waste. Police probing of the seemingly routine transport accident, at Kwardmeechelen, began to uncover problems that revealed serious mismanagement in the cross-border road transport of LLW sludges and liquids.[98] The sorting of LLW was also TN's responsibility at Mol, but in practice the job was subcontracted to the local Belgium firm SMET-JET.[99] Subsequent investigations revealed that instead of LLW being returned by TN to the FRG, consignments containing traces of radioactive cobalt and plutonium had been substitued in the 321 drums of LLW initially discovered with irregularities, a total which grew to over 2400 after subsequent revelations from Belgium.[100] Both consignment notes for international declaration, and the packaging drums were discovered to be inappropriate.[101]

The FRG had sent LLW to Mol (and Studsvik in Sweden) for treatment because, firstly, the capacity of the Karlsruhe centre was insufficient to handle and store the waste. The FRG total annual capacity was 270 tonnes at Karlsruhe and 200 at the nuclear centre at Jülich.[102] But the annual production in the FRG of such wastes was 650 tonnes. Mol and Studsvik were therefore essential until a new 700 t/per annum plant at Karlstein opened to reduce the load.[103] A second reason was that Mol offered very favourable prices for treatment, and thirdly, Mol could use boric acid in its outfall, authorization for which could not be obtained in the FRG, making its plant less efficient.[104]

Early in 1988 details began to appear in the press of the complicated nature of the TN scandal. It was revealed that nuclear functionaries were making dishonest gains and there was even the possibility that weapons grade nuclear materials had been illicitly sold to Pakistan and Libya from the FRG.[105] The corporate links between TN and other nuclear companies, particularly Alkem (fuel fabrication) and its sister company Nukem (fuel cycle services) based at Hanau in Hesse (Figure 6.1) provided a focus for political alliance on the nuclear issue between the Greens and the SPD. In the FRG where the nuclear issue is more closely bound up with mainstream politics than in any other country the TN affair provided a major setback for the nuclear industry.[106]

The SPD federal energy spokesperson, Volker Hauff, proclaimed 'we are faced with a moral catastrophe when not one firm, but an entire branch of industry has displayed a considerable degree of criminal energy in violating regulations'. The CDU federal environment minister Dr Klaus Töpfer commented, 'Whoever deals with nuclear fuels in the way they were dealt with at Hanau is not committing a minor offence, but a crime'.[107] The SPD supported a proposal in the Bundestag by the Greens to create a special committee to investigate the revelations. By the second week of January 1988, two implicated employees had apparently committed suicide. One, Hans Holtz, was a business broker for TN, who secured contracts for his firm by taking managers and operators of prospective client firms to sex and drink orgies at clubs, and proferring material bribes such as videos, cars and family holidays. In all illegal payments to the value of DM21 million (£7m) were discovered. Following arrest in December 1987 by the Public Prosecutor's Office in Hanau, Holtz slashed his wrists in jail.[108] Over 50 employees of TN, Nukem or utilities with which they had dealings were sacked. These included Manfred Stephany who was both Chairman of TN's administrative board and one of Nukem's managing directors;[109] and Nukem's other two managing directors were suspended.[110]

Herr Töpfer suspended TN's operating licence on 17 December 1987 and on 18 March 1988 the Federal government announced that there was no future for TN.[111] Nukem's operating licence was suspended indefinitely on 14 January and partially re-instated on 11 April after the inquiries in the Bundestag, the State Prosecutor's Office in Hesse, the European Commission and European Parliament had begun to clarify the complexity and complicity of Nukem's involvement in TN's corrupt practices. Its management had been aware of violations of waste management rules at TN since 1982, but had done nothing to inform the regulatory authorities. The federal Ministry for the Environment, Nature Conservation and Nuclear

Safety (BMN) discovered[112] Nukem had previously been involved in dubious commercial practices in the export of nuclear materials[113] and in an effort to sort out the fiasco, the company was radically restructured.

Herr Töpfer had made it clear that a substantial reshuffling of ownership was required in the FRG nuclear industry.[114] The network of industry links were to be dismantled to make it easier for Federal and Länder regulators to monitor the industry.[115] When the enormity of the TN scandal became apparent,[116] Töpfer's initial reaction was to continue with his nuclear fuel cycle management plans for Gorleben and Konrad, with only minor changes made because of the TN revelations.[117] However, as the scope of the problem became apparent, with the potential non-profileration violations emerging and other countries becoming embroiled, he was forced to change his plan. 'The Nukem scandal is a severe blow to the European nuclear power industry and will no doubt provide ammunition for the Greens' was how one European Commission (EC) official put it.[118]

The TN affair underlined the international implications of the transportation of nuclear waste. The EC had begun its own investigation in early January, openly expressing its disquiet to Töpfer.[119] The European Parliament also set up its own special enquiry committee. The Commissioner for Transport and Environment matters, Stanley Clinton-Davis, told the European Parliament (EP) that: 'matters known to date in the Transnuklear affair would be given the close and undivided attention of the Commission as a whole'.[120]

When he addressed the EP enquiry committee in March 1988, he reported that, following a meeting convened on 15 February by national experts on radioactive waste transport and management, it became clear that the TN case was 'an untypical instance'; nonetheless, it had been concluded in the Commission that there was the absolute need for all nuclear waste to be labelled in such a way that it could be tracked from place of origin, through eventual reprocessing, treatment and temporary storage, to the place of final storage. However, the implementation of this monitoring was a matter for EC member state authorities, although the Commission was considering 'extending existing Community law on the trans-frontier movement of hazardous waste to cover nuclear waste'.[121] Mr Clinton-Davis summed up the TN affair as follows.

> Our investigations as well as those of the national authorities involved, have brought disturbing facts to light. With all that we

> know of the dangers it is certainly from now on unacceptable for national authorities not to be in a position to know the location, nature and source of all radioactive wastes in their territory. Within its competence under the Euratom Treaty the Commission has a clear duty to propose these measures which the Community must in all logic and common sense now take following the revelations of policy weakness which the investigations have disclosed.[122]

Mr Clinton-Davis stated that Commission officials had made five fact-finding visits to Mol, had visited a Swiss nuclear power plant, three nuclear plants in the FRG, as well as the nuclear research centre at Karlsruhe (KFK), and also paid several visits to government agencies in Belgium and the FRG.[123] In addition, the European Parliament, which had long put nuclear waste on the top of its nuclear energy agenda (for example 143 written questions were posed by MEPs on nuclear waste against 83 on nuclear plant safety and 11 on toxic waste management between 1984–88) also put considerable external pressure on the FRG government, and the nuclear industry, to come up with substantial organisational and regulatory reforms.[124] However, the EP's special enquiry committee through its rapporteur Dr Gerd Schmid, a West German, expressed its disquiet that the Commission had not kept the EP properly informed of its investigations, especially as the TN affair was a matter of considerable relevance to many EEC States.

One response of the FRG government was to reorganise the nuclear waste management and fuel cycle sector by banning companies that created or treated waste material from transporting it. This was a more radical initiative than that proposed by the utility association VDEW, which had suggested a data bank for all radioactive wastes be organised by DWK, and that a parallel control system be operated by all Länder. More conditioning and compaction at reactor sites by utilities themselves would cut down the volume of transported wastes.[125] But, it was considered that the nuclear industry's own reform plans did not go far enough. Apart from creating a new centralised oversight watchdog (BAS), Töpfer, basing his plans on recommendations made by Länder regulatory officials, proposed important changes in the 1959 Atomic Law. To try to ensure the fullest oversight from the earliest date in nuclear waste projects, public hearings were to be a mandatory part of the licensing procedure for LLW, reactor waste, or spent fuel storage or processing applications for Wackersdorf and its associated facilities.[126]

As we have seen, the specific requirement for Wackersdorf was itself soon superseded.

Moreover, international co-operation agreements were to be strengthened, because, as Töpfer's deputy, Walter Hohlefeder told the German Atomic Forum meeting one year after TN first shook the industry, 'the lack of national controls was a problem contributing to the TN affair'.[127] The failure of the federal authorities to monitor exports from the FRG also surfaced in the scandals of covert sales of chemical warfare agent technology to Libya, of ballistic missile guidance technology to Argentina, Egypt and Iraq and tritium to Pakistan and India in 1988–89.[128]

The Lessons

Gorleben and the TN affair illustrate how the different levels of institutional conflict influence nuclear waste management and policy. In the FRG's system planning and regulation of the nuclear industry is undertaken by both central and Lander governments. The consequent legal and political complexity can frustrate the nuclear industry and its opponents alike. But, as is clear from the Gorleben case, the federal structure provides opportunity for local communities to exploit divisions between central and regional tiers of government. This can be particularly effective against proposals for greenfield development of the nuclear industry even in rural, conservative and relatively remote areas like Gorleben.

The TN affair was a consequence of the weakness of the regulatory regime and its lack of control over the powerful and interlocking private companies which dominated the nuclear industry in FRG. But, once the abuses were exposed, the case for greater federal control and reappraisal of overall strategy was reinforced by international concern, with the EC and the European Parliament taking the unusual step of instigating special investigations. TN made clear that decisions on radioactive waste in Western Europe could no longer be taken – if they ever really had been – purely on the basis of national priorities and policies. The very fact that the FRG had to send abroad spent fuel to France, Britain and Sweden, and LLW to Belgium and Sweden for treatment shows how the nuclear waste management programme had been internationalised in any case by the industry. In fact, as a further development post-TN in the summer of 1988, the Soviet Union also made an offer to the FRG to take some of its nuclear waste, adding a new dimension to the new East–West relationship in the age of nuclear glasnost.[129]

SWEDEN – THE PROTRACTED POLITICS OF BURYING THE PROBLEM

Sweden has a long history as a trading nation. The port of Göteborg (Gothenburg) – the home city of Volvo on the west coast has exported for over 350 years, whilst the national capital Stockholm built on 15 islands between lake Mälaren and the Baltic Sea on the east coast has been established for over 800 years as a great sea port in Scandinavia. Sweden presently exports around 40 per cent of everything it produces, which includes woodpulp, heavy engineering and nuclear technology expertise.[130] It is Sweden's desire to make full use of its nuclear expertise that results in the energy technology centre at Studsvik (Studsvik Energiteknik AB) in Nyköping taking contracts to treat LLW from power plants and fuel cycle facilities in the EEC, especially the FRG.[131]

Studsvik has an oven for incineration of combustible LLW, and in 1983 expanded its trade with the FRG, signing a contract with TN, since which time around 300 tonnes of foreign wastes have been incinerated annually.[132] Allegations of irregularities with these contracts emerged in the summer of 1987, suggesting Studsvik officials had accepted bribes from TN.[133] Following investigations by the Swedish Nuclear Power Inspectorate (SKI) early in 1988, some minor infringements in record keeping and the acceptance of small gifts (such as bottles of whisky) by Studsvik officials were unveiled, but the study of the full record of nuclear materials transfer between the FRG and Sweden between 1980–87 otherwise gave the Swedish side a clean bill of business health.[134]

The TN affair therefore only proved a minor blip on the political landscape in Sweden, unlike the volcano of problems it posed in the FRG. Nevertheless, it is nuclear issues that have been substantial contributory causes to the greatest political upheavals in Sweden's modern political history. Most recently the Green Party became the first new political party to enter the Riksdag (Swedish Parliament) for over seventy years, at the same time making it the second largest parliamentary Green faction after the Grunen in the FRG.[135] The relative success of the Swedish Greens has been attributed both to the environmentalist philosophy preached striking a chord with materialist voters also worried about sea pollution, acid rain and above all the shock of Chernobyl's radioactive fallout, which has caused substantial dislocation in Sweden's rural farming areas in particular.[136] A decade before the 1988 election, the ruling Social Democrat Party were actually voted out of office in 1976 after governing without interruption for 44 years, primarily because of their very unpopular strong advocacy of nuclear power.[137] It is to the political background of nuclear power we now turn.

Geographical and Resource Determinants

Sweden is Europe's fourth largest country (450 000 square km), but only has a population of some 8.4 million, one seventh of the UK or FRG. Despite its vast land area, more than 7m Swedes live in the south of the country, 80 per cent in urban areas. Greater Stockholm has a population of around 1 430 000, making it five times larger than Bonn.[138] The country is governed from the 349 seat Riksdag in Stockholm, which apart from the period 1976–82, had a Social Democrat majority since 1932. Described as 'one of the world's most conformist and predictable societies (with) a well ordered welfare state',[139] the country is the second most prosperous in terms of GDP per head after Japan.[140] The nuclear issue apart, the only real turbulence in the political scene in recent years has been the assassination of Prime Minister Olof Palme in 1986, the subsequent emergence of the so-called Bofors scandal, involving covert arms dealing with India[141] and opposition to new motorway construction. As with other democratic states, Sweden also has regional (County) government and municipality councils, each of which has a role in planning. The latter has, in theory at least, a 'community veto', a residuary of Sweden's ancient decentralised democracy.[142]

Nuclear Decisions and Developments – the Background

Sweden's approach to radioactive waste management and disposal is best understood in the context of its political system and its geology, both of which have apparently proved favourable to producing what may be a definitive solution to a definable problem. But there have been some serious political consequences along this pathway to compromise.

Atomic research began in Sweden in 1945. Until 1959 when the Riksdag deferred debate on nuclear weapons, there was serious consideration of nuclear weapon development along with nuclear power research. Although in 1968 after a soul-searching public debate,[143] Parliament decided against nuclear weapons procurement, revelations in the publication *Ny Teknik* in April 1985 made clear that from 1960–72 the Swedish military had covertly, in conjunction with AB Atomenergi (later renamed Studsvik Energiteknik AB), done experiments on plutonium. It is not known what was done with the plutonium contaminated wastes.[144] As early as 1956 a government study reported that uranium in Sweden's shales was the only reliable indigenous fuel. By 1970 it had been decided to develop a complete indigenous fuel cycle: uranium mining; enrichment; fuel fabrication; spent fuel reprocessing; and ultimate disposal.[145]

According to the Ministry of Industry the issues of reactor safety and radioactive waste have been focused upon since the 1956 study, although it was not until the early 1970s that 'they came to play a more decisive role for the Swedish nuclear programme'.[146] But according to the Swedish national anti-nuclear movement Folkkampanjen mot Kärnkraft och Kärnvapen (FMKK), which opposed both civil and military uses of nuclear energy, 'military motives delayed the serious recognition of the waste problem for two decades'.[147]

The first Swedish reactor was the Ågesta heavy water plant opened in the mid-1960s for heat generation only. Professor Hannés Alfvén, the Nobel Prize-winning physicist, had helped design it. Alfvén also expressed early doubts about its safety, and his expert criticism became important for those opposed to Sweden developing nuclear energy.[148] The first public protest against nuclear development came in 1970 when the municipality of Sanås used the threat of its right of veto to block the planned reprocessing plant. Seven years later the small municipalities of Skövde and Falköping near the Ranstad uranium mine in south central Sweden did veto its continuance[149] (see Figure 6.3).

The first commercial reactor was opened in 1972, at Oskarshamn (Figure 6.3). There are now twelve reactors at four coastal sites: three at Oskarshamn, three at Forsmark, four at Ringhals and two at Barsebäck (the last of which are controversially sited just north of Malmo and directly opposite Denmark's capital, Copenhagen). They provide 9700MW of power, about 45–50 per cent of Sweden's electricity demands, equivalent to about 15 per cent of energy demand.[150] This could be raised to 60 per cent if the reactors were run at full capacity. Sweden enjoys the world's second cheapest electricity, after South Africa, because half its power is generated by

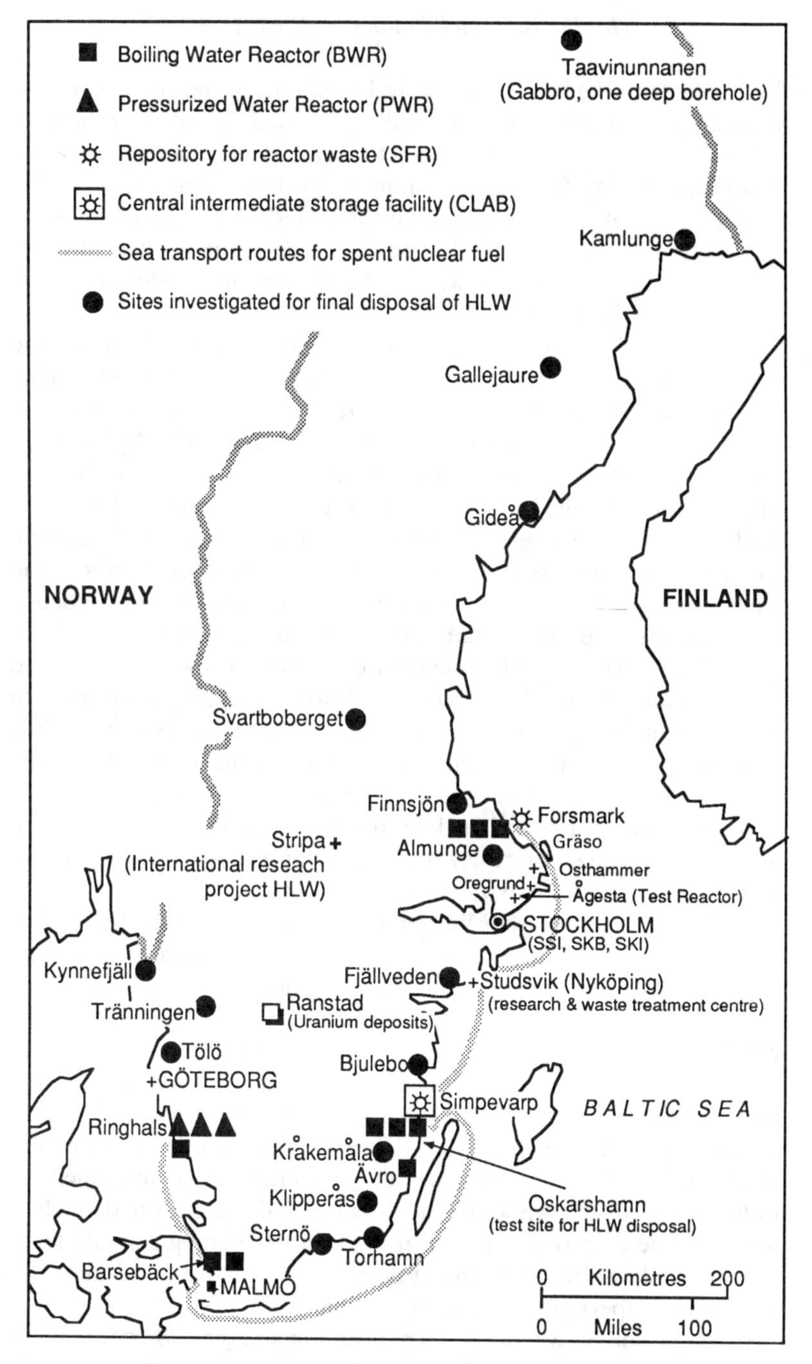

FIGURE 6.3 *Nuclear facilities and disposal sites in Sweden*

hydroelectric schemes. In the early 1970s, nuclear power was seen as a bonus to replace oil, which rapidly increased in cost after the 1973 oil crisis. However, in 1973 also, the Swedish Parliament passed a resolution calling for a moratorium on nuclear power until the full implications of the expanding nuclear industry, including waste disposal, were thoroughly studied. The controversy set off by this action by the Parliament was probably the most deep-seated in modern Swedish history.[151]

The spring 1973 decision in the Riksdag provoked widespread and acrimonious public debate. Two years later, Prime Minister Palme announced the go-ahead for a programme of twelve twin-reactor nuclear plants to ring Sweden's coast by the year 2000. Anti-nuclear groups supported by some eminent scientists such as Alfvén, had argued that technical and safety problems, particularly those connected with nuclear waste, had not been solved. Palme's decision overruled that of his own energy minister, Erik Grafström, who wanted to keep to the eleven reactors already in the planning process, until the safety problem had been satisfactorily solved.[152] Opinion polls also showed that 81 per cent of the population were prepared to reduce their standard of living by lowering energy consumption rather than accept the new nuclear programme. Despite public opposition, in May 1975 the Riksdag passed the first stage of the energy bill, by 192 votes, giving the go-ahead for thirteen reactors to be completed by 1985.[153] In 1976 the ruling Social Democrats were voted out of office after forty-four years in power, primarily because of their advocacy of expanding nuclear power.

In 1973, the Swedish Minister of Industry established a review committee (AKA-Utredningen) to evaluate strategies for handling HLW. Later AKA was also asked to consider LLW and ILW. AKA deliberated throughout the turbulent nuclear debate and produced its report just before the election in 1976, recommending that a central facility for the temporary storage of spent fuel should be established and that the final disposal of all categories of waste should be carried out in bedrock at the same site.[154] The AKA commission established the principles for both technical and organisational management of radioactive waste, whereby the private and state owned nuclear utilities take care of the practical work in developing suitable technology, and various state institutions play a complex and mutually supportive supervisory and licensing role.

In the period between AKA's publication and the 1980 referendum on nuclear power (discussed later), a variety of important developments took place in establishing the nuclear waste programme. A quasi-independent research council, Programrådet för radioaktivt avfall (PRAV), oversaw plans until disbanded in 1981 following criticism that it was insufficiently detached from the pro-nuclear interests. PRAV shared a switchboard with the Swedish nuclear fuel company Svensk Kärnbränslehantering AB (SKB),[155] which AKA had recommended should cover al the costs of spent fuel and radioactive waste. The Swedish Nuclear Power Inspectorate, Statens Kärnkraftsinspektion (SKI) established as an independent entity in 1974, after 18 years as an advisory and legislation-drafting ministerial subdivision, also created a special unit for radioactive waste facility oversight on AKA recommendations.[156] The Swedish National Institute of Radiation Protection, Statens Strålskyddsinstitut (SSI), which is overseen by the Ministry of Environment and Energy, was established in 1965 and also plays a role in supervising the implementation of the 1958 Radiation Protection Act and evaluating radioactive waste repository proposals.[157]

The most important direct policy development over the 1976–1980 period was the passing of the so-called Stipulation Act (Villkorslagen) in April 1977 by the Riksdag. The coalition government was led by the anti-nuclear Centre Party whose partners, the Liberals and Conservatives, were in favour of nuclear energy. The nuclear energy issue thus became part of political 'horse trading' and the Stipulation Act was a compromise.[158] It restricted the licensing of new reactors under the following conditions:

(A) either the utility must establish a contract for reprocessing of spent nuclear fuel, and can demonstrate how and where the final disposal of highly active nuclear waste will be handled; or

(B) the utility must show how and where unreprocessed radioactive waste will be stored with absolute safety.[159]

To meet the conditions of the Stipulation Act the utilities jointly formed the Projeckt Kärnbränslesäkerhet (KBS) nuclear fuel safety project. Towards the end of 1977 it submitted the five volume plan 'Handling of Spent Nuclear Fuel and Final Storage of Vitrified High Level Reprocessing Waste', known as the KBS-1 Report, produced by more than 400 scientists. KBS-1 itself had more than one hundred appendices in support. KBS-1 was devised to justify option A in the Stipulation Act. The usefulness of the report was immediately tested by Vattenfall, the Swedish state power board and a KBS partner who used it in support of an application to start up the Ringhals 3 reactor.

Under Swedish planning practice, when a complex government decision has to be made, the relevant documentation is circulated widely to various agencies, professional organisations, universities and trades unions for review under the 'remiss process'. As the review of the KBS-1 report was to be the first time the Stipulation Act was tested, two extra reviews were organised, undertaken by a group of foreign experts and by a working group of the Swedish Energy Commission (EK-A), specialising in safety and environmental issues. It was particularly important to involve foreign peer review, as practically all experts in Sweden had been involved in the preparation of the original KBS-1.

The KBS-1 report was evaluated by 24 foreign organisations or experts. Three endorsed it (the IAEA and geologists from the FRG and Finland) while the rest criticised it in varying degrees. Twenty five Swedish organisations were also involved in the review; the two given most weight by the ministry were SSI and SKI, both of which approved it.[160] The applications for fuelling licenses were considered by the Cabinet in October 1978 and granted in June 1979.[161] However, the Ministry of Industry accepted that the KBS-1 was an unusual remiss review: 'The status of the remiss organisation usually is, within the context of their own competence and interests, to comment on the desirability of proceeding with the described action. The strength of the remiss process is that it provides a formal mechanism for elements of society, holding very diverse opinions and values, to express their opinion as to whether or not a proposed action is acceptable, not whether it is technically possible ... it is obvious that what was hoped for in the KBS case was a technical review, not a remiss'.[162]

The anti-nuclear movement was extremely unhappy about the way the Centre Party Prime Minister Falldin had agreed to KBS-1 ('with minor exceptions'); so was his Party. The three party coalition government was dissolved, with the Liberals taking over control, after Falldin resigned in acrimony over the failure to compromise. For the second time in two years Sweden had lost a government over the nuclear question, on this occasion specifically over failure to consolidate a radioactive waste disposal plan. KBS-1 was followed in 1978 by KBS-2, which dealt with the direct disposal option for spent fuel. Of the 24 foreign authorities consulted on the first report, only six were approached to evaluate KBS-2. The FMKK suggested that

'It is no surprise that many of the most forceful critics were not consulted a second time'.[163] KBS-2 was never used in legal decision making, but was superseded by KBS-3 in 1983, discussed later.

The overall outcome of these complicated political and planning events was that the new Liberal-led government agreed to a referendum on nuclear power, with concern fuelled by the Three Mile Island accident in the USA in March 1979.

The 1980 Nuclear Referendum

The referendum, advisory in nature, took place on 23 March 1980, almost exactly a year after the Three Mile Island accident. Because of the wording of the referendum it is difficult to give a clear account of the outcome. Broadly, voting for the two options advocating the completion of the twelve reactor programme outweighed the single option to phase out the currently operating reactors within ten years by 58 per cent to 38.6 per cent, in a 74 per cent turnout. One commentator summed up the result as 'do away with nuclear power, Yes, but not yet'.[164] However, the 58 per cent vote was broken down into 39.1 per cent in favour of the softer 'continue' option that argued for completion of the programme and a phase out at the end of operating lifetime, which had Social Democrat support; and 18.9 per cent favouring the stronger pro-nuclear option, supported by the Conservatives. Later in 1980, the Parliament took the political decision to phase out the existing reactor programme by the end of the life of the newest reactors, calculated as the year 2010.[165] This decision went some way to defuse a political debate that had grown uncharacteristically divisive for a country used to political compromise.[166]

Inevitably, the Swedish programme for managing nuclear spent fuel and radioactive waste was greatly influenced by the referendum and subsequent Parliamentary decisions. But the KBS-1 and -2 remiss review process also highlighted how the formal planning system did not permit dealing with ethical issues or questions of value. The dispute, if there is to be one, must be between technical choices. This was exemplified by the transfer of decision making authority from the political arena to an agency. It exemplifies a state that promotes consensus decisions and technocratic order.[167] Despite criticism by FMKK that the referendum was biased towards the KBS-1 approach,[168] it nonetheless profoundly altered the nuclear industry's expansion plans. The decision determined the quantities of different types of waste to be disposed of in Sweden, and allowed for a timetable to be planned for the various facilities needed.[169] Whilst the Chernobyl accident caused a further re-evaluation, Birgitta Olsson, the adviser to the Swedish Energy and Environment Minister, stated in April 1987 that the detailed plan drawn up in 1985 to phase out nuclear power would not be seriously altered. The SKB estimated that by 2010 the nuclear waste inventory in Sweden would consist of:

> 7750 tonnes of spent fuel; 90 000m^3 of operating wastes LLW and ILW; and 115 000m^3 of decommissioning wastes. In addition there are expected to be around 10 000m^3 of very low activity waste created at each reactor site. [Despite the known closure date of the nuclear power programme it is noticeable from the official literature that even SKB gives varying estimates of final volumes in each category.][170]

Sweden has categorised the waste as follows:

- Long lived: wastes containing radionuclides with half-lives in excess of 30 years, such as spent fuel;
- Short-lived: wastes containing radionuclides with half-lives of less than 30 years, such as the bulk of operating wastes and decommissioning wastes;
- Very low-level: wastes with short half-lives and low activity.[171]

The Search for Acceptability: the HLW Drilling Impasse

As in most countries with extensive dependence on nuclear electricity, Sweden has sought locations and methods for final disposal (perhaps with retrievability) for HLW (Figure 6.3). Sweden is also the location for an international scientific research programme on HLW, the Stripa Mine project. The first test drilling was done in 1977 at Finnsjon, close to the Forsmark nuclear site; at Kråkemåla, near Oskarshamn; and at Sternö, close to Karlshamn. Despite the fierce nuclear debate in the mid 1970s, this work barely attracted attention. But with public sensitivities sharpened by the politics of the 1980 referendum, when test drilling was tried at Kynnefjäll in northern Bohuslän in April 1980, resistance by the local people was resolute. The local area had voted 60 per cent against nuclear power continuing, and soon set up a community defence group called 'Rädda (Save) Kynnefjäll'.[172] A guard hut – supplied with electricity from a

Swedish anti-nuclear protest symbols

wind generator and solar cells – was erected at a strategic position on the hillside, which had to be passed by any motor vehicle wanting to climb the mountain. Ever since the test drilling plans were publicised for the area, a 24 hour vigil has been kept at the guard hut, making it probably the longest running single anti-nuclear protest in the world. Granted a telephone by the local municipality, which threatened to use the communal veto against the test drilling, the Rädda Kynnefjäll has been able to alert its members of any potential transgression of its area.

Other protests have taken place in the areas of Svartboberget, in Ovanåker in 1981; in Klipperås in 1983–4; and in Almunge, east of Uppsala, in 1985–6. On each occasion, when the nuclear or geological authorities tried to test drill, the communities rebelled. Arrests, court appearances and fines followed. But the national media gave extensive coverage and SKB were reprimanded by the Minister for Energy and Environment in the Almunge case for creating a confrontation and not properly informing the community in advance of their plans.[173] These clashes between local Rädda groups and the authorities over test drilling are important parts of the planning process, yet receive no mention in official articles on the site investigation programme.[174]

The Ultimate Solution?

Whilst the various regional attempts to conduct test drilling were being conducted, and failing, in the early 1980s, another part of the long term strategy for radioactive waste management was being initiated. This was the KBS-3 study 'Final Storage of Spent Nuclear Fuel', which was submitted in May 1983 in support of a licensing application to fuel the reactors at Forsmark 3 and Oskarshamn 3 under the terms of the 1977 Stipulation Act. The KBS-3 report has 450 pages, supported by 140 background reports and references to a further 250 technical articles.[175] The KBS-3 was required in order to open Sweden's final two planned reactors, and forms the technical and planning foundation of what the Swedish nuclear industry have persisted in describing as the ultimate solution to nuclear waste problems in that country. Copies of KBS-3 were sent for review to the US Nuclear Regulatory Commission (NRC) and the UK Department of the Environment (DoE), who sent comments, and the Ministry of the Interior in the FRG, who declined. Its production involved scientists in the US, Canada, Norway, and UK as well as Sweden.[176]

The Current Nuclear Waste Management Programme

The waste management programme is currently undertaken by SKB, which was formed in 1972 (as the Swedish Nuclear Fuel Supply Company) and is owned by the Swedish electrical utilities. The present waste management strategy is based on the 1984 Act on Nuclear Activities which entered into force on 1 February 1984. It was drafted by the Social Democrats, who returned to government in

1982. It documents the responsibilities of nuclear plant operators and waste generators.[177]

The 1984 Act also set up the National Board for Spent Nuclear Fuel, Statens Kärnbranslenämnd (SKN), a supervisory and reviewing authority, which is the central body concerned with the financing of future expenses for spent fuel and post-operational wastes. SKN is required to propose to government the appropriate fee to be levied on nuclear electricity to cover costs of waste management.[178] The Act, as with its predecessors, requires that permission for reactor commissioning and operating is only given when it can be demonstrated that methods for the handling and final disposal of wastes are acceptable in terms of safety and radiation protection. SKB consequently has an R&D role in addition. A detailed six year plan was submitted to SKN by SKB in September 1986.[179] Swedish disposal policy identifies four basic principles: very high safety requirements; the use of the best available technology; the highest degree of national independence; and no burdens on future generations. This programme is supervised by the ministries of agriculture and industry, the SSI and SKI.[180]

Resulting from KBS-1, some reprocessing contracts were signed in 1977 with Cogéma in France. The details remain commercially confidential.[181] Sweden has also sent 140 tonnes of spent fuel to the UK for storage and/or reprocessing.[182] Perhaps 1000 tonnes (out of 7750 tonnes) of spent fuel might thus be reprocessed,[183] although the Swedish Government has attempted to extricate the utilities from their contracts with Cogéma.[184] Japanese companies have taken over 25 per cent of the contracts and FRG companies an undisclosed further share.[185] Having relinquished the foreign reprocessing option, Sweden is now banking on its home-based solution for nuclear waste disposal. The science correspondent of *The Times* (London) summed up the policy thus: 'For some years the Swedes have been applying to the entombment of radioactive waste the sort of devotion that the ancient Egyptians gave to the burial of their Pharoahs. They are systematically protecting their waste against all the combined forces of man and nature, from the threat of nuclear war to the advent of the next Ice Age'.[186]

In summary, the present nuclear waste R&D, management and long term disposal facilities in Sweden are:

The Stripa Mine. Located in Central Sweden, near Guldsmedshyttan. This is an *in situ* experimental project, based on an old iron ore mine.

Sponsored by the OECD and managed by SKB, the project began in 1976 in the crystalline granite formations. No waste is planned to be stored at Stripa,[187] though the effects of radioactivity will be simulated. The Swedish government rejected the suggestion by IAEA Director-General Dr Hans Blix, a former Swedish Foreign Minister, in October 1983, that the Swedish (and Canadian) granite formations might make excellent locations for centralised international deep waste disposal sites.[188]

Very low activity (VLA) wastes are to be packaged and stored at an above-ground facility at the Forsmark site, and a shallow burial trench facility at the Oskarshamn nuclear site. The projected lifetimes of these facilities are 25 years. VLA wastes from the other two reactor sites might be transported to these facilities by road.[189] A further facility is planned at Studsvik.

Centrala Lagret för Anvant Kärnbransle (CLAB). Located on the Simpevarp peninsula, next to the Oskarshamn site, CLAB is an interim spent fuel storage facility, strategically located on the south east coast from which sea transport radiates north and south to other Swedish nuclear facilities. The CLAB project was begun in late 1977, given government licensing approval in August 1979, and construction began in May 1980. It emerged as part of the KBS process and the Stipulation Act.[190] It consists of an above ground receiving building and underground storage in rock with present capacity for 3000 tonnes. Expansion is planned from mid 1990s to cope with around 8000 tonnes.[191] The nuclear opposition is concerned it might be converted into a full final disposal facility.[192]

As each Swedish reactor needs to replace around 30 per cent of its fuel assembly core each year, around 250 tonnes of spent fuel is discharged nationally a year. After on-site pool storage for a year, the fuel is then packaged in casks and transported on the French-built (under special Swedish design) M/S *Sigyn* roll-on, roll-off/lift-on, lift-off ship to CLAB. CLAB will store the spent fuel for a period of 30–40 years, during which time the fuel's activity content and residual heat will decline by around 90 per cent.[193] CLAB was commissioned in 1985, and opened in April 1986, the day after the Chernobyl radiation was first detected in Sweden.[194] It will probably cost on completion close to £190m.

M/S *Sigyn*: In Nordic mythology, Sigyn is the dutiful wife whose vessel protects her husband Loki from drips of poison.[195] M/S *Sigyn* was launched in 1982 – it beached on its inaugural voyage – and was used in 1983 to ship spent fuel to France (from 1988 it also shipped operating waste to Forsmark). M/S *Sigyn* became the focus of opposition in autumn 1987, partly because of its use in shipping spent nuclear 'mox' fuel from the FRG to CLAB, in a deal that became part of the nuclear waste transport scandal in FRG, and partly because it was delivering the first consignment of LLW to the quayside at Forsmark, awaiting final disposal.[196]

Slutförvar för Langlivat Avfall (SFL) Final Repository: Based on the KBS-3 concept, the SFL comprises a system for the handling and encapsulation and final storage of spent fuel in crystalline rock, without reprocessing. It is planned to develop a deep facility by 2010–20, based on the research done at Stripa into the penetrability of fractures and the propensity of rock to retard radionuclide migration through sorption and the buffering qualities of bentonite clay. Various potential sites have been identified. A detailed R&D plan was presented by SKB to the government in 1986 and has been under review, by several national and international organisations.[197] The R&D plan for 1987–92 proposed a new *in-situ* underground laboratory. Preliminary work on this was begun at Oskarshamn in September 1986.[198]

Slutförvar för Reaktoravfall (SFR) Forsmark: This is a central final repository for low and intermediate level reactor operating waste, built adjacent to the Forsmark reactors under the seabed of the Baltic, 160km north of Stockholm. It is being built in two phases. The first was completed in 1988, when ion exchange resins from the Ringhals nuclear plant were accepted on 27 April, once the final licence was granted.[199] The second phase is planned for the end of the 1990s. The repository consists of rock caverns of a different design for each type of waste. Short lived LLW will be placed in long caverns, and for ILW in concrete silos. The complex is 50 metres below the seabed, and 90 000m^3 is to be accommodated. It is planned to have a lifetime of at least 500 years[200] (Figure 6.4).

The SFR project was licensed in June 1983, following SKB's site licence application in March 1982. Work began in August 1983.[201] The Swedes have taken full advantage of the country's favourable geology in their plans to bury ILW and LLW at SFR. Apart from the natural barriers to migration afforded by the crystalline rock, the 50m × 25m concrete silos will also be surrounded by compacted bentonite clay. The site ensures a slow hydraulic gradient, high dilution by seawater, and a lack of unintended human interference in the repository over the period during which the most important radionuclides decay.[202] The SFR is accessed by two tunnels stretching 1km under the Gulf of Bothnia.

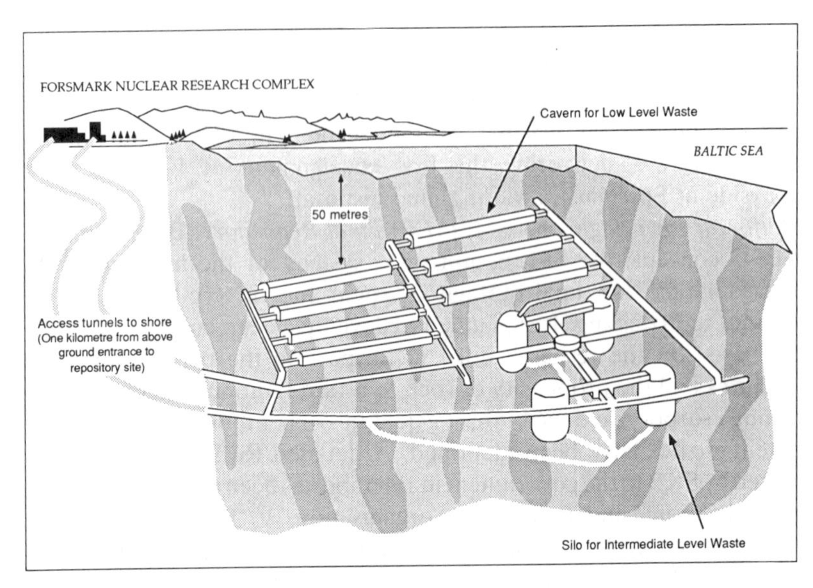

SOURCE Adapted from SKB original.
FIGURE 6.4 *The SFR-1 sub-seabed LLW and ILW repository at Forsmark, Sweden*

During a protracted R&D programme, followed by construction, the SFR project has had to meet many stringent licensing criteria for design, packaging integrity of waste, and safety (quality assurance) from the oversight authorities at SKI and SSI.[203] The SFR project is relatively expensive, double the costs for the comparative project in the French disposal programme for instance, but it is deemed a financial price worth paying to achieve public acceptability. As Sten Bjurström, SKB's President, put it, 'it has been described as the Rolls-Royce of waste disposal, but compared with the price of electricity it's just nothing and in relation to having a facility in operation that is looked on as being safe by the authorities and the people, I think it is not expensive'.[204]

Containing the Consensus – Selling the Solution

The SFR project is extremely important to Sweden for two main reasons. Firstly, it allows the government and nuclear industry to argue, with conviction, that a best possible solution to waste management and disposal has been meticulously designed and developed, although FMKK and other opponents do not agree. Secondly, it acts as a credible showpiece for the 'Swedish Way', which has attracted very positive international attention, carefully cultivated by an image-conscious Swedish industry. For example the major international conference on LLW and ILW, organised by the IAEA and European Communities, was located in Stockholm in May 1988, deliberately to coincide with the opening of the SFR first stage a few weeks earlier. One Swiss nuclear engineer noted during an underground tour of the SFR, 'It's a real repository. It's perfect, built by a tiny country with a small nuclear programme. The Swedes have demonstrated their resolve in cement and rock and not just in endless paperwork.'[205]

The Swedes have also gone abroad to discuss their showpiece,[206] as well as invite a series of foreign political groups and delegations to admire their success. The UK Parliamentary Environment Committee commented 'Of all the countries we visited, the most rigorous in its approach to radioactive waste is almost certainly Sweden'.[207] As shown in chapter 4, the UK now intends to adopt a modified version of the SFR, although ironically it is the Scandinavians who have most forcefully opposed the UK project.

Despite plaudits from abroad, it has still proved a protracted political struggle to win full acceptance for radioactive waste management policy in Sweden. An opinion poll taken in 1984 by the Swedish Institute of Public Opinion Research (SIFO) found that, despite SKB and government efforts, the public were still far from convinced that nuclear waste could safely be stored in bedrock. 64 per cent said they did not feel it could, 21 per cent believed a solution was possible, 15 per cent did not know.[208] The national Swedish anti-nuclear movement, FMKK, and the regional and smaller local campaign groups in the area of Östhammer, Oregrund, and the island of Gräso near Forsmark (Figure 6.3) and the SFR, have argued that neither SKB nor the government agencies have provided sufficient documentation, including technical appendices, to allow independent scientific assessment of the SFR project.[209] These complaints are rejected by SKB, who believe their various documents, appearances at licensing board hearings and the holding of 'open days' at SFR demonstrate their commitment to openness and access.[210] The SKI, moreover, usually permits its own staff, appearing in a personal capacity, to speak at meetings opposing the SFR and other waste management or disposal projects. The director of SKI's nuclear waste division asserts that whilst the planning system in Sweden may not be perfect, it is more open to review and re-evaluation than elsewhere.[211] Protestors such as the former school teacher Maj-Britt Andersson, a resident of Gräso, have argued that the complexity of the split licensing procedure, which takes questions of the 'installation' separately from 'safety', mean that responsibility is continuously passed around from one agency to another in local hearings.[212] Moreover, she argues that because the local area council felt incompetent to comment on the technical details it let the initial SKB application for the SFR go ahead unchallenged in 1982.[213] Other local campaigners complain that lack of resources have meant that insufficient pressure was put on local authorities to use the communal veto.[214].

However, some small scale protest was staged against the SFR. In July 1987, a group of twelve protestors refused to return to their sightseeing bus during a tour of SFR, claiming SKB literature was inaccurate and misleading. They sat-in for three hours before being arrested. The civil disobedience trial in November 1987 brought considerable national publicity to the opposition case on SFR.[215] Another disruptive protest by anti-nuclear activists, in pinstripe suits to symbolise being pillars of the local community, was held during the SFR open days on 10–11 October 1987[216] and also when the SSI granted its operating licence on 9 March 1988.

There have also been recent technical criticisms of SFR, from FMKK, which played a major part in the 1980 referendum. These criticisms were made public in April 1986, just prior to the Chernobyl accident. The FMKK questioned the hydrological, geological and dispersal predictions, and argued research on the long term future effects of sub-seabed disposal on the concentration of radioactivity in the food chain had not been undertaken, owing to inadequate legislation. They concluded that in the development of the SFR, the nuclear industry did not live up to its moral and legal responsibilities. They claimed that since the Baltic sea bed was rising the SFR was *de facto* a form of delayed sea 'dumping' – which was against Swedish atomic law – and contrary to the Baltic Sea Convention.[217] These concerns were not new. The SKI, in July 1984, had addressed both the worries over the appropriateness of the rock formation and the legality of the SFR within terms of the Baltic Sea Convention.[218] SKI went so far as to concede that

> in the opinion of the nuclear power inspectorate, with the proposed site at Forsmark, SKBF has not selected the best site for a final repository from the geological point of view. The bedrock is

> complex and certain major zones of weakness have been indicated by seismic measurements and borehole investigations. Situating the facility underneath the seabed also makes it more difficult to obtain information on the quality of the bedrock . . .

Nonetheless, SKI concluded that siting at Forsmark could be recommended and did not regard the SFR to be in conflict with the Baltic Convention.

The Swedish government has created a consultative committee for nuclear waste management, Samrådsnämnden för Kärnavfallsfrågor (KASAM), which reports annually to the Ministry of Environment and Energy. KASAM provides an opportunity for technical, political and ethical appraisal of the nuclear waste programme, by sponsoring seminars and forums, sometimes in conjunction with SSI, SKI and SKN. KASAM is a further example of the deep desire in Sweden's political establishment to keep the nuclear issue firmly on a consensual rather than confrontational basis.[219]

Conclusions

It may be judged that the bitter struggle over nuclear power development in Sweden in the 1970s has left a legacy of suspicion amongst some people that only time may allay. For the moment, at least compared to virtually any other nation, the Swedish political system has been more responsive in trying to accommodate antinuclear concerns within its democratic decision making system, on both the power reactor phase-out and the waste management programme itself. Indeed after Chernobyl, whose fallout was first discovered in the West at the Forsmark complex, the Swedish government undertook a special evaluation of the consequences for energy policy, nuclear safety, radiation and environmental protection, conducted by the Committee on Nuclear Safety and the Environment. This was done in case the phase-out programme needed to be speeded up. It seems this is not going to happen, but the Swedish government were clearly sensitive to the possible consequences of Chernobyl on their carefully laid plans.[220]

An alternative approach to adopting widespread democratic involvement in the formulation and execution of policy, is to decide from the political centre, in secret, with minimal consultation. This is possible in some political cultures, as the example of France next demonstrates.

FRANCE – L'ÉTAT NUCLÉAIRE

Background: State Sponsored Power

Of all the countries in Western Europe, France is both the most difficult to explain and the most simple to understand, with regard to the politics of radioactive waste (les déchets radioactifs). In short, the issue is practically de-politicised and has produced comparatively little public debate or power struggles between political groups in the society. This final country review therefore is necessarily rather different in approach than that taken for the other countries where radioactive waste has proved a contentious issue. Moreover, on the technical side, France has opted for a shallow burial for LLW whereas the UK, FRG and Sweden have chosen a variant of deep geological disposal for LLW.

France has both the largest nuclear energy programme in Europe, making it the most nuclear dependent of any country in the world, and has a substantial nuclear weapons programme and attendant infrastructure.[221] The reasons why this came about, are the key to understanding nuclear France in the 1990s.

The French State was in disarray and its people humiliated at the end of World War II. The provisional government was determined that France would never again face such a shattering disaster. One of the very earliest decisions of President Charles De Gaulle, in October 1945, was to establish an atomic energy commission, Commissariat à l'Energie Atomique (CEA).[222] Some of France's greatest scientists in the twentieth century had been key contributors to the Manhattan atomic bomb programme in Canada and the USA in the early 1940s, so France had a head start on other European countries, except the UK which was in the same position. But it was not just the existence of this group of atomic experts that explains the importance and power that the CEA very soon achieved. It was given a unique structure amongst French institutions – being almost a power unto itself – with an administrative autonomy, yet direct access to the French government executive.[223] The power of the CEA, and thus eventually the expansive political power of the nuclear establishment across France, was ensured by the rise of the engineer-administrators who displaced the original scientists who had set up the CEA's initial structure. The French administrative élite have created since the 1789 revolution, a culture of 'classic bureaucracy', a centralised rigidly hierarchical structure – 'les grands corps de l'Etat' – to produce, via the Ecole Polytechnique, a homogenous group of ambitious, rigorous and like-minded technocrats to control the state bureaucracies. One of these élite centres, the Corps des Mines, has operated with a sort of 'freemasonry of power', monopolizing access to a key range of positions in all sectors of central government and top positions in private industry.[224]

In 1951 the Corps des Mines, in effect, annexed the CEA. From then on, the CEA increasingly turned towards its secret military programme, and this centralised, secretive and increasingly unaccountable body created the political framework for the dominance of the nuclear industry within an accommodating political culture in post-war France. The French approach to atomic energy was similar to its technocratic approach to science in general, but the political approval of atomic energy offered, in perhaps a greater way than any other technology, a return to the 'gloire ancienne'.[225] The fact that nuclear technology was equated with the twin goals of national independence and economic progress enabled the nuclear establishment to blur the distinction between public and private interests and between political and administrative choice. So dominant has the nuclear sector become in France that some critics have dubbed it 'the nucleocratic state'.[226] The issue of radioactive waste management policy development in France must thus be understood against this background of élite, centralised and secretive decision-making.

Geography, Demography and the Nuclear Infrastructure

France has a population of around 55.4 million, which is virtually the same as the UK, but its area of almost exactly 550 000 square km makes it more than double the UK's 242 000 square km. It has a democratic centralised political system, with three main centre to extreme right parties (UDF, RPR, and National Front), two centre left or left parties (PS and PCF);[227] and in 1989 the French Green Party made its first substantial presence in the municipal and EEC elections, although the party remained comparatively small.[228] France has a strong agricultural sector, supported strongly by the EEC; a declining heavy industry, and an expanding high technology sector, especially in military and space development. France however is energy-poor, being able to produce domestically only around 3–4 per cent of its own oil demand, around 24 per cent of coal requirements and about 10 per cent of its natural gas needs.[229] In addition, about 20 per cent of French primary energy demand is met by indigenous hydroelectric power. This paucity of indigenous energy resources was a key factor in France's decision to establish a substantial nuclear industry. Although France does have considerable resources of uranium, estimated at 90 000 tonnes of economically recoverable reserves,[230] it has in practice obtained most of its uranium from abroad.

France is second only to the USA in installed nuclear generating capacity with over 51 000MW at the beginning of 1990,[231] and rising to perhaps 60 000MW by the late 1990s, around a third of total energy demand.[232] The 1973 oil crisis precipitated a policy of 'tout électrique, tout nucléaire' designed to secure French energy independence and boost French technology. In 1973 oil accounted for 40 per cent of electricity production (and two-thirds of total energy consumption) and nuclear for less than 10 per cent. By 1989, around 70 per cent of France's electricity was from nuclear and virtually none from oil[233] amounting to about a third of the total energy needs of France. In 1989 there were available four gas-graphite reactors (built before 1973); 34 900MW PWRs; 14 1300MW PWRs; and the prototype Phénix and Superphénix fast breeder reactors, or FBRs. Eighteen reactors were also under construction. In addition to nuclear reactors France has involvement in all stages of the nuclear fuel cycle from uranium to mining and milling, through enrichment, fuel fabrication, reprocessing to waste management. The major reprocessing plants are at Cap de La Hague near Cherbourg in the north and at Marcôule on the Rhône in the south (see Figure 6.5).[234] In addition to its civil nuclear programme France has a major nuclear defence commitment, involving nuclear warhead construction and nuclear submarine reactor and fuel development.[235]

The Politicisation of Energy: from Conflict to Consensus

The massive nuclear infrastructure developed by the CEA, and the national electricity utility Electricité de France (EdF) in the 1950s and 1960s was created with virtually no public involvement or objection. At the end of the 1960s the Consultative Committee set up in 1957 by the government (the Commission Consultative pour la Production d'Electricité d'Origine Nucléaire, or PEON), to coordinate CEA and EdF development, was asked to produce a new nuclear expansion plan for the 1970 and 1980s.[236] The PEON commission's original proposals in 1969 were expanded in 1973 and 1974 in response to the oil crisis.[237] These massive projections for

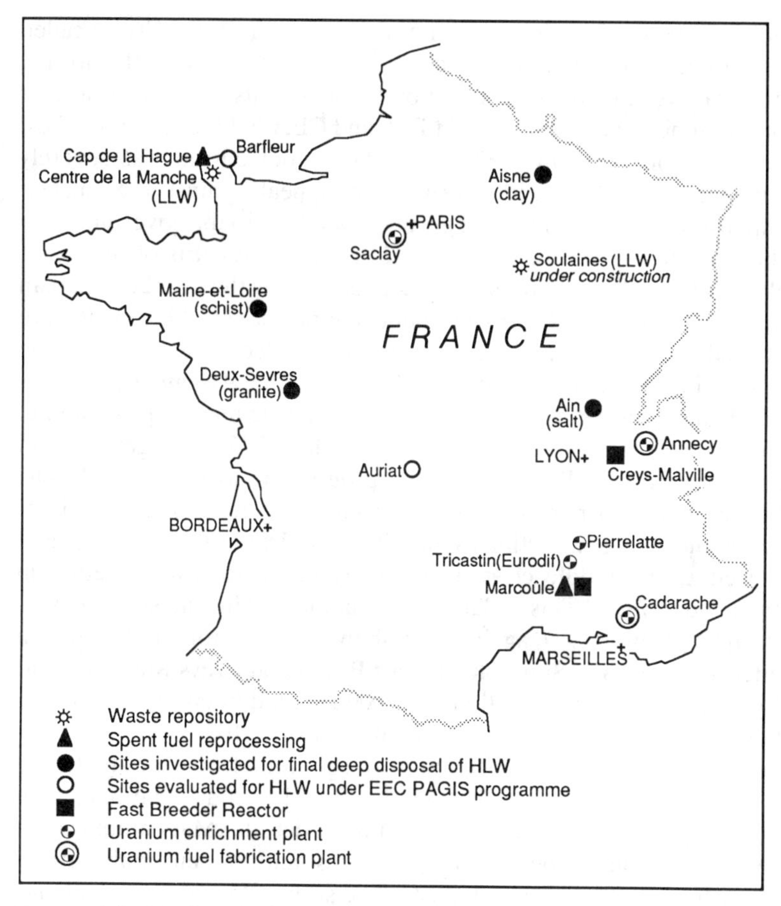

FIGURE 6.5 *Nuclear fuel facilities and disposal sites in France*

expansion created the conditions for the emergence of France's anti-nuclear movement in 1973. It was to have a relatively short life, compared to other countries and it focused almost entirely upon reactors, not radioactive waste.

In 1972 Regional Assemblies were created in France to act as mediating bodies between the local government 'communes' and central government. For the first time Paris recognised some form of regional involvement was required in planning policy. The Regional Assemblies were asked to comment on EdF's list of 34 potential reactor sites submitted in 1975. But the Assemblies were only expected to comment on the acceptability of specific sites, not nuclear policy in general, which was set at the centre, in Paris. Critics called the process 'a phantom exercise'.[238] Frustrations with the formal political system led to direct action by opponents of the vast nuclear programme. The technocrats at EdF and CEA held to the view 'C'est décidé. Donc ce sera fait'.[239] But such aloof élitism was severely challenged in the middle 1970s. An appeal against the nuclear expansion, on environmental grounds, was sent to the government by the College of France, signed by 420 nuclear physicists of repute.[240] With EdF confidently requesting 200 reactors by the year 2000 (a plan actually rejected by the government) the nuclear establishment were abused by the expanding opposition as 'electro-fascists'.[241] The opposition were frustrated that *députés* in the *Assemblé Nationale* (Parliament) were resolutely in favour of the big nuclear programme. In a rare debate on nuclear power, in May 1975, both government and opposition MPs endorsed the programme as essential.[242] The French and German activists, both emerging following periods of economic change affecting especially the agricultural community and the educated middle class, increasingly began to work together in 1976 and 1977.[243] This collaboration culminated in a massive demonstration in July 1977 against the showpiece reactor of the French nuclear future, the Superphénix Fast Breeder at Creys-Malville, near Lyon. A mass demonstration of 80 000 resulted in one demonstrator being killed and five police injured with over 100 badly hurt anti-nuclear activists, as police blamed agitators from the FRG for creating a violent confrontation.[244] This also ended national anti-nuclear protest in France for more than a decade. Mass mobilization ceased, although local ecology groups and 'green' publications flourished.[245] The nuclear industry had survived its biggest challenge and was *de facto* left in peace to develop its somewhat reduced reactor programme (despite small-scale and intermittent regional protest), and its nuclear waste projects for the next decade.[246]

Radioactive Waste – Out of Sight, Out of Mind

One of the consequences of the large nuclear programme was inevitably a major waste management problem. Like the UK, France reprocesses its spent fuel thus reducing the volumes of HLW that must be managed but correspondingly increasing the amounts of ILW and LLW. It is estimated that by the year 2000 the volume of LLW produced will amount to 800 000m^3 there will be 45 000m^3 of alpha wastes, long-lived but of low or medium activity; and 3000m^3 of high-level wastes (fission products and transuranic wastes in the form of vitrified wastes) containing around 99 per cent of the total radioactivity but only about 3 per cent of the total volume of radioactive wastes.[247]

France, initially through the CEA, and then its offshoot nuclear fuel company created in August 1975 – Compagnie Générale des Matières Nucléaires, or Cogéma – does considerable business in reprocessing spent PWR fuel from abroad. This itself is adding to France's indigenous radioactive waste burden. Some of the LLW and ILW, (and possibly even some HLW) arising from such contracts is not intended to be returned to the country of origin. Up to mid-1989 Cogéma had signed contracts with 17 utilities in FRG (9), Switzerland (2), and Belgium, Japan, Netherlands and Sweden (1 each, Sweden later cancelled) in addition to two French contracts. Cogéma's main operations are at Cap de la Hague, where they have one plant, UP2 (*usine de plutonium*, or plutonium factory) which they hope to double in capacity to 800 tonnes of heavy metal (h/m) per year throughput by 1991.[250] A second plant, UP3, with an intended final annual capacity of 800 th/m per year, is planned to open in 1990, for which contracts for some 7000 th/m per year have already been signed.[251] Cogéma also operates a smaller 600 th/m per year capacity plant at Marcôule, for gas-graphite fuel from the early reactors.[252] The reprocessing situation in France was complicated by the new agreements signed with the FRG, in the wake of the economic collapse of the Wackersdorf plant.[253] The memorandum of understanding signed between Cogéma and VEBA AG stipulates that over a period of 15 years, from the year 2000, VEBA would have access to 400 mt/h/m pa of La Hague's reprocessing capacity with an option for a future 200 metric tons. The designation plant, UP-3, is fully booked with foreign contracts up to 1999.[254] For this requirement to be met, some existing EdF spent fuel already contracted for reprocessing at La Hague will have to be displaced and stored longer. So, if Cogéma operates La Hague to its full intended capacity, around 7400 mt spent fuel from EdF reactors will be reprocessed by 1999. By the end of the same period, according to Jean Beaufrère, a senior manager in EdF's fuel division,[255] at least 13 700 mt/h/m would have accumulated, leaving around 6300 mt spent fuel in cooling ponds. Thus the new agreements with the FRG, which have assisted the FRG utilities in their spent fuel management problem, may create a new problem for Cogéma and France.[256]

Until the new FRG-Cogéma contracts were signed France had

developed a clear waste management strategy recognising that the future of the civilian nuclear industry depends on the implementation of satisfactory means of processing radioactive wastes. The intention is to vitrify all HLW and store them pending deep geological disposal after about year 2010. At present reprocessed HLW from the gas-graphite reactors are vitrified and stored at Marcôule which has a capacity for storing 330m^3. The development of vitrification mainly for PWRs and FBRs at La Hague will provide capacity sufficient to deal with accumulated and current production of HLW from reprocessing. Vitrification is managed by Cogéma, which is also responsible for storage of wastes before and after reprocessing and of vitrified wastes pending final disposal. Cogéma is also mainly responsible for on-site storage of alpha wastes which will also increase in volume as new reprocessing facilities come on stream at La Hague. Storage capacity for vitrified and alpha wastes is limited.[257] Radioactive wastes are also generated elsewhere in the fuel cycle and at research plants at Tricastin and Pierrelatte (uranium enrichment and fuel fabrication) and Cadarache (Figure 6.5).

The agency responsible for the long-term management of all radioactive wastes is Agence Nationale pour la Gestion des Déchetes Radioactifs (ANDRA) which, like Cogéma, is controlled by the state under the CEA. ANDRA was created in November 1979 by the Ministries of Industry, Economy and the Budget.[258] France is also a participant in the international project at the Stripa mine in Sweden. But France is pressing ahead with its own plans for a deep disposal site to take alpha and vitrified wastes.

Site Choice for HLW Repositories

The HLW research programme is guided by the so-called Goguel Panel Report produced for the Ministry of Industry.[259] The earlier Castaing Commission (1982–84) on the nuclear fuel cycle had recommended two research sites be chosen, in case one proved unsuitable after investigation. The Castaing Commission, set up by the Ministry of Industry in December 1981 following the debate on nuclear energy in the Assemblé, which had been promised by the new Socialist government during their election campaign, contained twelve senior scientists from the nuclear industry and research institutes including members of the Trades Union CFDT and two independent critical scientists. It held 45 meetings before completing its first report in November 1982. Considering its composition and the 'technocratic brotherhood' from which most of its members came, it produced a surprisingly critical appraisal of the French policy of reprocessing and consequent waste management. Despite its force of argument, it was never really used politically in France.[260] It was agreed that if the findings were positive then an application for construction would be made. The aim was to have the deep disposal facility ready to receive alpha wastes early in the next century and vitrified wastes from about year 2010 by which time they will have cooled sufficiently.[261]

In early 1987 four sites in different geological settings were announced as potential candidates for a deep repository. This was done without prior consultation with the communities picked. Referendums were organised by local mayors in October 1987. Votes of between 75–95 per cent were registered against ANDRA's proposals. The programme nonetheless continued. The sites are: a clay formation in the department of Aisne near Laon in the north east; a granite massif in Deux-Sevres and a schist formation in Maine-et-Loire both in the west central part of the country; and a salt site in Ain in the east central area north east of Lyon (see Figure 6.5). Geological investigation will take place at each site for the selection of one as the site for an underground laboratory for conducting more detailed studies. These plans went ahead practically unhindered until early in 1988 the regional opposition groups, supported by local mayors and prefectures, formed a national anti-dumping campaign which although nowhere near as coherent as national opposition coalitions in other European countries or the USA, was a new development in France.[262]

The Disposal of LLW and ILW: Shallow Burial

LLW and ILW with short half-lives (less than 30 years) are at present disposed of at ANDRA's Centre de la Manche next to the reprocessing plant at La Hague (see Figure 6.5). The facility, opened in 1969 and covering 32 acres, is designed as a permanent surface disposal site. The wastes from power plants, reprocessing and other nuclear sites across the country are compacted and solidified using bitumen and concrete, and packaged prior to disposal. Those wastes which require further shielding are placed in excavated concrete trenches called 'monoliths' and backfilled. These monoliths form the base for 'tumuli' or mounds in which low activity wastes are stored in drums surrounded by concrete blocks. Water from run-off is separated and monitored before discharge into the sea and the wastes are

SOURCE WISE, Paris.

intended to be shielded from ground water. Once completed the tumuli are covered and left to form shallow grassy mounds merging with the surrounding 'bocage' landscape.[263] Once the site's capacity of 480 000m^3 is full in around 1991, it will be closed but will remain under state supervision and surveillance for 200 to 300 years until it is safe for unrestricted access. In 1986 ANDRA announced that Soulaines, east of Troyes in the department of Aube had been selected as the site for the second shallow disposal facility for short-lived wastes (Figure 6.5). This Centre del'Aube is expected to open in 1990–91. The choice of site was challenged in the Supreme Court by local opposition activists, attempting to reverse the Decree of Public Utility issued by ANDRA.[264]

France intends to spend the majority of its waste disposal budget (80 per cent) on treatment and packaging, leaving only 20 per cent for disposal.[265] This policy has directed effort towards volume reduction by chemical and compaction means.[266] In terms of disposal routes France has developed a system of surface disposal for LLW which has been operating since 1969, and is likely to have a replacement facility ready within a few years. Although a specific site for a deep repository for high activity wastes has not yet been selected there appears to be every prospect that France will achieve the deadlines for selection and construction enabling a repository to be opened around the turn of the century. France is almost unique among western countries in its ability to develop and implement civil nuclear programmes and policies without the kind of political hesitation and conflict endemic elsewhere. Even involved activists and researchers do not see the French anti-nuclear movement being resurrected to confront the ANDRA plans.[267]

Conclusions

The lack of conflict over radioactive waste management policy can be explained partly by the features of the French political culture and decision making system. There is, first, a strong partnership between state and industry which acts to minimise public debate or dissent, although evidence of re-emerging opposition suggests this may change. In the waste management field ANDRA is the archetype of the corporatist state agency serving the needs of industry, in this case an industry dominated by state control. This massive state involvement strengthens the second feature, the central domination of decision making. The central state effectively controls the nuclear industry and ensures decisions are effected with only token local participation. Although there is, formally, a right to be consulted at regional level, in practice siting decisions are vested in the industry and government and local interests are excluded from the decision making process. Official disdain for public anxiety about the dangers of nuclear power was aptly expressed by Remy Carle of Electricité de France who commented on French nuclear planning procedure, 'You don't ask the frogs when you drain the marsh'.[268] This exclusion of public opinion is encouraged by a third feature, the technocratic élitism of French political culture and the general confidence placed in the bureaucratic regime. There is a tendency 'to exploit this élitism in order to maximise economic efficiency and minimise political debate, impasse, and indecision in policy implementation'.[269]

The strength of the nuclear industry is also to be explained by a deliberate policy to minimise opposition. The French nuclear industry has capitalised its political advantage by a 'skilful blend of public relations and tangible inducements'.[270] Among the latter have been preference to local firms during construction, tax and rating levies on the nuclear industry to provide funds for host departments and localities, and other forms of compensation such as provision of community facilities. Most controversial was the tariff reductions on electricity, subsequently declared illegal. Above all the nuclear industry provides employment, a potent lever with which to ensure support especially in remote 'nuclear oases' such as La Hague where around 13 000 workers are directly dependent on the industry. Trade unions have had little desire to develop opposition to an industry which provides substantial employment. In addition careful and sustained publicity for the presumed benefits of nuclear power serves to entrench the French nuclear culture. The 'capture' of the trade unions, after earlier dissent, has proved crucially important.

The weakness of the anti-nuclear opposition is not simply to be explained in terms of the centralised French political culture and the careful and selective use of appropriate inducements. It has also been achieved by the willingness of the central state to use force as most clearly shown at Creys-Malville. The use of violence by some anti-nuclear protestors in France has tended to divide and thereby weaken the opposition. Despite this, opposition has not been entirely eliminated. On occasion it has been able to prevent nuclear development such as the proposed reactors at Plogoff in Brittany in 1981.[271] On his election, President Mitterand redeemed his campaign promise to cancel the proposal[272] but reneged on a commitment to have a full Parliamentary debate on the nuclear future.[273]

Although in the post-Chernobyl period almost 50 per cent of the population indicated opposition to nuclear power (59 per cent opposed new reactors[274]) and 69 per cent expressed their concerns over nuclear waste disposal,[275] in practice the opposition remains passive, cowed and ineffective. While France appears to have made more progress in establishing its nuclear waste management regime she still has to establish the suitability and safety of HLW sites. Despite centralised power and support for the nuclear industry,

France may encounter the determined opposition to any greenfield location that has been experienced in other countries.

France presents the extreme case of close cooperative relationships beween central political power and the civil-military industrial nexus with a concomitantly feeble opposition. The government counters any political wavering with the reminder that France not only depends on nuclear electricity but relies on its nuclear weapons programme: and that the two are linked. This ensures that French commitment to the nuclear option is able to withstand the commercial and political pressures which have weakened the nuclear lobby in other European countries.

RADIOACTIVE WASTE – THE POLITICAL OUTLOOK

The differences among the three countries considered in this chapter may be explained partly in terms of geological and geographical conditions. Sweden has an abundance of crystalline formations, a low population density and a coastal distribution of power stations. The FRG possesses a number of abandoned mine workings in salt and other suitably dry formations which are considered suitable for radioactive waste disposal. France has selected four sites in different geological formations for HLW disposal and the method of disposing of other radioactive wastes does not depend on the underlying geology.

The problems are also technically different. France has developed civil reprocessing facilities enabling HLW to be reduced in volume and vitrified prior to final disposal. In the meantime such wastes are stored at the reprocessing plant. But, reprocessing immensely increases the volumes of ILW and LLW which may be further increased by wastes created from imported wastes which are not returned to the exporting country. In this way Sweden and the FRG have reduced their waste problems by exporting part of it abroad to France and the UK.

The contrasts of policy and progress towards solutions are primarily politically determined. Sweden and France appear to have made most progress in gaining public acceptance (or quiescence in the case of France) and in achieving implementation. They are the countries with the greatest dependence on nuclear energy yet they are also the countries which show the greatest political contrasts. In Sweden nuclear power was a major political issue at national level until the government accepted the public wish to phase out nuclear energy, thus establishing a broad consensus. France has adopted a vigorously pro-nuclear policy. Here consensus has been based on national independence and technical development and subsequently ensured through the use of inducements and the systematic exercise of centralised power. In the FRG the nuclear issue, military and civil, has been a major source of conflict. The federal political structure provides opportunities for the anti-nuclear opposition to challenge proposals such as reactors and repositories. The lack of effective control over the private sector revealed by the TN affair has also strengthened anti-nuclear opposition.

The future credibility, progress (or in the case of the FRG even existence) of the nuclear industry in all three countries depends crucially on politically acceptable solutions to the radioactive waste management problem. In Sweden and the FRG it is clear that it is possible to mobilise broad alliances able to influence policy at national level as well as to deploy effective opposition to specific siting proposals. In France opposition has been coopted or circumvented or dealt with by a summary exercise of central power. Although each country has plans and policies for nuclear waste management in none of them has a site for final disposal of HLW been identified.

In Western Europe, nuclear waste has remained a problem for individual countries to solve. So far international cooperation has been largely confined to developing such projects as fusion and breeder reactors, and the French and British fuel fabrication and reprocessing operations. As for international agreement on radioactive waste disposal the London Dumping Convention on disposal at sea and the Paris Commission on controlling pipeline emissions into the sea have met with limited success as we saw in chapter 2. International research projects such as the Stripa mine in Sweden are intended to find solutions that satisfy technical criteria and thus clear the way for future expansion of the nuclear industry. For the anti-nuclear opposition the problem of radioactive waste is the Achilles heel which will fatally weaken the industry. Both supporters and opponents recognise nuclear waste as the issue which will determine the way forward.

NOTES

1. Brown, S., 'Anti Nuclear feeling grows in GDR', *END Journal*, No. 15 December, 1986, p. 14. Reuter, 'Polish nuclear marchers gassed', *Guardian*, 7 September 1987. Glenny, M., 'Czech plans for nuclear reactor stun scientists', *Guardian*, 30 October 1987. Ravasz, K., 'PAKS radwaste site opponents becoming more vocal in Hungary', *Nucleonics Week*, 22 September 1988. Keller, B., 'Soviet Public warn of nuclear power', *International Herald Tribune*, 14 October 1988. Cornwell, R., 'Green movement threatens Soviet nuclear dump plan', *Independent*, 29 June 1989. Smiley, X., 'Kremlin admits more mishaps at Chernobyl', *Daily Telegraph*, 16 July 1989. Schneider, K., 'The Soviets show scars from nuclear arms production', *New York Times*, 16 July 1989.
2. Mackay, L., and Thompson, M. (eds), *Something in the Wind: politics after Chernobyl* (Pluto Press, London, 1988).
3. Thames Television, 'Brothels, bribes and nuclear wastes', *This Week* Transcript, 10 March 1988.
4. Evidence to European Parliament by EC Environment and Transport Commissioner, S. Clinton-Davis, 10 March 1988, PE 121.303; statement from the European Commission IP (88) 432, 5 July 1988, Strasbourg; statement by EC Energy Commissioner, N. Mosar, 26 October 1988, Verbatim Report of European Parliament proceedings, pp. 208–10.
5. Schmid, G., Report drawn up on behalf of the Committee of Enquiry on the handling and transport of Nuclear Material, on the results of the Inquiry Parts A and B, European Parliament Doc. A 2–120/88. PE 123.491/fin. 24 June, 27 June (1988a) respectively.
6. Hibbs, M., 'West German official urges caution on European Radwaste Regime', *Nuclear Fuel*, 23 January 1989a, p. 4.
7. Hibbs, M., 'SKI concludes TN safety probe, but Swedish prosecutors continue', *Nucleonics Week*, 14 April 1988b.
8. Hülsberg, W., 'The German Greens. A social and political profile', esp. Chapters 2 and 3, (London, VERSO Press).
9. Goodhard, D., 'System without a centre – but with history on its side', *Financial Times*, 21 October 1988.
10. Dangelmayer, D., 'West Germany's nuclear dilemma', *New Scientist*, 13 July 1978, pp. 103–5.
11. Marsh, D., 'West Germany finds its nuclear legacy still handicapped by history', *Financial Times*, 3 December 1986.
12. Timm, M., 'Significance of Nuclear for Germany's Energy Supply', *Nuclear Europe*, January 1987, pp. 13–15.
13. Ibid.
14. Weinländer, W., 'Reprocessing and Waste Management in the Federal Republic of Germany', paper presented at the International Conference on Nuclear Fuel Reprocessing and Waste Management, RECOD, Paris, 23–7 August 1987.
15. Marsh, D., 'Nuclear Power in West Germany: the consensus lies in tatters', *Financial Times*, 14 January 1987a.

16. Reuters, 'Kohl Doubt on N-power', *Financial Times*, 25 January 1988.
17. Hibbs, M., 'West German Fabricators Alkem, R.B.U. consolidated under Siemens-KWU Division', *Nuclear Fuel*, 23 January 1989b, pp. 2–3. Jelinek-Fink, P., Memorandum submitted 27 February 1986 by Nukem GmbH to House of Lords, Select Committee on the European Communities, Session 1985–86, published in *Nuclear Power in Europe*, Vol. II (Evidence) 227–II, 15 July 1986a, pp. 143–9; and: Meysenburg, H. H., 'The Fuel Cycle Industry in the Federal Republic of Germany', *Nuclear Europe*, No. 4, April 1983, pp. 27–30.
18. Duxbury, L., 'Nuclear Power in the Federal Republic of Germany', *ATOM*, No. 384, October 1988, pp. 12–13.
19. Hibbs, M., 'Wackersdorf cost estimate rises to DM7.7 billion', *Nuclear Fuel*, 6 February 1989c, p. 10.
20. Thym, R., 'Ende der Bauarbeiten an der Wiederautarbeitungsanlage', *Suddeutsche Zeitung*, 31 May 1989. Marsh, D., 'Bavarian N-plant construction to halt', *Financial Times*, 31 May 1989a.
21. Hibbs, M., 'Bayernwerk calls Wackersdorf essential to future of German nuclear energy', *Nuclear Fuel*, 15 June 1987a.
22. Hirsch, H., Helmut Hirsch was interviewed by D. Lowry in Hanover, Lower Saxony in November 1987a.
23. Agence France-Presse, 'Paris, Bonn in nuclear waste pact', *International Herald Tribune*, 7 June 1989. Hibbs, M., 'UK-German fuel pact expected soon: Germans look at Swedish UP-3 contracts', *Nuclear Fuel*, 10 July 1989d, pp. 1–2.
24. Hibbs, M., 'Hanau firms tangled in a web of legal charges over plant changes', *Nuclear Fuel*, 10 August 1987b. Marsh, D., 'Scandal put West German nuclear industry under a cloud', *Financial Times*, 29 April 1987b.
25. Marsh, D., 'A delicate balancing act'. Interview with Klaus Töpfer, the West German Environment Minister, *Financial Times*, 12 January 1989.
26. Hübenthal, K. H., 'The policy of the Federal Republic of Germany concerning the management of low-and-intermediate-level waste', BMFT, Bonn. Paper to *The International Symposium on Management of Low-and-Intermediate-Level Wastes*, Stockholm, 16–20 May 1988, IAEA–SM–303/157.
27. Hirsch, H., Interviewed by D. Lowry, Hanover, November 1987b.
28. PTB Info-Blatt, 12 November 1985.
29. Brennecke, P., Griller, H., Marlens, B. R., Schumacher, J., Warnecke, E. and Odoj, R., 'Waste Acceptability and Quality Control for the Planned Konrad Repository', PTB, Braunchweig, KJ GmbH, Jülich. Paper to Stockholm Conference, 16–20 May, 1988, IAEA–SM–303/139.
30. Salander, C., Proske, R., and Albrecht, E., 'The Asse Salt Mine. The World's only test facility for the disposal of radioactive waste', *Interdisciplinary Science Reviews*, Vol. 5, 1980, No. 4, pp. 292–303.
31. Wollny, Lilo, Interviewed by D. Lowry, Shetland Islands in July 1987.
32. DBE, 'The Final Disposal and Related Waste Management', Peine 1987.

33. Olivier, J. P., 'Disposal Policies in Other Countries', OECD Paper to the 1987 European Summer School of Radioactive Waste Management, 6–9 July, Christ College, Cambridge.
34. Marsh. D., 'East and West Germany to improve links', *Financial Times*, 9 September 1987c. TAZ, Cited in 'Gorleben "a wobbly pudding" – local response to collapse', *WISE News Communiqué*, 278.2382, 14 August 1987.
35. Marsh, D., '"French Connection" boosts German N-plant', *Financial Times*, 16 March 1987d.
36. Tomforde, A., 'Celebration as Bavaria's Sellafield site goes green', *Guardian*, 8 August 1989.
37. Hibbs, M., MacLachlan, A., 'Official asks utilities whether they want to reprocess at Wackerdorf', *Nuclear Fuel*, 15 May 1989, pp. 4–5.
38. Hibbs, M., 'Abandoning Wackersdorf will cost German utilities another DM1–billion', *Nuclear Fuel*, 10 July 1989e, pp. 12–13.
39. Goodhart, D., 'UK undercuts French in nuclear reprocessing, *Financial Times*, 26 July 1989.
40. Clough, P., 'German nuclear fuel may come to Britain', *Independent*, 26 July 1989.
    Brown, P., 'Nuclear agreement brings £1.6bn deal', *Guardian*, 25 July 1989. Carr, J., 'Bonn may rethink N-Waste Plan', *Financial Times*, 22 May 1985.
41. Catteral, T., 'N-Waste Store for Gobi', *Observer*, 16 June 1985; 'China-West Germans near agreement on spent fuel', *WISE News Communiqué*, 209.1519, June 1984. Perera, J., 'China and Sudan want Germany's Nuclear Waste', *New Scientist*, 5 September 1985. 'Gobi Grave for Nuclear Waste', *New Scientist*, 23 July 1987.
42. Petroll, M., 'Taking the "worry" out of the back end of the Fuel Cycle', *Nuclear Europe*, Vol. 7, No. 1–2, January/February 1987a, pp. 25–7.
43. OECD (1987), op. cit.
44. ERL (1987), op. cit.
45. *Power Europe*, 18 June 1987.
46. *Wise News Communiqué*, 278.2382, 14 August 1987.
47. Salander, K., Proske, R., and Albrecht, E., op. cit., 1980, pp. 292–303.
48. Hübenthal, K-H., (1988), op. cit.
    OECD (1987), op. cit.
49. Scheffler, K., Strilzke, D., and Tittman, E., 'Pamela demonstrates the vitrification of high level waste', *Nuclear Engineering International*, September 1983, pp. 39–42. Körting, K., 'Karlsruhe begins cold trials of prototype melting facility for vitrification of reprocessing high-level waste', *Nuclear Europe*, August 1987.
50. Petroll, M., op. cit.
    DBE (undated), op. cit.
51. Nau, H. R., *National Politics and International Technology; nuclear reactor development in Western Europe* (Baltimore, Johns Hopkins University Press, 1974).
52. Hülsberg, W., op. cit., 1988, pp. 42–3, 54–76.

53. Nelkin, D., and Pollak, M., *The Atom Besieged: extraparliamentary dissent in France and Germany* (London, MIT Press, 1981a).
54. Marsh, D., 'The consensus lies in tatters', *Financial Times*, 14 January 1987e.
55. Cervenka, Z., and Rogers, B., *The Nuclear Axis: secret collaboration between West Germany and South Africa* (London, Julia Friedman Books, 1978).
56. Nelkin & Pollak (1981b), op. cit., p. 28.
57. Nelkin & Pollack (1981c), op. cit.
58. Palmer, J., *Europe Without America? The crisis in Atlantic Relations* (Oxford, Oxford University Press, 1988).
'A Campaign Against a Model West Germany' (Bochum, 1978).
59. Kelly, P., *Fighting for Hope* (London, Chatto and Windus, Hogarth Press, 1984).
60. Petroll, M., op. cit., 1987c.
61. Jelinek-Fink (1986b). See reference 17, p. 146.
62. Hirsch, H., interview by D. Lowry, in Hanover, November 1987c.
63. Piontek, N., interview by D. Lowry, in New York, October 1987.
64. Goodhart, D., 'Fresh Setback for West German N-Power', *Financial Times*, 13 September 1988.
65. Hirsch, H., 'Nuclear Waste Management in West Germany: the battle continues', *The Ecologist*, Vol. 13, No. 1, January 1983.
66. Salander et al., op. cit., 1980.
67. Berkhout, F., Boehmer-Christiansen, S., Skea, J., 'Deposits and repositories: Electricity wastes in the UK and West Germany', *Energy Policy*, April 1989, pp. 109–115.
68. Colchester, N., 'Bonn rejects atomic waste site near East German border', *Financial Times*, 23 February 1977.
69. Buschschluter, S., 'Site Soon for Nuclear Waste', *Guardian*, 12 November 1976.
70. Fritzen, Marianne, interview by D. Lowry, in Lüchow, November 1987.
71. Wollny, L., interview by D. Lowry, in New York, September 1987.
72. Fritzen, M., op. cit., 1987.
73. Barnaby, F., 'An unclear decision for Germany', *New Scientist*, 26 April 1979, pp. 266–7.
74. Pesel, Astrid, interview by D. Lowry in Lüchow, in November 1987.
75. Schaarschmidt, Dieter, interview by D. Lowry in Lüchow, in November 1987.
76. Zint, G., *Gegen den Atomstaat*, Zweitauseudeius, Frankfurt, 1979.
77. Steudel, Evelyn, interview by D. Lowry in Lüchow, in November 1987.
78. Hirsch, H., interview by D. Lowry in Hanover, in November 1987d.
79. Bonn Correspondent, 'Salting it away', *Economist*, 31 March 1979, p. 34.
80. Patterson, W., 'Grim Legacy of Meltdown', *Guardian* 28 March 1989.
81. Dicks, A., 'Bonn may modify nuclear disposal plans', *Financial Times*, 4 May 1979.

82. Buschschluter, S., 'Giant atom waste plant scrapped', *Guardian*, 17 May 1979.
83. Hirsch, H., op. cit., 1983.
84. Buschschluter, S., 'Protest Villagers evicted', *Guardian*, 5 June 1980.
85. Harms, R., interview by D. Lowry, in Grabow, November 1987.
86. Hirsch, H., 'Gorleben: winning the battle, losing the war?, *Nature*, 24 May 1979, Vol. 279, p. 283.
87. Hopfner, K., 'Germany to decentralise nuclear waste treatment', *Nature*, 1 October 1979, Vol. 281, p. 419.
88. Hirsch, H., op. cit., 1987e.
89. *WISE News Communiqué*, 205.1438, 'Large demonstration at Gorleben', 27 March 1984a. *WISE News Communiqué*, 210.1523, 'Hearing on Waste Repository in Salt Domes FRG', 3 July 1984b.
90. Mayer, R., and Spannbruker, K., PTB officials, interviewed by D. Lowry in Gartow, in November 1987.
91. Hibbs, M., 'Geologists Probing Problems at Gorleben Test Shaft', *Nuclear Fuel*, 13 July 1987c.
92. *WISE News Communiqué*, 187–1246 'Hearings on temporary spent fuel storage facility in Ahaus', 12 July 1983, pp. 4–5. *WISE News Communiqué*, 184.12.15 'Ahaus waste Dump Protested', 14 June 1983, p. 1.
93. *WISE Communiqué*, 207.1471, 'Successful Blockade in Wendland', 15 May 1984c.
94. Hirsch, H., op. cit., 1987f.
95. Ibid.
96. Letter from Transnuclear, Inc (USA), to Charles R. Head, Operations Division of US Office of Civilian Radioactive Waste Management, dated 5 July 1984. Reprinted in *Public Comments on the Draft Mission Plan for the Civilian Radioactive Waste Management Program*, Vol. III, June 1985, DOE/RW–0005, p. 249, USDOE, Washington DC.
97. Clough, P., 'Germany unearths nuclear scandal', *Independent*, 30 April 1987a.
98. Hibbs, M., 'Transnuklear Affair widens to include foreign clients', *Nucleonics Week*, 24 September 1987d.
99. Schmid, op. cit., 1988B, p. 27.
100. Schmid, op. cit., 1988A, p. 9.
101. *Der Spiegel*, 'Der Atom Müll-Schwindel', 11 January 1988, pp. 20–32.
102. Dietrich, G. and Gussman, H., 'Solid Radwaste Treatment: incineration, compaction, and decontamination', *Nuclear Europe*, March 1986, pp. 20–21.
103. Lowry, D., 'Corrupt to the Core: the hidden agenda for nuclear waste', *Environment Now*, August 1988, p. 27–9.
104. Schmid, op. cit., 1988B, p. 24.
105. Hibbs, M., 'US repeatedly warned Germany on nuclear exports to Pakistan', *Nuclear Fuel*, 6 March 1989d, pp. 12–13. Schutze, C., 'Nuclear waste scandal "no evidence" that suspended firms sent material to Pakistan, Libya', *The German Tribune*, No. 1307, 24 January 1988, p. 3.

106. MacShane, D., 'Nuclear Waste panic hits Germany', *New Statesman*, 22 January 1988.
107. Tomforde, A., 'W. Germans concede crisis of confidence in nuclear industry', *Guardian*, 14 January 1988.
108. See Ref. 3.
109. Hibbs, M., 'German Parliament orders inquiry into National Fuel Cycle Plan', *Nucleonics Week*, special issue, 15 January 1988c.
110. Hibbs, M., 'Nukem re-instates two directors', *Nucleonics Week*, 2 June 1988d, p. 6.
111. Hibbs, M., 'No Place for Transnuklear in German Waste Management Future', *Nucleonics Week*, 24 March 1988e.
112. Hibbs, M., 'Nukem said to agree to market French research reactor fuel', *Nucleonics Week*, 18 April 1988, p. 5.
113. Hawkes, N., 'Alarm over plutonium firm', *Observer*, 11 December 1977.
114. MacLachlan, A., Hibbs, M., 'European Nuclear players altered as West German firms consolidate', *Nucleonics Week*, 7 April 1988.
115. Roth, T., 'Bonn moves to break up Nuclear Firm's Equity Links', *Wall Street Journal*, 21 March 1988.
116. *Der Spiegel*, 'Nukem: Ausser Kontrolle', 18 January 1988, pp. 18–30. *Der Speigel*, 'Uran-Schwindel: Atome tragen keine Flagge', 15 February 1988, pp. 51–66.
117. Hibbs, M., 'TN scandal reverberates in more countries, top management', *Nucleonics Week*, 14 January 1988, p. 10.
118. Greenhouse, S., 'Safety issues test Europe's faith in nuclear power', *New York Times*, 31 January 1988.
119. Dickson, T., 'Commission steps into row over nuclear waste', *Financial Times*, 7 January 1988.
120. Commissioner S. Clinton-Davis, Address to European Parliament, verbatim reports, 19 January 1988, pp. 109–10.
121. Commissioner S. Clinton-Davis, Statement to the European Parliament Special Enquiry Committee, 10 March 1988, PE121.303 (Annex).
122. Commissioner S. Clinton-Davis, Address to European Parliament, verbatim reports, 5 July 1988 pp. 109.
123. Ibid.
124. Bangeman, Reply from the European Commission to L. Smith MEP, 17 February 1989; QXW 1979/88 En.
125. Hibbs, see note 111.
126. Hibbs, M., 'West German Official urges caution on European radwaste regime', *Nuclear Fuel*, 23 January 1989, p. 4.
127. Hibbs, M., 'Bonn announces plans for new controls on nuclear exports from West Germany', *Nuclear Fuel*, 23 January 1989e, p. 3.
128. Reported by Jane Corbyn, 'The Condor Connection', Panorama BBC-1 Television, April 1989. AP/Agence France-Press, 'W. German firms linked to Pakistan bomb', *Independent*, 21 April 1989.
129. Dickman, S., 'Nuclear Waste may go East', *Nature*, vol. 334, 18 August 1988b, p. 557.

130. Stockholm, Sweden, Stockholm site and Development Company, SML, 1984, p. 7.
131. 'Nuclear Sweden VI', Swedish Atomic Forum, Stockholm 1986, pp. 14–16.
132. Schmid, G., op. cit., 1988b.
133. Hibbs, M., 'Transnuklear Affair widens to include Foreign Clients', *Nucleonics Week*, 24 September 1987b, pp. 4–5.
134. Hibbs, M., see note 7.
135. Ezard, J., 'Greens fall short of holding balance of power: Carlsson returns to power in Sweden', *Guardian*, 19 September 1988.
136. Taylor, R., 'The Greens may end an idyll', *Financial Times*, 16 September 1988.
137. Abrams, N., 'Nuclear Politics in Sweden', *Environment*, May 1979, Vol. 21 (4), pp. 6–11, 39–40.
138. Energy in Sweden, 1987, National Energy Administration, Stockholm.
139. Mosey, C., 'Green Revolution poses threat of political upheaval', *The Times*, 16 September 1988.
140. Nordic Correspondent, 'Sweden Still Shining', *Economist*, 24 September 1988, p. 62.
141. Persson, L., and Duckert, H., 'Swedes vote on return to innocence', *Sunday Times*, 18 September 1988.
142. Goldstick, M., interview by D. Lowry in Yttersby, December 1987.
143. Olsson, B., 'The feasibility of non-nuclear strategies'. Paper to Conference on Nuclear or Non-nuclear Futures, London April 28–30.
144. Åhåll, K-I., Lindström, M., Holmstrand, O., Helander, B., and Goldstick, M., 'Nuclear Waste in Sweden: the problem is not solved' (Uppsala, 1988).
145. Cross, M., 'Nuclear Sweden's final melt down', *New Scientist*, 22 May 1986, pp. 34–6.
146. Ministry of Industry, 'New Swedish Nuclear Legislation', DSI 1984: 18, Stockholm, part 2, p. 1.
147. See note 144, p. 20.
148. Joss-Liljegren, E., interview by D. Lowry in Sweden, December 1987.
149. See note 144, p. 57.
150. Olsson, B., op. cit., 1987.
151. Abrams, N., op. cit., 1979.
152. Choate, R., 'Sweden to go ahead with vast nuclear programme', *The Times*, 19 February 1975.
153. Walker, J., 'Swedes approve N-plant increase', *Financial Times*, 28 May 1975.
154. Larsson, A. H., 'Regulatory aspects of radioactive waste management in Sweden and its implementation'. Paper for Radioactive Waste Management Conference, BNES, London, 1984.
155. See note 144, p. 21.
156. SKI (1987), 'Swedish Nuclear Power Inspectorate: a presentation of our activities'.

157. Bergman, C., Boge, R., Johansson, G., and Snibs, J. O., 'Radiation Protection Aspects of Waste Acceptance Criteria', SSI-rapport 85–15. Paper to seminar on radioactive waste products, Jülich, 10–13 June 1985.
158. See note 144, p. 23.
159. Abrams, N., op. cit., 1979, p. 7.
160. Ibid., p. 10.
161. See note 146.
162. Abrams, N., op. cit., 1979, p. 10.
163. See note 144, p. 26.
164. Barnaby, W., 'The Swedish Referendum: Do away with it but not yet', *Bulletin of the Atomic Scientists*, Vol. 36, No. 6, June 1980, pp. 58–9.
165. Olsson, B., op. cit., 1987.
166. Fishlock, D., 'Sweden's nuclear programme: buried at sea for up to 500 years', *Financial Times*, 23 January 1985.
167. Abrahamson, D., 'Governments fall as consensus gives way to debate', *Bulletin of the Atomic Scientists*, Vol. 35, No. 11 1979, pp. 30–7.
168. See note 144, p. 32.
169. Bjurström, S., and Sverke, E., 'The Swedish Program for Spent Fuel and Waste', *Nuclear Europe*, October 1984, pp. 26–8.
170. ERL, 'Disposal of Low and Intermediate Level Radioactive Waste in Sweden', Working paper No. 12. Prepared for Bedfordshire County Council by Environmental Resources Ltd, London. July 1987.
171. Environmental Resources Ltd, 'Disposal of Radioactive Waste in Sweden, West Germany and France', prepared by ERL for the County Councils Coalition, January 1987.
172. Joss-Liljegren, E., Interviewed by D. Lowry, Goteborg, December 1987.
173. See note 144, pp. 43–50.
174. Nilsson, L. B., and Carlssen, H. S., 'Site investigations for a Swedish repository for HLW', *Nuclear Europe*, February 1985, pp. 19–21.
175. SKI, 'Review of Final Storage of Spent Nuclear Fuel – KBS-3', SKI, 84, 5. Technical Report Series, Stockholm 1984.
176. Ibid., pp. 12–13.
177. Andersson, K., Larssen, A., and Wingefors, S., 'Sweden: policy and licensing – an update of projects and the new framework for regulation', *IAEA Bulletin*, 1986, Vol. 28, No. 1, p. 41.
178. Hedman, T., Aronsson, I., and Petterssen, S., 'Design, Construction and Safety Assessments for a repository in bedrock', SKB and Vattenfall, joint paper to IAEA/CEC Symposium, Stockholm, May 16–20 1988. IAEA–SM–303/138.
179. SKB, SKB Annual Report 1986, Stockholm, May 1987.
180. SKB, 'Swedish Nuclear Fuel and Waste Management Company, Activities', Stockholm 1987, p. 12.
181. Barnaby, W., 'Nuclear Waste Problem Solved, claims Sweden's nuclear industry', *Nature*, 6 July 1978.
182. OKG Aktiebolag press officer, Telephone interview with D. Lowry in Stockholm, December 1987.

183. Fishlock, D., 'Swedish plan to bury nuclear waste rather than reprocess it', *Financial Times*, 18 August 1983.
184. Marsh, D., 'Sweden ends N-Fuel deal with French', *Financial Times*, 9 October 1984.
185. Done, K., 'Sweden's nuclear reaction cools', *Financial Times* 2 August 1985.
186. Prentice, T., 'Burying the future in a nuclear tomb', *The Times*, 20 May 1986.
187. OECD, 'The International Stripa Project. Background and Research', OECD Nuclear Energy Agency, Paris, March 1983.
188. *WISE News Communiqué*, 195.1332. 'Swedish Energy Minister rejects proposal for centralised waste dump in Sweden', 7 November 1983, p. 13.
189. See note 170.
190. Gustafsson, B., 'Sweden goes for central temporary storage facility', *Nuclear Engineering International*, October 1980, pp. 51–2.
191. SKB, 'Central Interim Storage Facility for spent nuclear fuel – CLAB', SKB, Stockholm 1986.
192. FMKK National Committee, Interviewed by D. Lowry in Stockholm, December 1987.
193. See note 191.
194. See note 186.
195. See note 166.
196. Söderström, S., and Karlsson, K., 'Protest mot Avfallslagret', *Norrtelje Tidning*, 12 October 1987.
197. See note 33.
198. Reuter, 'Rocky Burial', *Guardian*, 30 September 1986.
199. Serber, M., 'Swedish Nuclear Industry puts final touches on Forsmark', *Nucleonics Week*, 26 May 1988, pp. 9–10.
200. Dalglish, J., 'Swedish subseabed store – phase 1 nears completion', *Atom*, No. 363, January 1987.
201. SKB, 'Final Repository for Reactor Waste', SKB, Stockholm, 1985.
202. SKI, 'Review of Final Repository for Reactor Waste', SKI TR 88:5, March 1988.
203. Norrby, S., and Bergman, C., 'Licensing of the Swedish Final Repository for Reactor Waste', IAEA/CEC Symposium, Stockholm, May 16–20 1988, IAEA–SM–303/18.
204. Tansey, G., 'Watery grave for Sweden's nuclear waste', *Financial Times*, 26 July 1989.
205. See note 199.
206. For instance, 'Briefing on the Swedish Programme on Nuclear Waste Management', by SKB before US Congress Committee on Energy and Natural Resources, US Senate 100th Congress. US GPO, Washington, January 1987.
207. For instance, The UK House of Commons Environment Committee visited Sweden in August 1985 during its examination of radioactive waste management. 1st Report, 1985–86 HC–191, para 105, pl iii and appendix 3c.

208. Wikdahl, C-E., 'The Future for Nuclear Power in Sweden', *Nuclear Europe*, October 1984, pp. 15–16.
209. Godstick, M., Törnqvist, M., and Andersson, M-B., interviewed by D. Lowry in Yttersby and Grasö, respectively, December 1987.
210. Ungemark, S., Public Information Director of SKB interviewed by D. Lowry, Stockholm, December 1987.
211. Norrby, S., Director of SKI nuclear waste division interviewed by D. Lowry, Stockholm, December 1987.
212. *WISE News Communiqué*, 175.1142 'Swedes Push Through Plans for Waste Disposal Facility', 31 March 1983, p. 7.
213. See note 209.
214. Ibid.
215. Ohman, A., 'Glömd klump i havet', *Dagens Nyheter*, 25 November 1987.
216. FMKK, 'Pin stripe suit action at Swedish nuclear PR show', press release, Soderboda, 12 October 1987.
217. *WISE News Communiqué*, 275.2340 'SFR-1 project is no more than slow ocean dumping', Stockholm, 12 June 1987.
218. SKI, 'Licensing of Final Repository for Reactor Waste – SFR-1', Technical Report SKI.84:2, Stockholm, July 1984, pp. 10–11.
219. SKN, 'Etikoch Kärnavfall', SKN Rapport 28, March 1988.
220. Ahlberg, E., 'Swedes' atom-free aim', *Daily Telegraph*, 18 January 1989.
221. Sanders, N., and Bolt, R., 'France: the Nuclear Renegade', *Australian Democrat*, Tasmania/Melbourne, December 1987.
222. Scheinman, L., 'Atomic Energy Policy in France under the Fourth Republic' (Princeton University Press, 1965), pp. 4–5.
223. Ibid., pp. 12–13.
224. Pringle, P., and Spigelman, J., *The Nuclear Barons*, Chapter 8, 'The Power of les X' (Michael Joseph, London, 1982).
225. Nau, op. cit., 1974, p. 68.
226. Nelkin and Pollak, op. cit., 1981, p. 11.
227. *Financial Times*, 'France' (Survey), 29 September 1988.
228. *Examiner News Service*, 'Greens make gains in French elections', *San Francisco Examiner*, 13 March 1989.
229. Sweet, C., 'Some Lessons from the French Nuclear Power Experience that the UK might Learn'. Proof of Evidence to Sizewell B Public Inquiry. Centre for Energy Studies, London, October 1983, p. 3.
230. Sweet, C., op. cit., 1983, p. 4.
231. *Nuclear News*, 'World List of Nuclear Power Plants', Vol. 33, No. 2, February 1990, pp. 63–82.
232. Friends of the Earth, 'Nuclear France; Power at any price?' FoE, London, 1986.
233. MacLachlan, A., 'French PWR Health', *Nucleonics Week*, Vol. 30, No. 9, 2 March 1989, p. 12.
234. Davis, M., 'The Civilitary Atom', Wise, Paris, 1988.
235. See note 221.
236. See note 53, pp. 12–13.

237. Pecqueur, M., 'How one organisation runs the whole nuclear industry', *Nuclear Engineering International*, December 1976, pp. 45–447.
238. See note 53, pp. 29–30.
239. Mortimer, E., 'French mount a nuclear attack on their oil import bill', *The Times*, 10 October 1974.
240. *New Scientist*, 'France's Nuclear Controversy Grows', 13 March 1975.
241. Leigh, V. W., 'Professors attack "electro-fascism"', *Sunday Times*, 20 April 1975.
242. *Nuclear Engineering International*, 'No opposition heard in French Parliament', June/July 1975, p. 499.
243. Anti-Superphénix action group members, interviewed by D. Lowry, Lyons, December 1983.
244. Murray, I., 'German-led group blamed for starting battle at French nuclear reactor site', *The Times*, 2 August 1977.
245. See note 53, p. 79.
246. Touraine, A., 'Political ecology: a demand to live differently – now', *New Society*, 8 November 1979, pp. 307–9.
247. Barthoux, A., and Faussat, A., 'Low and Medium Waste Storage in France', *Nuclear Europe*, No. 3, March 1986, pp. 7–9.
248. Cogéma, 'The Nuclear Fuel Cycle Back-End: A Cogéma View', Velizy-Villacoublay, 1989.
249. Schneider, M., interviewed by D. Lowry in Strasbourg, October 1988.
250. *Nuclear Engineering International*, 'La Hague extension behind schedule', February 1989.
251. Delange, M., 'Over 2000t of LWR Fuel Reprocessed at La Hague', *Nuclear Europe*, No. 9, August/September 1988, pp. 9–10.
252. Michallet, F., 'France's Nuclear Fuel Cycle Industry', *Nuclear Europe*, No. 10, October 1987, pp. 25–6.
253. Eisenhammer, J., 'Kohl seeks French help on nuclear treatment plant', *Independent*, 21 April 1989.
254. Delange, M., M. Delange, the head of Cogéma's reprocessing division, correspondence with Mycle Schneider, Paris, April 1989.
255. Beaufrère, J., M. Beaufrère, director adjoint, Service Combustible, EdF, interviewed by Mycle Schneider, Paris, 11 May 1989.
256. Schneider, M., letter to D. Lowry, Paris, 22 May 1989.
257. *Revue Générale Nucléaire*, 'Les déchets radioactifs et leur gestion', N4, Juillet–Août 1988, pp. 289–339.
258. ANDRA, 'ANDRA; a government agency for safe radioactive waste management', CEA, Paris, 1983.
259. MacLachlan, A., 'French Government Publishes Report on Waste Site Technical Criteria', *Nuclear Fuel*, 30 November 1987a, pp. 19–20.
260. Report of the Castaing Commission, (in translation). 1 December 1982. Sizewell Inquiry document SSBA/5/76A (revised).
261. Olivier, J. P., op. cit., 1987.
262. *Nuclear Fuel*, 'Candidate HLW Site Protested in France', Vol. 12, No. 22, 2 November 1987, pp. 15–16. Schneider, M., interviewed by D. Lowry in Strasbourg, October 1988.

263. CEA, 'The "Centre de la Manche"', Andra, Paris, 1985.
264. MacLachlan, 'The French Government issues a Permit for a Low and Medium level Waste Burial', *Nucleonics Week*, 6 August 1987b, p. 8.
264. MacLachlan, 'Waste decree challenged', *Nucleonics Week*, 13 August 1987c, p. 18.
265. Erickson, T., 'Civilian Nuclear Programs: the United States, France and Japan', Washington State Institute for Public Policy, Olympia, January 1988, p. 11.
266. Celeri, J. J., 'Etude Techno-économique des procédés industriels de réduction de volume des déchets technologiques f.a. provenant de l'exploitation des centrales de puissance EDF', IAEA/CEC Symposium, Stockholm, 16–20 May 1988, IAEA–SM–303/80.
267. Schneider, M., op. cit.
268. Wavell, S., *Guardian*, Diary, June 1986.
269. Feldman, D. L., 'Public Choice Theory Applied to National Energy Policy: The Case of France' in *Journal of Public Policy*, Vol. 6, No. 2, 1986, pp. 137–58, p. 139.
270. Boyle, M., and Robinson, M., 'Nuclear Energy in France: a foretaste of the future', in Blowers, A. and Pepper, D. (eds), *Nuclear Power in Crisis* (Croom Helm, 1987) pp. 55–84.
271. Dodsworth, T., 'Intermission time for France's nuclear dream', *Financial Times*, 30 May 1981.
272. *New Scientist*, 'France under Mitterrand thinks again about nuclear policy', 4 June 1981, p. 608.
273. Gee, J., 'France cuts nuclear programme, but political battle lies ahead', *Electrical Review*, 9 October 1981.
274. Clefs CEA, Juillet, No. 2, 1986.
275. *Le Monde*, 'Les Francais craignent davantage la drogne que l'accident nucleaire', 23 February 1989.

7 The Way Forward

> There is little doubt that any development of this magnitude will have a substantial impact on the local community. Because of this, there is a need for open discussion and feedback from a wide audience on the various aspects of site selection, investigations and development, environmental protection, monitoring and control of repository operations. It is important, in particular, that the local community is represented at every stage of the development so that benefits to local people can be maximised and any disadvantages minimised.[1]

SUMMING UP – LEARNING FROM PAST MISTAKES

The disposal of radioactive wastes has become a critical problem for many governments and the nuclear industry. Nuclear wastes provide fewer local benefits than do nuclear power stations but impose major costs on individual communities. Indeed, these wastes are perceived by many people as posing intolerable health and environmental risks, whether it be LLW, ILW or HLW, and irrespective of the disposal technology. The nuclear industry and governments have a credibility problem in making the case for radioactive waste disposal, as long as communities perceive a proposed facility as being in their 'backyard'. Opponents of nuclear waste sites in many countries have been able to build coalitions on a broader front covering a wide range of nuclear issues, including nuclear power and armaments. Such coalitions are by nature politically fragile and liable to fragmentation, but they can command considerable popular support using the media and by lobbying their elected officials.

In Chapter 1 we suggested that the outcomes of conflicts over nuclear waste between competing interests were determined by the political power of the protagonists. Furthermore, we suggested that the context within which power is exercised is determined largely by three factors: political institutions (that is, the degree to which the government system is centralised), political participation (namely, the degree of openness of the political system), and the government policy (that is, the extent to which policy is comprehensive rather than incremental). A fourth factor, geography (by which we mean the availability of technically suitable sites) operates as a constraint. It is the combination of geographical and political factors that determines outcomes.

Political institutions were identified in chapter 1 as the major factor of a nation's political context that usually influences outcomes on nuclear waste disposal. Perhaps surprisingly, we have seen that the centralised systems of Britain and France have made less discernable progress on waste *disposal* than have the federalist systems of the USA and the FRG. The latter are two of the four countries that have identified a preferred site for HLW disposal (the others being Argentina and South Africa). Britain and France, on the other hand, have cancelled what were thought to be required sites for LLW facilities, without backup sites, and are farther away from decisions on HLW (though France is seeking an underground research laboratory that could be converted into an MRS). But our major conclusion on institutions is that the centralised societies cannot be seen as having an urgent need to identify new nuclear waste sites, since these countries are also firmly committed to reprocessing. Thus, they can utilise their own massive warehouses for reprocessing wastes (UK, France, Belgium, USSR), and nearby LLW landfills, or reprocessing centres of other nations (as is the case for Japan). The centralised country that is an exception to this pattern and has moved much more quickly to define its radioactive waste management strategy is Sweden. Sweden does not reprocess spent fuel and differs from Britain and France on the two other main characteristics comprising the political context: *political participation and waste management policy*.

Sweden is very open to *political participation* by its citizens on nuclear waste issues. Nuclear waste disposal, in fact, was the major issue influencing the nuclear power debate in Sweden in the late 1970s and the debate on the 1980 referendum. A similar debate in France was promised by President François Mitterrand, but has never occurred. The federalist systems, however, have *not* been fully open to public participation on nuclear power and waste management matters, and have often been secretive with facts considered sensitive by the central government. This has certainly been true in the USA and FRG, just as it has been in France, the UK and USSR. Indeed, there appears to be an endemic secrecy among the nuclear states, though perhaps less so in Sweden. The problem is that it is probably harder to keep secrets for long in the decentralised societies, with access to government greatly increased by the added layers of government. The result, in the USA and FRG, is a familiar one: public participation fuels distrust of government and a relatively weak nuclear industry, and encourages strong and active opposition to nuclear waste sites using both administrative and judicial channels (of course, this may ultimately lead to politically-acceptable solutions).

Waste management policy (or its absence) is also crucial in setting the political context for nuclear waste proposals. Britain in particular, among the major nuclear states, has had the least clearly defined waste management policy. Incremental waste policy coupled with secretiveness in the UK has now led to excessive suspicion and distrust of nuclear programmes in a country that was once apathetic on all nuclear matters. In contrast, waste policy is much more comprehensive in Sweden, France, FRG and the USA. While there are vocal critics of nuclear waste programmes in these countries, especially the USA and FRG, most opposition focuses on specific sites rather than policy. Clearly, comprehensive or synoptic waste policy (as in Sweden) has provided greater legitimacy to government proposals on nuclear waste, though the plans are still hotly debated.

Political geography of a country is a final though usually less important factor in determining political context. It is less important because while geographical factors such as land area and geology of a country obviously matter in waste facility siting, there are no perfect sites (if there were they would assuredly be quickly chosen) and each nation has a range of alternatives to consider. Political factors determine specific sites among the geographically and technically available options. It is this political geography that we have illustrated with the opposition to NIREX from rural conservative strongholds (UK), East–West conflict (US), the existence of nuclear oases, and so on. Obviously, large nations such as the USA and Canada have more potential disposal locations than does Britain. Yet, even in large nations political opposition severely limits the availability of nuclear waste sites. Inevitably each major nuclear state will need multiple sites and will have to face the conflict with its critics.

We have shown *political power* to be the key dimension of conflict over nuclear waste. There is evidence from our study that power has shifted a little to the nuclear opposition wherever it is able to achieve alliances combining various diversified interests cutting across party, social class and geography. Diversified interests in Britain coalesced into the BANDs and overcame powerful odds and contributed to the defeat of four government proposals for LLW sites. In the USA,

cross-cutting interests in powerful eastern states convinced the Reagan administration that it should cancel its plans for an eastern HLW site, and instead focus attention on a site in sparsely-populated Nevada. And pressure groups in the FRG forced the central government to move its proposed reprocessing plant from Gorleben to Wackersdorf, which was finally cancelled in 1989, while the planned repository at Gorleben has been delayed and could still be cancelled as well. Other nuclear waste facilities have been recently completed. The USA's WIPP site in southeast New Mexico has been built, though it is in nuclear-friendly territory and its military nature has 'freed' it from civil licensing procedures. Even so, opposition to the WIPP has been strong and its opening has been delayed several times. Sweden, of course, has effectively side-stepped anti-nuclear critics by constructing the SFR beneath the Baltic seabed, in no (human) community's back yard. We would be foolish to try and predict the outcome of nuclear waste conflicts in the future, and to project the location of new disposal sites. Yet it is clear that the interaction of political institutions, political participation, waste management policy and geography will continue to set the stage for political conflict over radioactive waste siting decisions.

From the analysis in earlier chapters it appears that 'progress' is possible under both centralised and decentralised regimes, political characteristics which in any case are slow to change. Political participation or the degree of openness, however, is subject to greater change (consider the USSR) and is vitally important in building public confidence and trust in governmental strategy. But participation must go hand-in-hand with a rational policy for waste management, that has been openly debated in affected communities. To do otherwise would lead to precisely some of the problems and the impasse that we have observed in the UK and US, the 'parents' of the nuclear age. Finally, political geography cannot be ignored, for in the final analysis, remote and nuclear-friendly sites may prove in some cases to be the best resting places for nuclear wastes.

WHY DO ANYTHING? THE ETHICAL DIMENSIONS

Virtually everyone agrees that 'something' should be done about nuclear wastes: the debate is often as much about procedures and technical options as it is about specific sites. Is it too soon to proceed with deep geological disposal? Should surface waste storage facilities be relied on, and if so, for how long? And last but not least, should we even continue to produce nuclear power and reprocess spent nuclear fuel while these questions are unresolved? These issues come down to the accountability question raised in chapter 1 – whose problem is nuclear waste anyway, and what is the most responsible thing to do with it, and when?

While radioactive wastes are generated by the nuclear industry (broadly defined), it is unrealistic to expect the industry to solve this problem on its own. Some environmentalist groups, such as Friends of the Earth, advocate that the nuclear industry should find solutions without government aid. This would probably never work because not only are there government-generated wastes, but multiple private utilities would probably not agree on all the required disposal sites and the public would not believe that their health and safety were given utmost consideration. Past disposal sites sponsored by the industry, such as the USA's Maxey Flats facility for LLW disposal (chapter 5), have closed early for both environmental and commercial reasons.[2] Other major nuclear states, such as Japan, have yet to open a commercial facility for LLW disposal. Thus, there is an obvious need for government to step in and represent the 'national interest', democratically decided upon. The central government, after all, in the guise of the UK Atomic Energy Authority (and the US Atomic Energy Commission) made a civil nuclear industry possible. On the other hand, the state certainly doesn't have a monopoly on knowledge of waste disposal technology and proper siting procedures, so all the affected interests have an important role to play in dealing with nuclear waste.

It could be argued that long-term storage is a practical way to deal with nuclear waste until the technological uncertainties and political problems surrounding permanent disposal have been resolved. The problem with this view is that it is not clear if the problems and uncertainties involving nuclear waste disposal, given the longevity of many of the radioisotopes, will ever be resolved;[3] nor do they necessarily need to be since we do not live in a risk-free world. All of the major nuclear states that we have examined in fact favour disposal over long-term storage. Existing disposal techniques can apparently minimise safety risks, as is being tried in Sweden.[4] Moreover, an ethical case can be made that the current nuclear waste problem should be 'solved' by those who created it (which is strongly argued in Sweden), without burdening future generations. In any event, the choice on whether to produce more nuclear power and thus more radioactive waste, should be left to the latter. Surely we have a better legacy to leave for the future than irreversible decisions on nuclear power development and radioactive waste disposal, and foreclosed energy and land use options!

The case for long-term storage of nuclear waste at a reprocessing plant or monitored retrievable storage (MRS) facility is also questionable on other grounds. For example, it has not been clearly established that such a facility makes final disposal any easier or more economical, while the opposite may be true. Certainly an MRS plant, without a waste disposal respository, in countries lacking reprocessing centres would drastically increase the frequency of waste shipments and the risk of transport accidents. Similarly, pre-disposal processing, packaging and management of wastes can be done at nuclear power plants just as easily as at an MRS plant, especially by using dry storage casks.[5] The best argument for an MRS (which was influential in Sweden but has been successfully challenged in the FRG) may well be that additional storage space is urgently needed to 'cool down' spent fuel. But some spent fuel may remain at older nuclear plants indefinitely, depending on the scheduling of nuclear reactor decommissionings.

Another reason to thus proceed expeditiously with the permanent disposal of nuclear waste is that at least one to two dozen 'first generation' commercial nuclear plants worldwide will be decommissioned (that is, taken out of service) in the next 20 years. At that point the plants themselves become radioactive waste. One ethical argument here again favours near-term plant dismantlement and disposal (that is, the decommissioning problem should be solved by those who created it), but in all likelihood a delay of at least 20 years will be required to lower the radiation hazard to the decommissioning workforce. In the meantime, the closed nuclear power plants will have to be monitored for safety and security, and can store and cool spent fuel until deep disposal facilities are ready. Countries like Britain with reprocessing centres, of course, can continue to use these facilities for interim waste storage. Over 80 per cent of the volume of decommissioning wastes is low-level, however, and existing disposal capacity for such wastes is woefully inadequate.[6]

Permanent disposal of nuclear waste by the current generation of people within 20–30 years may be desirable on ethical grounds. It would relieve future generations from dealing with this political nightmare, presuming of course that waste disposal is accomplished soundly. Geological uncertainties regarding the chosen disposal medium, however, would assuredly remain, requiring technical monitoring and surveillance of disposal sites for a century or longer. If environmental problems are eventually uncovered (such as groundwater contamination by radioactivity), it would be imperative that the waste canisters were accessible from the surface. Consequently, it is essential that if deep geological repositories are further developed for radioactive waste disposal that the wastes be retrievable for a period matching their short-term geological uncertainty, perhaps a hundred years. Future generations a hundred years hence would still retain the option of maintaining retrievability, if it was deemed desirable. This viewpoint effectively rules out salt (such as at the WIPP) as a permanent disposal medium for HLW, since repositories in salt would eventually hermetically seal themselves when subjected to heat-emitting waste canisters, in a time period probably shorter than that of the short-term geological uncertainty surrounding salt.

In the meantime nuclear wastes keep accumulating while plans for disposal facilities are inadequate. Until such time that a successful waste disposal system is in hand, a powerful case exists for stopping nuclear power production and spent fuel reprocessing. Voters in Sweden recognised this in 1980 when they decided to end their nuclear power programme by 2010 (see chapter 6). Lack of waste disposal plans also contributed to the decision to decommission the Zwentendorf reactor in Austria before it even operated. Similarly, various laws link waste disposal (or reprocessing) plans to nuclear reactor operating licences in FRG, Switzerland, Japan and several American states. In our view, it would be prudent to strengthen these provisions now by setting strict though reasonable timetables for progress on developing LLW *and* HLW disposal facilities, with a penalty of a nuclear power phase-out (as in Sweden). A phased shut-down of the industry would also allow time for the development of alternative, environmentally-sound power sources. While a nuclear power moratorium and thus lowered waste generation does not solve the waste disposal problem, the Swedish experience suggests that this path could lower public anxieties over waste disposal programmes, although the political culture also plays an important role.

TECHNICAL AND GEOGRAPHICAL OPTIONS: THE HOW AND THE WHERE

Our analysis suggests that political solutions to the problem of nuclear waste management might be achieved either by exercising central power and coercion or through a process of decentralised and democratic decision-making. Yet while politics is the key, technical aspects of potential waste facilities cannot be ignored. To take two extreme examples, reconsider Sweden's SFR repository (chapter 6) and the USA's WIPP (chapter 5). The SFR, while in an apparently politically-acceptable offshore location, beneath the Baltic seabed, has also been widely praised on technical grounds. This repository was designed to ensure a seaward-dipping water gradient in the event of contamination, and uses the 'multi-barrier' approach of waste containment with metal canisters, rock and soil, in order to minimise the risk of accident. The WIPP, in contrast, has been delayed repeatedly by technical questions raised by independent reviews of the project, though it is located in nuclear-friendly New Mexico. If the WIPP opens (as is highly probable), future geological problems with the salt medium could lead to even greater political setbacks at other proposed sites for waste disposal.

Irrespective of other developments, volume reduction of nuclear wastes will come to be seen as increasingly necessary. This should be done for all levels of nuclear waste. The existing LLW quantities and those following reactor decommissionings can be greatly reduced through waste compaction and/or incineration, thereby lessening the need for increasing disposal requirements. While reprocessing reduces HLW volumes by recycling the spent fuel, it also produces very large quantities of LLW and ILW.[7] Remaining liquid HLW quantities will be lowered through vitrification, calcination, or other techniques of waste solidification. Of course, the most effective overall means of volume reduction of nuclear waste is to stop making it.

Waste minimisation can ultimately only go so far. In tackling the problem of radioactive wastes, decision makers must first confront whether permanent disposal is necessary. If permanent waste disposal is the agreed solution, questions of timing, location and depth must then be considered. The timing of nuclear waste disposal must be dictated by the suitability and safety of temporary storage measures (which all nuclear states have), as well as by the availability of permanent repositories. Dry storage casks, popularised in the FRG, have the advantage of avoiding the irradiation of cooling water. Interim waste storage can be done either above-ground or below-ground; the latter has been demonstrated at the CLAB in Sweden (chapter 6), and has obvious security advantages. The location of nuclear waste facilities should be and is influenced by hydrology, seismicity, population distribution and politics.[8] Thus, the geography of a nuclear state limits the range of siting options, and historically pushed densely-peopled island nations such as the UK and Japan toward favouring ocean disposal of wastes. Disposal of radioactive wastes beneath the deep ocean floors, however, is not currently technically acceptable[9] while cruder, ocean dumping has been banned by international agreement (chapter 2). Only an SFR-like tunnel into a sub-seabed repository can be realistically considered in countries short of land-based sites. But the superiority of one geological medium or location over others has not been proven, and many alternative rock types may eventually be shown to be technically suitable for nuclear waste disposal. Similarly, the 'proper' depth for geological disposal is debatable and depends on local geological factors such as rock strata, tendency for gas build-up, and so on.

Lingering concerns about the safety of deep disposal of nuclear waste can be lessened by the multi-barrier approach, again most typified by Sweden. This policy requires decision-makers to minimise safety risks by using extensive engineered and geological barriers to contain wastes. For example, thick copper canisters could be emplaced in a suitable mine shaft, surrounded by crystalline rock and then back-filled with other rock or soil and sealed. Each layer of containment is thereby added to the next to minimise the chance of release of radionuclides into the environment. If radioactivity does escape, its chance of migrating very far is also minimised by these precautions. A given proposal for nuclear waste disposal should be analysed in this light – are safety standards 'acceptably' high? And if not, how can they be improved?

Suitable geology for LLW disposal is widespread, though severely constrained by political feasibility.[10] For HLW disposal, however, the best or more favourable sites are limited to a few regions on each continent. Geologically speaking, the best rocks for disposing of these nuclear wastes are in Canada, USSR and most of Scandinavia, which are underlain with stable, Precambrian crystalline rock.[11] Poor geological areas for HLW disposal include the nuclear states of Italy, France, Switzerland, Japan, South Korea and Taiwan; the suitability of several other countries is highly questionable. Given the shortage of technically acceptable sites and the international 'proliferation' of nuclear power, it is apparent that many countries (especially in southern Europe and eastern Asia) will be hard pressed to find suitable sites for HLW disposal. Consequently, it can be argued that it is imperative that all nuclear states accelerate investigations for technically suitable waste sites, and pursue the 'best' ones. Failing this, the other geographical options for radioactive waste disposal include sending the wastes to existing nuclear oases, and negotiating agreements for international disposal sites in other countries.

Potentially-suitable sites for disposing of nuclear wastes can be thwarted by political opposition. Examples of this abound in the USA. The cancellation of investigations in the eastern US removed potentially-suitable sites in granite from detailed consideration. Similarly, salt domes in Mississippi and Louisiana (setting aside the retrievability problem) were believed to have superior hydrogeological characteristics, until political opposition ended the broad-based site search. The preferred HLW site in Yucca Mountain, Nevada, though potentially suitable, also has faced substantial political opposition that may yet turn the search to other sites. Richard Bryan, a Democrat from Nevada and former Governor, was elected to the US Senate in November 1988 in large part because he was seen as a more effective opponent to the Yucca Mountain proposal than was the incumbent Senator.

The nuclear industry and government always have the option of retreating to nuclear oases for disposal of nuclear wastes. Examples from our study include Sellafield in the UK, La Hague in France, and Hanford in the USA. Communities at these and similar sites may welcome nuclear waste, since they have grown up in a 'nuclear culture'.[12] There are three problems with this option. First, nuclear oases are few in number, and these sites may also face stiff political opposition, especially from non-residents and those along transport routes. Second, such sites may have technical or geological problems not fully appreciated, because of the historical lack of concern for local environmental problems.[13] Third, nuclear oases may not be able to accept new wastes when there are serious problems with existing facilities or wastes. The best example is Hanford, where clean-up of existing liquid HLW was estimated by several studies in 1988 to cost tens of billions of dollars.

International sites for nuclear waste disposal present a final geographical option. European nations such as FRG, Switzerland, Austria and Sweden have approached other countries about taking foreign nuclear wastes, and in a few cases preliminary agreements were signed. Potential recipient nations have included China, USSR, Iran, Egypt, Sudan and Namibia. Indeed, the Soviet Union regularly takes spent reactor fuel from East European countries (being their supplier of uranium oxide), and China has considered accepting West German and Swiss HLW for disposal in the Gobi Desert.[14] The USA has also been proposed as a host for an international repository for commercial spent fuel.[15] Of course, deals such as these may always be tenuous, though many nuclear states would gladly ship their waste problem abroad. In a sense, this is what Belgium, Switzerland and the Netherlands used to do by contracting with Britain for LLW dumping in the northeast Atlantic Ocean (chapters 2 and 3), and many nations still do who have Britain and France reprocess their spent fuel and have yet to take back the radioactive wastes. International agreement for geological disposal of nuclear wastes may be desirable, though serious political, transport and security obstacles will have to be overcome to make it a reality. Can the Soviet Union be trusted with foreign nuclear wastes, in the light of the 1957 accident in the Urals, and Chernobyl? And what about developing countries with lower scientific and engineering capabilities for managing radioactive materials, where some leaders have been attracted by financial and trade incentives for accepting foreign nuclear wastes?

POLITICAL OPTIONS: SOME PROPOSALS FOR BUILDING REAL AND LASTING PUBLIC TRUST

In the present absence of development of international agreements on multinational sites for radioactive waste, the nuclear industry and governments have three management options in their search for sites for the wastes. First, they can send their wastes to the existing nuclear oases where further facilities may be welcome. Second, they can store the wastes at existing nuclear plant sites for the foreseeable future. Third, they can attempt to secure new waste facilities by a combination of positive incentives or compensation, and greater consultation and involvement of local communities. This might include open government, equality of access to decision makers, impartiality, greater local control and consideration of the widest range of options for nuclear waste management.[16] The prospects for achieving a political settlement will ultimately depend on societal values and attitudes toward the future of nuclear power. Where these reforms are not yet instituted (as in the UK, FRG and most of the US), radioactive waste assuredly will continue to be the focus of political conflict with uncertain outcomes. Overcoming uncertainty will require policy-makers to engender real and lasting public trust as a means for achieving acceptability of nuclear waste management.

Indeed, public *mistrust* of the nuclear state in most countries is so high that its reversal can be argued to be the most critical task at present for nuclear waste policy makers.[17] This mistrust, debate and subsequent conflict stems from many factors: the connection between military and civil applications of nuclear power, past blunders of nuclear agencies, nuclear accidents and Windscale/Sellafield, Chernobyl and Three Mile Island, secret nuclear decisions, environmental worries, and *ad hoc* policies on nuclear power and radioactive wastes. Therefore, to capture the public trust it appears that governments must 'clean the slate' on the nuclear waste issue. This would require the creation of a new and representative siting body with democratic procedures, possible selection of some fresh new sites based on open and credible siting criteria, and most of all an open debate on radioactive waste disposal and the future of nuclear power. While this course of action could move the waste issue forward in the USA, UK and FRG, it is probably too much to ask of France and may not be required in Sweden.

In the present political climate, indefinite storage of radioactive waste would do little to establish the credibility of the nuclear waste management authority. Critics could always argue that a waste storage site will become a *de facto* disposal site, thereby 'demonstrating' the inadequacy of the government's policy and plans for long-term management of nuclear waste, further fuelling public mistrust. While such a scenario could potentially help anti-nuclear activists in their quest to shut down the nuclear industry permanently, it would assuredly be counterproductive to government efforts to establish comprehensive policy for nuclear waste.

Incentives for community acceptance of a nuclear waste site can be economic, environmental, socio-cultural, or public safety and medical, although the most important ones are economic.[18] Economic incentives take several forms, and have been seriously considered by NIREX and the USDOE. A common type is a fund to deal with the fiscal consequences of a facility, which can be derived through a surcharge on the waste generators; or from a direct monetary payment.[19] Property tax levies from a respository (or equivalent grants) could be set at an appropriately high rate, and the revenues used to mitigate the impact of the repository and to develop the local economy. Job opportunities might be enhanced, too, especially if developers committed themselves to on-site training and employment of a minimum number of local residents for the project. Finally, the government could offer to commence other, more desirable projects in the locality once a community agrees to host a nuclear waste facility.

Environmental incentives might involve protection of coastal access or landscape improvement; farmland soil erosion control and game preserves could be arranged by the government on surrounding lands. Socio-cultural incentives could tie in with fiscal incentives, since improved infrastructure (such as land assembly for local industry, water and sewerage services, roads, schools) could be developed. Recreation and arts facilities might also be established, with funds generated from the nuclear waste facility or associated grants. The last type of incentive, for public health, safety or medical care, may be required because of the influx of people needed to operate the nuclear waste facility, and may also involve specialised training courses for local residents.[20]

Carefully crafted economic incentives will be helpful in eliciting local acceptance of nuclear waste facilities, but will not deal with all of the public's worries. Other studies have found that two additional concerns in this area are of utmost importance to the public: the integrity of public health, safety and the environment; and shared political power in decision making over nuclear waste sites.[21] The latter requirement can be satisfied if significant local control in the waste repository siting and licensing process is granted, and negotiations are recognised as a viable mechanism for reconciling differences and building public confidence and trust. Indeed, Bord[22] found in his study of LLW siting in Pennsylvania (USA) that power sharing options are more important in promoting local cooperation than are economic incentives. Such incentives that may be offered should be, clearly, specific to the local political culture.

A final consideration for building lasting public trust in nuclear waste policy and programmes is that incentive packages and power sharing options do not lead to technically-suboptimal sites. Such a result would inevitably backfire and erode the precarious public confidence garnered by decision makers, since environmental problems would eventually 'surface'. Public opposition to the WIPP project has grown as geological problems have become more apparent, and will hopefully ensure that a technically sound course is followed. Thus, it is important that nuclear waste programmes ensure that only the most technically-qualified sites (including nuclear oases) will be subject to detailed investigation or site characterisation. Only after that point is reached should incentives, compensation and reformed decision making procedures enter into play to deliver final sites for nuclear waste disposal. As we saw in chapter 5, such a procedure was distorted in the selection of HLW sites in the USA by political bias of the federal government.

The radioactive waste problem is here to stay and we do not expect it to be 'solved' in our lifetimes. Moreover, it could get much worse if the nuclear industry has its way in achieving a massive expansion in world nuclear power capacity, in response to scientific and public concern about potential global warming resulting from carbon dioxide (CO_2) buildup in the atmosphere (the 'greenhouse effect').[33] But nuclear power cannot solve this problem for three reasons. First, nuclear power can displace only fossil-generated electricity, which accounts for just a third of fossil fuel CO_2 emissions, with most of the remainder coming from space and water heating, transport, and industrial process steam. Forest fires, deforestation and cement manufacture are also major CO_2 sources. Second, other greenhouse

SOURCE *ECOnews*, December 1989, p. 1.

gases, such as methane, nitrous oxide, chlorofluorocarbons and halons, which come primarily from non-energy sources, contribute in the aggregate as much to global warming as does CO_2.[24] Third, the massive capital investment required for nuclear power to have much effect on global warming would slow efforts at reforestation and investment in energy efficiency, far more cost effective, quicker and therefore more likely answers to global warming.[25]

Recent history in several countries has taught us to expect frequent nuclear policy shifts and reversals, accidents and other surprises. Disjointed and incremental policy for nuclear waste management, as we have shown in this book, is incompatible with the need for the development of public acceptability of proposed solutions.[26] The public expects and deserves better and much more can be done. We hope that our political analysis of the problem has shed sufficient light on a major Achilles heel of the nuclear industry. It is up to all sides of the debate to come together and be part of the solution rather than part of the problem. Obstacles to a real solution however are very high, and Amory Lovins's[27] belief that the problem is insoluble may yet be proved correct. While government adoption of our proposed reforms will help to chart 'the way forward', we remain sceptical about the outcome and also encourage a continual vigilance by nuclear critics.

NOTES

1. NIREX, *The Way Forward: a discussion document*, Harwell, UK NIREX Ltd., November 1987.
2. Shrader-Frechette, K. S., 'Values and hydrogeological method: how not to site the world's largest nuclear dump', In Byrne, J., and Rich, D., eds., *Planning Under Changing Energy Conditions*, (New Brunswick, NJ., Transaction Books, 1988), pp. 101–37;
   Resnikoff, M., *Living Without Landfills* (New York, Radioactive Waste Campaign, 1987).
3. Lovins, A. B. (1978), 'Comments on the 10/78 draft of the IRG report to the President (TID–28817 Draft)', Unpublished memorandum to the Interagency Review Group on Nuclear Waste Management, available from the Rocky Mountain Institute, Old Snowmass, Colorado.
4. Blowers, A., 'Sweden buries its radioactive waste problems', *Geography*, Vol. 71, June 1986, pp. 260–3.
5. U.S. Department of Energy, *Initial Version Dry Cask Storage Study*, DOE/RW–0196, Oak Ridge, Tennessee, US Department of Energy, August 1988.
6. Solomon, B. D, and Cameron, D. M., 'The impact of nuclear power plant dismantlement on radioactive waste disposal', *Man, Environment, Space and Time*, Vol. 4, Spring 1984, pp. 39–60.
7. Carter, L. J., *Nuclear Imperatives and Public Trust: Dealing with Radioactive Waste* (Washington, D.C., Resources for the Future, 1987) p. 30.
8. Solomon, B. D., and Shelley, F. M., 'Siting patterns of nuclear waste repositories', *Journal of Geography*, Vol. 87, March/April 1988, pp. 59–71.
9. Parker, F. L., Kasperson, R. E., Andersson, T. L., and Parker, S. A., *Technical and Sociopolitical Issues in Radioactive Waste Disposal, 1986*, Vol. 2, Stockholm, 1987, Report to the Beijer Institute of the Royal Swedish Academy of Sciences.
10. Milnes, A. G., *Geology and Radwaste* (London, Academic Press, 1985).
11. Carter, op. cit., p. 370.
12. Loeb, P. R., *Nuclear Culture: Living and Working in the World's Largest Atomic Complex* (New York, Coward, McCann & Geoghegan, 1982).
13. Historically, nuclear oases were established in areas remote from human populations because the technical enterprise was both dangerous and a military secret.

14. Carter, op. cit., pp. 380–2.
15. Barkenbus, J. N., Weinberg, A. M., and Alonso, M., 'Storing the world's spent nuclear fuel', *Bulletin of the Atomic Scientists*, Vol. 41, No. 10, 1985, pp. 38–42.
16. Solomon, B. D., and Cameron, D. M., 'Nuclear waste repository siting: an alternative approach', *Energy Policy*, Vol. 13, No. 6, December 1985, pp. 564–80.
17. Kemp, R., and O'Riordan, T., 'Planning for radioactive waste disposal: some central considerations', *Land Use Policy*, Vol. 5, January 1988, pp. 37–44; Carter, 1987, op. cit.
18. Solomon and Cameron, 1985, op. cit.
19. Kunreuther, H., Kleindorfer, P. R., Knez, P. J., and Yaksick, R., 'A compensation mechanism for siting noxious facilities: theory and experimental design', *Journal of Environmental Economics and Management*, Vol. 14, 1987, pp. 371–83.
20. O'Connor, W., 'Incentives for the construction of low level nuclear waste facilities', in Energy and Natural Resources Program, *Low-Level Waste: a Program for Action*, National Governors' Association, Washington, D.C., 1978, Appendix II.
21. Carnes, S. A., Copenhaver, E. D., Sorensen, J. H., Soderstrom, E. J., Reed, J. H., Bjornstad, D. J., and Peele, E., 'Incentives and nuclear waste siting: prospects and constraints', *Energy Systems and Policy*, Vol. 7, 1983, pp. 323–51. Bord, R. J., 'Judgments of policies designed to elicit local cooperation on LLRW disposal siting: comparing the public and decisionmakers', *Nuclear and Chemical Waste Management*, Vol. 7, 1987, pp. 99–105; Kozak, R. M., 'Public perception of low-level radioactive waste disposal issues', unpublished PhD dissertation, Texas Tech University, 1987.
22. Bord, R. J., op. cit. See n. 21.
23. See, for instance, Blix, H., Presentation in Helsinki for 'Earth Day '88', International Atomic Energy Agency, Press Release 88/55, 'Nuclear energy and the greenhouse effect', Vienna, 1988.
24. Lashof, D. A., and Tirpak, D. A., eds., *Policy Options for Stabilizing Global Climate*, Draft Report by the US Environmental Protection Agency to the US Congress, 2 vols., 1989.
25. Keepin, W., and Kats, G., 'Greenhouse warming: comparative analysis of nuclear and efficiency abatement strategies', *Energy Policy*, Vol. 16, No. 6, December 1988, pp. 538–61; Chandler, W. U., Geller, H. S., and Ledbetter, M. R., *Energy Efficiency: a New Agenda* (Washington, DC, American Council for an Energy-Efficient Economy, 1988).
26. Cf. Kemp and O'Riordan, 1988, op. cit.
27. See n. 3.

Epilogue

As the world entered the 1990s, the theme of political change through peaceful and violent revolution, especially in regard to Eastern Europe, dominated politics, at least in the industrialised world. These changes showed how political institutions can indeed be frail: what seemed immutable one day was swept aside the next. In general terms, this requires us to reflect upon the stability and predictable permanence of many of the institutions presently established to oversee radioactive waste management, perhaps across many generations into the future. In particular, it highlights perhaps the most important change in political arrangements since the authors agreed upon the final draft of this book in early November 1989, that of the inexorable developments towards a unified Germany.

None of the political changes are likely, we believe, to disturb the general analysis of institutional arrangements and political factors governing the progress of policies for the management of radioactive wastes. But there are developments that have occurred since this book was written which should be mentioned to bring our record up to date.

One of the most interesting consequences of Germany reunification is likely to be on the management of the significant nuclear waste projects which straddle the present border between the two countries (see Figure 6.2). An extremely worrying feature of the East European regimes, and a factor in their demise, has been the environmental despoliation and mismanagement that have been allowed to develop.

With the lifting of official secrecy and raising of political consciousness of the people on environmental matters, no longer will there be a quiescent acceptance in Eastern Germany of large volumes of toxic wastes from the FRG for disposal. At the same time the East German authorities will be requiring nuclear safety and waste management expertise from the FRG.[1] It seems likely that 'common cause' will develop between citizens of the GDR who live close to Gorleben, and the Wendlanders to oppose the nuclear waste projects located there, including the new PKA treatment facility. Moreover, the anti-nuclear SPD won the Lower Saxony Lande elections in May 1990, and ended the 14 year reign of the pro-nuclear CDU who had sponsored the nuclear projects in the state. Seventy per cent of those voters polled at the election put the environment ahead of economic prosperity or German unification as their priority, which gave a strong indication of the ranking of voter concerns in the region.[2] The struggle over Gorleben and Konrad seems set to take a new twist as a result of these political changes.

In Sweden, the political pressures arising from the growing awareness of global warming caused the government to review its timetable for the phasing out of nuclear power by 2010. This was the key factor that had persuaded the Swedish population to give support to the waste management programme, as finite volumes of waste arisings could be calculated. The ministerial portfolio for energy and environmental matters was split in January 1990, with energy being absorbed into the Industry Ministry and Environmental Affairs retaining its own identity. The governing Social Democrats produced an equivocal policy review on 19 April 1990, but a new Energy bill was promised before the end of the year. With the next general election due in September 1991, it seemed nothing would be finally resolved before then.[3]

Some significant developments have taken place in France. At the end of 1989 and in early 1990, two technical reports were released, produced by senior officials of EdF and the CEA respectively, called the Tanguy and Rouvillois reports, after their respective authors. Each was critical of nuclear safety in France. The powerful trades union, the CFDT, called for a revision of the waste management programme and a review of reprocessing following its assessment of the reports. In February 1990 two review committees were given the task of evaluating France's nuclear future; one, made up of members of the Assemblé Nationale, was charged with reviewing risk assessment; the other, a technical committee of scientists, was asked by the Industry Minister, Roger Fauroux, to reassess the HLW programme.[4] Prime Minister Michel Rocard meanwhile suspended the test drilling at the four prospective HLW sites for a period of at least a year whilst the reviews were undertaken.[5]

Notwithstanding the doubts that have crept into the orderly French nuclear waste programme, in May 1990 the go-ahead was given to a new MOX facility at Marcoûle, which will produce a new type of 'Melox' waste. In December 1989 French Environment Minister Brice Lalonde, former president of French Friends of the Earth and a 1981 Presidential candidate for the Green Party, had objected to the new MOX-Melox project, on the grounds of uncertainty over the capability of safely handling the waste arisings. After six months political and technical debate between ministers, Lalonde acceded to pressure from Fauroux and Prime Minister Rocard[6] and the go-ahead for the Melox programme was granted.

Developments in Britain took place in three main areas. In November 1989 the Energy Secretary announced that following an extensive review of the economics of nuclear power, the nuclear industry was being withdrawn entirely from the privatisation programme. Thus the AGR plants, and the one PWR under construction, joined the Magnox plants in remaining in public ownership. This meant the entire waste management programme remained in the public sector, to be paid for by tax payers.

The second important development was the announcement by the Secretary of State for Scotland in May 1990, following intense media speculation, that test bore drilling at Dounreay could go ahead.[7] This delighted NIREX, but caused dismay amongst many in Scotland, including a substantial number of people living near to the Dounreay site. The issue dominated the regional press, with letters columns filled with objections, and provoked headlines such as 'Riots fear at nuclear plant'[8] and 'Orkney's Greens promise "direct action" on NIREX'.[9] Dounreay therefore joined Sellafield as a site with permission granted for drilling.

The prospective construction of a repository at Sellafield was also affected by a third development, the agreement by BNFL with an FRG utility, for the additional import of spent fuel for reprocessing at THORP, in a deal worth some £225m.[10] BNFL's plans to expand the throughput at THORP, with the concomitant increase in LLW and ILW arisings becoming an additional problem for NIREX, led to the creation in spring 1990 of a coalition of environmentalists opposed to the opening of THORP, co-ordinated by the Cumbrian group CORE. *Greenpeace* published a detailed report on the potential risks of the increase in rail transport of irradiated fuel to Sellafield.[11] In both Dounreay and Sellafield, as chapter 4 indicated, the nuclear industry was encountering predictable opposition, but also enjoyed some local support from its dependent workforce. It hoped for wider support when it came to the ultimate decision.

In the United States a well developed tendency for political vacillation and policy drift identified in chapter 5 continued and affected both civil and military nuclear waste programmes. The various interests were encouraged to participate as responsibility for guidance was devolved to a series of committees and specialist panels.

Energy Secretary Admiral James D. Watkins established two expert panels to review how the impasse met by the military waste programme could be resolved. Problems with the WIPP were examined by committees of the House and Senate in the Congress, and Watkins told the House Armed Services hearings that he was 'confident the pre-operational activities' had been completed for the WIPP. Given past delays, time would tell if confidence was well founded.[12]

The civilian waste programmes were the subject of two key reports, one on the MRS[13] and the other on Yucca Mountain.[14] The key conclusions of the MRS commission, which reported in November 1989, were that it did not recommend a linked MRS, as required by the current law and as proposed by the DOE. This means that the construction of the MRS and its opening schedules need to be linked to those of the HLW repository. This would prevent the MRS becoming a *de facto* permanent repository. The Commission recognised, however, that some linkages were justified, but not on technical grounds; and that some interim storage facilities, substantially more limited in capacity and built under different conditions than the DOE-proposed MRS, would be in the national interest to provide for emergencies and other contingencies. Any development of such facilities would not be linked to the schedule of the HLW repository.

The report's main recommendations were that the Congress should authorise construction of a Federal Emergency Storage (FES) facility with a capacity limit of 2000 te of uranium (that is, to deal with nuclear accidents) and that the Congress should authorise construction of a User-Funded Interim Storage (UFIS) facility with a capacity limit of 5000 tons. This would provide facilities for those utilities that are short of at-reactor-storage, or which cannot obtain a license for additional storage. The DOE announced plans to work with the Congress to modify the current linkages between the MRS and the HLW repository and to embark on an aggressive programme to develop an integrated MRS facility for spent fuel. As things stand the DOE will not meet the 1998 disposal date set out in the 1982 law, and will not be able to accept the utility waste by then at an MRS (as originally planned), unless the linkages are modified.

The future of the deep repository continued to be in doubt. In November 1989 the DOE reported to the Congress on its review of the HLW programme, after being requested by Secretary Watkins to conduct the study. The Office of Civil Radioactive Waste Management report recognised the delays at Yucca Mountain, such as the State of Nevada's refusal to issue the environmental permits that the

DOE requires to complete its site characterisation/suitability activities. Consequently the Justice Department sued the State. Nevertheless, the DOE has put the expected opening back to 2010. This appears to be still somewhat speculative, however. The DOE in the meantime plans to take advantage of some early surface-based tests in advance of the ability to construct the exploratory shaft. Major site-specific design activities have been deferred for the time being.[15]

Environmentalists meantime have continued to oppose many sites identified for LLW. Confrontation over the choice of potential sites in Allegany and Cortland counties in New York State provoked particularly powerful opposition.[16] Additionally the issue of Below Regulatory Concern (BRC) wastes became further politicised, as citizen action groups such as NIRS in Washington DC claimed BRC definitions were being used to 'de-regulate' radioactive waste problems,[17] and the whole issue became embroiled in the politics of the Clean Air Act, passed through Congress in spring of 1990.

It is perhaps apposite to end this review of changes and developments in radioactive waste programmes, dominated by politics and committees, with the fact that the detailed plans for radioactive waste disposal in the US may yet flounder on the future of the desert tortoise after it was declared an endangered species.[18] Its desert habitat at Yucca Mountain and Nettles, a proposed site for the California LLW, is claimed by environmentalists to be threatened by the nuclear waste projects. Decisions and progress with nuclear waste have something in common with the tortoise: they move ahead ever so slowly.

NOTES

1. Hibbs, M., 'Acceptance of Nuclear Power rising in Germany, East-West meeting told', *Nucleonics Week*, 17 May 1990a.
2. Hibbs, M., 'Pronuclear Christian Democrats Suffer set back in West Germany', *Nucleonics Week*, 17 May 1990b.
3. Wikdahl, C-E., 'Premiers Statement Leaves Swedish Energy Future Unsure', *Nuclear Europe*, Nos. 5–6, May–June 1990, p. 43.
4. Schneider, M., communication with D. Lowry, June 1990.
5. 'French Farce', *SCRAM Energy Bulletin*, No. 76, April–May 1990, p. 7.
6. MacLachlan, A., 'Melox License issued with Prime Minister's signature', *Nucleonics Week*, 24 May 1990, pp. 3–4.

7. Edwards, R., 'Dounreay dump tests to go ahead', *Guardian*, 16 May 1990.
8. Porter, M., *Scottish Sunday Express*, 27 May 1990.
9. *Caithness Courier*, 30 May 1990.
10. Brown, P., 'Dustbin fury over BNFL £225m deal', *Guardian*, 25 April 1990.
11. Ghazi, P., 'UK becomes nuclear dump of Europe as waste imports soar', *Observer*, 10 June 1990.
12. *In Committee*, Weekly Bulletin, 2 April 190, B13–14, Washington DC.
13. MRS Commission, 1989, *Nuclear Waste: Is There A Need For Federal Interim Storage?* Report of the Monitored Retrievable Storage Commission Washington DC. US Government Printing Office, 1 November 1989.
14. US Department of Energy, 1989, *Report to Congress on Reassessment of the Civilian Radioactive Waste Management Program*. DOE/RW–0247, Washington DC. Office of Civilian Radioactive Waste Management, November.
15. Details of further planning problems at Yucca Mountain are contained in two other studies: the *First Report to the US Congress and US Secretary of Energy from the Nuclear Waste Technical Review Board*, Washington DC, March 1990. The review board was established in 1987, under NWPAA, Public Law 100–203; and Malone, Charles R., 'Geologic and Hydrologic Issues Related to Siting a Repository for High-Level Nuclear Waste at Yucca Mountain, Nevada, USA', *Journal of Environmental Management*, Vol. 30, 1990, pp. 914–30.
16. Radioactive Waste Campaign, news update, 11 May 1990, New York.
17. D'Arrigo, D. interviewed by D. Lowry, Washington DC, March 1990.
18. *Horizon*, 'The Ten Thousand Year Test'. Transcript, BSS/BBC, March 1990.

Sources and Bibliography

The material gathered, sifted and synthesised in the preparation of this book was drawn from a wide variety of published and unpublished sources. These vary from primary documents published by the key institutional actors that make up the study, to anonymous tip-offs received, or documents sent to one or other of the authors from sources unknown. By definition, such sources have to remain unacknowledged by name; notwithstanding this, on occasions they provided important heuristic tools for the comprehension of often complex inter-twinned accounts of a not always transparent political process.

This book could not have been produced in the final form in which it appears without the access one author (Andrew Blowers) had to internal reports and meetings of Bedfordshire County Council. Chapters 3 and 4 especially are indebted to this information. The footnotes to the chapters have identified where materials have been drawn upon directly, but our access to reports, debates and conversations also informed the perception of the subtle changes that developed in the political process.

In general terms the book has drawn upon many official reports produced by governments, parliaments, nuclear industry bodies and international agencies. Reports of special commissions of inquiry, such as those chaired by Lord Flowers and Professor Holliday in Britain, Professor Castaing in France, or inter-agency committees in the US have also complemented these official publications. We also drew upon annual reports of bodies such as NIREX, RWMAC, BNFL and the UKAEA in the UK.

We have drawn extensively from debates and government answers in the UK Parliament (recorded in the *Official Report, Hansard*), and to a lesser extent from specialist sub-committees of the US Congress. No direct reference is made to the proceedings of the national parliaments or regional governmental assemblies in Sweden or Germany, as the authors speak neither language. However, in both countries a substantial literature based on parliamentary decisions has been translated into English in full or summary form. This was of great assistance. The proceedings of the French Parliament (the Assemblée Nationale) do not enter into the account in any substantive way, because of the limited role played by this institution in political decision-making in regard to nuclear matters in France.

Information was also gathered from an extensive list of transient and ephemeral publications, which captured the 'flavour of the moment' but were not intended for deeper analysis. These include local and national newspapers and news magazines. Some more detailed information was obtained from specialist magazines and journals such as the *SCRAM Energy Journal, WISE News Communiqué, New Scientist, Nature, Science, Atom, Nucleonics Week* and *Nuclear Fuel*, as well as press releases from government departments and agencies, such as the US Nuclear Regulatory Commission's regular *News Releases*.

Our information is biased towards printed sources, but on occasions we have had recourse to material broadcast on television or radio stations, both local and national.

Each of the authors interviewed many 'actors' in the countries discussed: these included national and local government officials and civil servants, scientists and policy makers in the nuclear industry, politicians at local and national levels, professional environmentalists, and participants in national, regional and local resistance and citizen action groups. The differing insights provided by this diverse array of participants in the policy making process for radioactive waste management have assisted us enormously in our comprehension of the issues. We have listed many by name in the preface; but there were many more, particularly officials from the nuclear industry and its regulatory authorities, who gave of their time and help.

Interviews with such actors in our account were supplemented with reference to the partial literature published by the opposing parties, such as *BandWagon* (BAND) and *Plaintalk* (NIREX) in the UK, and *The Waste Paper* in the US. Summary translations were also made of similar publications in Sweden and the FRG; very little material of this sort exists in France.

Other sources on which we have drawn include the extensive transcripts of the Sizewell 'B' and Hinkley 'C' public inquiries (340 and 182 days respectively), each of which included extended analysis of the nuclear waste situation in Britain.

We have listed our sources in two categories: primary sources to which we make direct reference and other support sources that aided our interpretation of events. We have added a third list of sources we suggest for further exploration of this fascinating subject.

PRIMARY

Abbotts, J. (1979), 'Radioactive Waste: A Technical Solution'. *Bulletin of the Atomic Scientists* Vol. 35, No. 8, October, pp. 12–18.

Abrahamson, D. (1979), 'Governments Falls As Consensus Gives Way To Debate', *Bulletin of the Atomic Scientists*, Vol. 35, No. 8, pp. 30–7.

Abrams, N., 'Nuclear Politics in Sweden', *Environment*, May 1979, Vol. 21, No. 4, pp. 6–11, 39–40.

Ahall, K-I., Lindstrom, M., Holmstrend, O., Helander, B. and Goldstick, M., *Nuclear Waste in Sweden: The Problem is not Solved*, Uppsala, 1988.

Anderson, R. Y. (1988), *Open letter to the US Congress on Radioactive Waste Disposal at WIPP*, University of New Mexico, Department of Geology.

Andersson, K., Larssen, A. and Wingefors, S. (1986), 'Sweden: Policy and Licensing – an Update of Projects and the New Framework for Regulation', *IAEA Bulletin*, Vol. 28, No. 1, Spring, pp. 41–4.

ANDRA. (1983), *ANDRA: A Government Agency for Safe Radioactive Waste Management*, CEA, Paris.

Arnott, D. G. (1957), *Our Nuclear Adventure: Its Possibilities and Perils*. London, Lawrence and Wishart.

Ashworth, P. (1984), *Judgement of Mr Piers Ashworth QC, in the High Court of Justice, 20 June*, London, Royal Courts of Justice.

Aspinwall and Company (1981), *Possible Sites for Shallow Burial of Radioactive Waste: Report on Programme of Research*, Shrewsbury, Aspinwall & Co.

Barkenbus, J. N., Weinberg, A. M. and Alonso, M. (1985), 'Storing the World's Spent Nuclear Fuel', *Bulletin of the Atomic Scientists*, Vol. 41, No. 10, pp. 38–42.

Barlett, S. and Steele, J. B. (1985), *Forevermore – Nuclear Waste in America*, New York, Norton.

Barnaby, W. (1980), 'The Swedish Referendum: Do Away With It, But Not Yet', *Bulletin of the Atomic Scientists*, Vol. 36, No. 6, June pp. 58–9.

Barthoux, A. and Faussat, A. (1986), 'Low and Medium Waste Storage in France', *Nuclear Europe*, March, pp. 7–9.

Bedfordshire Against Nuclear Dumping (1986), *Evidence Against a Shallow Land Burial Nuclear Waste Repository at Elstow in Bedfordshire, or any Similar Oxford Clay Based Repository*, 20 January, Bedford, Band.

Bedfordshire County Council (1980), *County Structure Plan*, County Hall, Bedford.

Bedfordshire County Council (1984a), *Oxford Clay Subject Plan*, County Hall, Bedford.

Bedfordshire County Council (1984b), *County Structure Plan, Proposed Alterations to Policies, Policy 97, Nuclear Waste*, County Hall, Bedford.

Bedfordshire County Council (1985), *A Review of Research and Development in the Field of Intermediate and Low-Level Radioactive Waste Management*. Additional written evidence to the House of Commons Environment Committee. July.

Bergman, C., Boge, R., Johansson, G. and Snibs, J. O. (1985), *Radiation Protection Aspects of Waste Acceptance Criteria*. Paper to a Seminar on Radioactive Waste Products, Jülich, 10–13 June. SS1–RAPPORT–85–15.

Berkhout, F., Boehmer-Christiansen, S., Skea, J., 'Deposits and Repositories: Electricity Wastes in the UK and West Germany' (1989), *Energy Policy*, April, pp. 109–15.

Bernstein, B. J. (1985), 'Radiological Warfare: The Path Not Taken', *Bulletin of the Atomic Scientists*, August, pp. 44–9.

Bertin, L. E. (1957), *Atom Harvest*, London, Scientific Book Club.

Bjurstrom, S. and Sverke, E. (1984), 'The Swedish Programme for Spent Fuel and Waste', *Nuclear Europe*, October, pp. 26–8.

Black, D. (1984), *Investigation of Possible Increased Incidence of Cancer in West Cumbria*. Report of Independent Advisory Group Chaired by Sir Douglas Black. London, HMSO.

Blackett, P. M. S. (1948), *Military and Political Consequences of Atomic Energy*, London, Turnstile Press.

Blix, H. (1988), *Nuclear Energy and the Greenhouse Effect*. Presentation in Helsinki for 'Earthday '88'. Press Release 88/55. IAEA, Vienna.

Blowers, A. (1986), 'Sweden Buries its Radioactive Waste Problems', *Geography*, Vol. 71, June, pp. 260–3.

Blowers, A. and Pepper, D. (eds) (1987), *Nuclear Power in Crisis: Politics and Planning for the Nuclear State*, London, Croom Helm.

Blowers, A. (1988), 'Radioactive Waste in the United States – Will New

Mexico Draw the Short Straw?', *Environment Now*, Vol. 9, pp. 26–7, October.

Bord, R. J. (1987), 'Judgements of Policies Designed to Elicit Local Cooperation on LLRW Disposal Siting: Comparing the Public and Decision Makers', *Nuclear and Chemical Waste Management*, Vol. 7, pp. 99–105.

Brennecke, P., Griller, H., Marlens, B. R., Schumacher, J., Warnecke, E. and Odoj, R., *Waste Acceptability and Quality Control for the Planned Konrad Repository*. PTB, Braunchweig, KJ GmbH, Jülich. Paper to Stockholm Conference, 16–20 May 1988, IAEA–SM–303/139.

Brown, J. (1980), 'Kyshtym Whitewash', *Undercurrents*, No. 42, October–November, pp. 24–6.

Bryan, R. H. (1987), 'The Politics and Promise of Nuclear Waste Disposal: The View from Nevada', *Environment*, Vol. 29, No. 8, pp. 14–17, 32–7.

Cameron, D. M. and Solomon, B. D. (1990), 'Nuclear Waste Landscapes: How Permanent?' pp. 137–86 in Cullingworth J. B. (ed.), *Energy, Land and Public Policy*, New Brunswick N.J., Transaction Books.

Cannell, W. and Chudleigh, R. (1984), *The Gravedigger's Dilemma – A Guide for the Curious, the Perplexed and the Irate*, London, Friends of the Earth.

Carnes, S. A., Copenhaver, E. D., Sorensen, J. H., Soderstrom, E. J., Reed, J. H., Bjoornstad, D. J. and Peele, E. (1983), 'Incentives and Nuclear Waste Siting: Prospects and Constraints', *Energy Systems and Policy*, Vol. 7, pp. 323–51.

Carter, L. J. (1987), 'Nuclear Imperatives and Public Trust: Dealing with Radioactive Waste', *Resources for the Future*, Washington, DC.

Caufield, C. (1989), *Multiple Exposures: Chronicles of the Radiation Age*, Harmondsworth, Penguin Books.

Chandler, W. U., Geller, H. S. and Ledbetter, M. R. (1988), *Energy Efficiency: A New Agenda*, American Council for an Energy-Efficient Economy, Washington, DC.

Clark, R. W. (1981), *The Greatest Power on Earth: The Story of Nuclear Fission*, London, Book Club Associates.

Cochran, T. *et al.* (1987), *US Nuclear Warhead Facility Profiles: Nuclear Weapons Databook*, Vol. 3, Cambridge, MA, Ballinger.

Commissariat à L'Energie Atomique (1985), *The Centre de la Manche*, Paris, ANDRA/CEA.

County Councils Coalition (1986a), *Radioactive Waste Disposal – Policy Statement.*

County Councils Coalition (1986b), *County Councils Response to Government's White Paper on Radioactive Waste*, 29 July.

Coyle, D. *et al.* (1988), *Deadly Defense – Military Radioactive Landfills. A Citizen Guide*, New York, Radioactive Waste Campaign.

Cumbrians Opposed to a Radioactive Environment (with Greenpeace and Friends of the Earth) (1987), *Radioactive Waste Management, The Environmental Approach*. Briefing Paper, London, November.

Cutler, J. and Edwards, R. (1988), *Britain's Nuclear Nightmare*, London. Sphere Books.

Cutter, S. L. (1984), 'Emergency Preparedness and Planning for Nuclear Power Accidents', *Applied Geography*, Vol. 4, pp. 235–45.

Davis, M. (1988), *The Civilitary Atom*, Paris, Wise.

Deese, D. A. (1982), 'A Cross-National Perspective on the Politics of Nuclear Waste', pp. 63–97 in Colglazier E. W. (ed.), *The Politics of Nuclear Waste*, New York, Pergamon.

Delange, M. (1988), 'Over 2000T. of LWR Fuel Reprocessed at La Hague', *Nuclear Europe*, August/September, pp. 9–10.

DiMento, J. F., Lambert, W., Suarez-Villa, L. and Tripodes, J. (1985–6), 'Siting Low-Level Radioactive Waste Facilities', *Journal of Environmental Systems*, Vol. 15, No. 1, pp. 19–43.

DoE *et al.* (1977), *Nuclear Power and the Environment, the Government's Response to the Sixth Report of the Royal Commission on Environmental Pollution (cmnd 6618) cmnd 6280*, London, HMSO.

DoE (1982), *Radioactive Waste Management cmnd 8607 – White Paper*, July, London, HMSO.

DoE (1983), *Disposal Facilities on Land for Low and Intermediate-level Radioactive Wastes: Principles for the Protection of the Human Environment*. Draft, October. Department of the Environment.

DoE (1984a), *Radioactive Waste: The National Strategy*, July, London, HMSO.

DoE (1984b), *Disposal Facilities on Land for Low and Intermediate-level Radioactive Wastes: Principles for the Protection of the Human Environment*, December, London, HMSO.

DoE (1986a), *Assessment of Best Practicable Environmental Options (BPEOs) for Management of Low- and Intermediate Solid Radioactive Wastes*, February, Final Report, prepared by Professional Division, Department of the Environment, London.

DoE *et al.* (1986b), *The Government's First Stage Response to the Environment Committee Report on Radioactive Waste*, 2 May, London, HMSO.

DoE *et al.* (1986c), *Radioactive Waste: The Government's Response to the Environment Committee's Report*, Cmnd 9852, London, HMSO.

Downey, G. L. (1985), 'Politics and Technology in Repository Siting: Military Verses Commercial Nuclear Wastes at WIPP 1972–1985', *Technology in Society*, Vol. 7, pp. 47–75.

Duncan, A. B. and Brown, S. R. A. (1982), 'Quantities of Waste and a Strategy for Treatment and Disposal', *Nuclear Energy*, vol. 21, No. 3, June, pp. 161–6.

Dunster, H. J. (1958), 'Waste Disposal in Coastal Waters', pp. 390–9. In *Proceedings of the Second United Nations Conference on the Peaceful Uses of Atomic Energy*, Geneva.

Energy Committee (1989), *British Nuclear Fuels PLC: Report and Accounts 1987–88*. Third Report of the House of Commons Energy Committee, HC-50. 5 April, London, HMSO.

Environment Committee (1986), *Radioactive Waste (3 Volumes)*. Report of the House of Commons Environment Committee, February, London, HMSO.

Environmental Risk Assessment Unit (1988), *Responses to the Way Forward*, November, University of East Anglia, Norwich, ERAU.

Environmental Resources Ltd (undated), *Comments on DoE's Radioactive Waste Management – The National Strategy*, London, ERL.

Environmental Resources Ltd (1984a), *Suitability of the Elstow Site for Radioactive Waste Disposal Compared with other Options in the UK*, Working Paper No. 3, London, ERL.
Environmental Resources Ltd (1984b), *Preliminary Hydrogeological Appraisal of Elstow as a Disposal Site for Radioactive Wastes*, Working Paper No. 4, London, ERL.
Environmental Resources Ltd (1984c), *A Review and Technical Evaluation of the Case Against NIREX Proposals for a Radioactive Waste Repository at Elstow*, June, London, ERL.
Environmental Resources Ltd (1984d), *The Status of Research and Policies on Intermediate Waste Management in the UK and other OECD Countries*, Working Paper No. 1, London, ERL.
Environmental Resources Ltd (1984e), *The Future Disposal of Low and Intermediate-level Radioactive Wastes in the UK*, November, London, ERL.
Environmental Resources Ltd (1985a), *Integrity of Containment Systems for Low and Intermediate-level Radioactive Wastes*, Working Paper No. 5, November, London, ERL.
Environmental Resources Ltd (1985b), *BPEO Exercise*, Report prepared for Oxford Clay Sub-committee of Bedfordshire County Council, November, London, ERL.
Environmental Resources Ltd (1986a), *Review of the Capacity of the LLW Disposal Site at Drigg*, Working Paper No. 9, February, London, ERL.
Environmental Resources Ltd (1986b), *Review of the BPEO*, Report prepared for Bedfordshire County Council. March, London, ERL.
Environmental Resources Ltd (1987), *The Disposal of Radioactive Waste in Sweden, West Germany and France*. Prepared for County Councils Coalition. January, London, ERL.
Environmental Resources Ltd (1987), *The Disposal of Low and Intermediate Level Radioactive Waste in Sweden*, July, Report prepared for Bedfordshire County Council, Working Paper No. 12, London, ERL.
Erickson, T. (1988), *Civilian Nuclear Programs: The United States, France and Japan*, July, Olympia, Wa., Washington State Institute for Public Policy.
Falk, J. (1982), *Global Fission: The Battle Over Nuclear Power*, Oxford, Oxford University Press.
Feldman, D. L. (1986), 'Public Choice Theory Applied to National Energy Policy: the case of France', *Journal of Public Policy*, Vol. 6, No. 2, pp. 137–58.
Flavin, C. (1987), *Reassessing Nuclear Power: The Fallout from Chernobyl*, Paper No. 75, Washington, DC, Worldwatch Institute.
Flowers, R. (1982), *Proof of Evidence on the Disposal of Low-level and Intermediate-level Solid Wastes*, November, London, CEGB.
Freudenburg, W. R. (1987), 'Rationality and Irrationality in Estimating the Risks of Nuclear Waste Disposal'. In Post, R. (ed.), *Waste Management 87*, Vol. 2, pp. 109–15. University of Arizona, Tucson, AZ, College of Engineering and Mines.
Fuerst, I. (1979), 'New Mexico's Radiation River: Aftermath of the Church Rock Spill', *Not Man Apart*, Vol. 9, December, San Francisco, Friends of the Earth.

Galpin, F. C. and Meyer, G. C., *Overview of EPAs Low-level Radioactive Waste Standards Development Program*. Paper presented at the DOE Low-level Radioactive Waste Participants Meeting, Denver, Colorado, 22–26 September 1986.
Ginniff, M. E. (1984), *Implementation of UK Policy on Radioactive Waste Disposal*. Paper to British Nuclear Energy Society Conference, London, 26–29 November.
Goodin, R. (1978), 'Uncertainty as an Excuse for Cheating our Children: The Case of Nuclear Waste', *Policy Sciences*, Vol. 10, pp. 25–43.
Goodin, R. (1980), 'No Moral Nukes', *Ethics*, Vol. 90, April, pp. 417–49.
Gowing, M. (1974), *Independence and Deterrence: Britain and Atomic Energy, 1945–1952*, London, Macmillan.
Grover, J. R. (1958), 'Disposal of Long-lived Fission Products', *Journal of British Nuclear Energy Conference*, January, pp. 80–5.
Hall, T. (1986), *Nuclear Politics*, Harmondsworth, Penguin.
Hansen, C. (1988), *US Nuclear Weapons: The Secret History*, New York, Orion Books.
Hartley, H. (1950), 'Man's Use of Energy', *Bulletin of the Atomic Scientists*, November, pp. 322–4.
Hemming, C. R., Hill, M. D. and Pinner, A. V. (1984), *An Assessment of the Radiological Protection Aspects of Shallow Land Burial of Radioactive Wastes*, April, Didcot, National Radiological Protection Board.
Herrington, J. S., *Testimony before the Sub-committee on Interior and Related Agencies of the Committee on Appropriations for FY 1989*, Part VII, US House of Representatives, Washington, DC, 23 February 1988.
Hill, M. D., Mobb, S. F. and White, I. F. (1981), *An Assessment of the Radiological Consequences of Intermediate-level Wastes in Argillaceous Rock Formations*, Didcot, National Radiological Protection Board.
Hill, M. D. and Pinner, A. V. (1982), *Radiological Protection Aspects of Shallow Land Burial of PWR Operating Wastes*, October, Didcot, National Radiological Protection Board.
Hirsch, H. 'Nuclear Waste Management in West Germany: The Battle Continues', *The Ecologist*, Vol. 13, No. 1, January 1983.
Hiruo, E. (1980), *A Background Report for the Formerly Utilized Manhattan Engineer District/Atomic Energy Commission Sites Program*, September, Washington, DC, US Department of Energy.
Holliday, F. G. T. (1984), *Report of the Independent Review of Disposal of Radioactive Waste in the Northeast Atlantic Ocean*, Chairman, Professor Holliday, London, HMSO.
House of Lords, Select Committee on the European Communities (1988), *Radioactive Waste Management Report and Evidence, Session 1987–88*, London, HMSO.
Hubenthal, K. H. (1988), *'The Policy of the Federal Republic of German concerning the Management of Low- and Intermediate-level Waste'*, BMFT, Bonn. Paper to The International Symposium on Management of Low- and Intermediate-level Wastes, Stockholm, 16–20 May, IAEA–SM–303/157.
Hülsberg, W. (1988), *The German Greens: A Social and Political Profile*, London, VERSO Press.

Humberside County Council (1986), *Humberside Structure Plan, Proposed New Policy – 1985.* Grimsby, Humberside County Council.

IAEA-CEC (1989), *Management of Low- and Intermediate-level Radioactive Wastes*. Proceedings of a Symposium, Stockholm, 16–20 May 1988. Volumes 1 and 2. Vienna/Brussels, International Atomic Energy Agency and Commission of the European Communities.

Jacob, G. (1988), *Conflict, Politics and Location: Siting a Nuclear Waste Repository*, University of Colorado, Unpublished PhD Dissertation.

Jelinek-Fink, P., *Memorandum submitted 27 February 1986 by Nukem GmbH to House of Lords Select Committee on the European Communities, Session 1985–86.* Published in Nuclear Power in Europe, Vol. II (Evidence) 227-II, 15 July 1986a, pp. 143–9.

Jungk, R. (1964), *Brighter Than A Thousand Suns*, Harmondsworth, Pelican.

Kasperson, R. E., Berk, G., Pijawka, K. D., Sharaf, A. B. and Wood, J. (1980), 'Public Opposition to Nuclear Energy: Retrospect and Prospect', *Science, Technology and Human Values*, Vol. 5, Spring, pp. 11–23.

Kasperson, R. E., Derr, P. and Kates, R. W. (1983), 'Confronting Equity in Radioactive Waste Management: Modest Proposals for a Socially Just and Acceptable Program', pp. 331–68, in Kasperson R. E. (ed.), *Equity Issues in Radioactive Waste Management*, Cambridge, Mass., Oelgeschlager, Gunn and Hain.

Kearney, R. C. and Stucker, J. J. (1985), 'Interstate Compacts and the Management of Low Level Radioactive Wastes', *Public Administration Review*, Vol. 45, January/February, pp. 210–20.

Keepin, W. and Kats, G. (1988), 'Greenhouse Warming: Comparative Analysis of Nuclear and Efficiency Abatement Strategies', *Energy Policy*, Vol. 16, No. 6, December, pp. 538–61.

Kemp, R. and O'Riordan, T. (1988), 'Planning for Radioactive Waste Disposal: Some Central Considerations', *Land Use Policy*, Vol. 5, No. 1, January, pp. 37–44.

Kemp, R. (1989), *The Politics of Siting Radioactive Wastes*. Paper to Conference on Radioactive Waste Management, London, 22/23, February.

Kirby, A. and Jacob, G. (1986), 'The Politics of Transportation and Disposal: Hazardous and Nuclear Waste Issues in Colorado, US', *Policy and Politics*, Vol. 14, pp. 27–42.

Korting, K. (1987), 'Karlsruhe Begins Cold Trials of Prototype Melting Facility for Vitrification of Reprocessing High-level Waste', *Nuclear Europe*, August.

Kozak, R. M. (1987), *Public Perception of Low-Level Radioactive Waste Disposal Issues*, Texas Tech University, Unpublished PhD Dissertation.

Kunreuther, H., Kleindorfer, P. R., Knez, P. J. and Yaksick, R. (1987), 'A Compensation Mechanism for Siting Noxious Facilities: Theory and Experimental Design', *Journal of Environmental Economics and Management*, Vol. 14, pp. 371–83.

Laporte, T. R. (1978), 'Nuclear Wastes: Increasing Scale and Sociopolitical Impacts', *Science*, Vol. 201, 7 July, pp. 22–8.

Larsson, A. H. (1984), *Regulatory Aspects of Radioactive Waste Management in Sweden and its Implementation*, Paper for Radioactive Waste Management Conference, London, British Nuclear Energy Society.

Lash, T. (1979), 'Radioactive Waste: Nuclear Energy's Dilemma', *Amicus Journal*, Vol. 1, No. 2, Fall, pp. 24–34, New York, Natural Resources Defense Council.

Lashof, D. A. and Tirpak, D. A., (eds) (1989), *Policy Options for Stabilizing Global Climate Change*, Draft Report by the US Environmental Protection Agency to the US Congress, 2 Vols.

Layfield, F. (1987), *Sizewell 'B' Public Inquiry Report by Sir Frank Layfield (9 Volumes)*, January, Department of Energy, London, HMSO.

Leigh, I. W. (1988), *International Nuclear Fuel Cycle Fact Book, PNL-3594 (Rev. 8)*, prepared by the Batelle Pacific Northwest Laboratory for the USDOE, Richland, WA.

Lipschutz, R. D. (1980), *Radioactive Waste*, Cambridge, Mass., Ballinger.

Loeb, P. R. (1982), *Nuclear Culture: Living and Working in the World's Largest Atomic Complex*, New York, Coward, McCann and Geoghegan.

Lovins, A. B. (1978), *Comments on the 10/78 Draft of the I.R.G. Report to the President (TID-28817, Draft), Unpublished Memorandum to the Interagency Review Group on Nuclear Waste Management*, Snowmass, Colorado, Rocky Mountain Institute.

Lovins, A. B. and Lovins, L. H. (1980), *Energy/War: Breaking the Nuclear Link*, New York, Harper & Row.

Lowry, D. (1986), 'Disasters in the Dark', *Times Higher Education Supplement*, 5 September, p. 13.

Lowry, D. (1988), 'Corrupt to the Core: The Hidden Agenda for Nuclear Waste', *Environment Now*, August, pp. 27–9.

Mackay, L. and Thompson, M. (eds) (1988), *Something in the Wind: Politics After Chernobyl*, London, Pluto Press.

Maxwell, A. (1958), 'The Accident at Windscale', *Contemporary Issues*, Vol. 9, No. 33, April–May, pp. 1–49.

McLean, A. S. and Marley, W. G. (1956), 'Health and Safety in a Nuclear Power Industry', *Journal of the British Nuclear Energy Conference*, January, pp. 26–7.

Medvedev, Z. (1979), *Disaster in the Urals*, London, Angus and Robertson.

Meysenburg, H. H. (1983), 'The Fuel Cycle Industry in the Federal Republic of Germany', *Nuclear Europe*, No. 4, April pp. 27–30.

Michallet, F. (1987), 'France's Nuclear Fuel Cycle Industry', *Nuclear Europe*, October, pp. 25–6.

Miller, C. (1987), 'Efficiency, Equity and Pollution: The Case of Radioactive Waste', *Environment and Planning*, Vol. 19, pp. 913–24.

Milnes, A. G. (1985), *Geology and Radwaste*, London, Academic Press.

Mitchell, N. T. and Shepherd, J. G. (1988), 'The UK Disposal of Solid Radioactive Waste into the Atlantic Ocean and its Environmental Impact', pp. 119–54, in *Symposium on Environmental Impact of Nuclear Power*, London, British Nuclear Energy Society.

Murauskas, G. T. and Shelley, F. M. (1986), 'Local Political Responses to Nuclear Waste Disposal', *Cities*, Vol. 3, May, pp. 157–62.

National Academy of Sciences (1988), *Report on Brine Accumulation in the WIPP Facility*, Washington, DC, WIPP Panel of Board of NAS Board on Radioactive Waste Management.

National Research Council (1987), *Safety Issues at the Defense Production Reactors*, Washington, DC, National Academy Press.

Nau, H. R. (1974), *National Politics and International Technology: Nuclear Reactor Development in Western Europe*, Baltimore, Johns Hopkins University Press.

Neill, R. H., Channell, J. K., Chaturvedi, L., Little, M. S., Rehfeldt, K. and Spiegler, P. (1983), *Evaluation of the Suitability of the WIPP Site*, EEG-23, Santa Fe, New Mexico, Environmental Evaluation Group of the State of New Mexico.

Nelkin, D. and Pollak, M. (1981a), *The Atom Besieged: Extraparliamentary Dissent in France and Germany*, London, MIT Press.

Nilsson, L. B. and Carlssen, M. S. (1985), 'Site Investigations for a Swedish Repository for HLW', *Nuclear Europe*, February, pp. 19–21.

NIREX, *Fact Sheet 4 – What is a Radioactive Waste Repository?*

NIREX, *Fact Sheet – How Long Can Concrete Last?*

NIREX (1983a), *The Disposal of Low and Intermediate-level Radioactive Wastes: The Elstow Storage Depot, A Preliminary Project Statement*, October, Harwell, NIREX.

NIREX (1983b), *The Disposal of Low and Intermediate-level Radioactive Wastes: The Billingham Anhydrite Mine, A Preliminary Project Statement*, October, Harwell, NIREX.

NIREX (1986), *Disposal of Low and Intermediate-level Radioactive Wastes, A Preliminary Project Statement*, February, Harwell, NIREX.

NIREX (1987a), *Response to the County Councils Coalition Report on the Disposal of Radioactive Waste in Sweden, West Germany and France*, February, Harwell, UK NIREX, Ltd.

NIREX (1987b), *The Way Forward: The Development of a Repository for the Disposal of Low and Intermediate-level Radioactive Waste*, a discussion document, November, Harwell, UK NIREX, Ltd.

NIREX (1989a), *Going Forward: The Development of a National Disposal Centre for Low and Intermediate-level Radioactive Waste*, March, Harwell, UK NIREX, Ltd.

Norrby, S. and Bergman, C. (1988), *Licensing of the Swedish Final Repository Waste*, AEA–SM–303/18. BMFT, Bonn. Paper to The International Symposium on Management of Low- and Intermediate-level Wastes, Stockholm, 16–20 May, IAEA–SM–303/18.

Nuclear Energy Agency (1983), *The International Stripa Project. Background and Research*, March, Paris, NEA/OECD.

Numark, N. N. (1987), *Analysis of Factors Influencing National Spent Fuel Management Strategies*, International Conference on Nuclear Power Performance and Safety, October, Vienna, International Atomic Energy Agency.

NUS (1980), *The 1979 State-by-State Assessment of Low-level Radioactive Waste Shipped to Commercial Burial Grounds*. Prepared by the NUS Corp for the US Department of Energy, Washington, DC, November.

O'Connor, W. (1978), 'Incentives for the Construction of Low Level Nuclear Waste Facilities, Appendix II' in *Low Level Waste: Program for Action*, Energy and Natural Resources Program, Washington DC, National Governors Association.

Olivier, J. P. (1987), *Disposal Policies in Other Countries*. OECD Paper to the European Summer School of Radioactive Waste Management, 6–9 July, Christ College, Cambridge.

Olsson, B. (1987), *The Feasibility of Non-nuclear Strategies*. Paper to Conference on Nuclear or Non-nuclear Futures, London, April 28–30.

O'Riordan, T. (1988), 'The Prodigal Technology: Nuclear Power and Political Controversy', *Political Quarterly*, Vol. 59, No. 2, April–June.

Parker, F. L., Kasperson, R. E., Andersson, T. L. and Parker, S. A. (1986), Technical and Sociopolitical Issues in Radioactive Waste Disposal. Report to *The Beijer Institute of the Royal Swedish Academy of Sciences*, Vol. 2, Stockholm.

Pasqualetti, M. J. (1983), 'Nuclear Power Impacts: A Convergence/Divergence Schema', *The Professional Geographer*, Vol. 35, pp. 427–36.

Patterson, W. C. (1983), *Nuclear Power* (2nd edn), London, Penguin.

Patterson, W. C. (1984), *The Plutonium Business and the Spread of the Bomb*, London, Paladin.

Petroll, M. (1987a), 'Taking the "Worry" Out of the Back End of the Fuel Cycle', *Nuclear Europe*, Vol. 7, No. 1–2, January/February, pp. 25–7.

Pocock, R. F. (1977), *Nuclear Power: Its Development in the United Kingdom*, Old Woking, Unwin Press.

Pollock, C. (1986), *Decommissioning: Nuclear Power's Missing Link*. Paper No. 69. Washington, DC, Worldwatch Institute.

Pollution Prevention Consultants (1982), *Feasibility Study of Shallow Land Burial of Specified Intermediate and Low-level Radioactive Wastes*. Prepared by PPC Ltd for the CEGB, April.

Reader, M. (ed.) (1980), *Atom's Eve: Ending the Nuclear Age, An Anthology*, New York, McGraw Hill.

Resnikoff, M. (1983), *The Next Nuclear Gamble: Transportation and Disposal of Nuclear Waste*, New York, Council on Economic Priorities.

Resnikoff, M. (1987), *Living Without Landfills*, New York, Radioactive Waste Campaign.

Revue Générale Nucléaire (1988), *Les Déchets Radioactifs et Leur Gestion*, No. 4, Juillet-Août, pp. 289–339, Paris.

Richardson, P. J. (1989), *Exposing the Faults: The Geological Case Against the Plans by UK NIREX to Dispose of Radioactive Waste*, March, London, Greenpeace/Friends of the Earth.

Salander, C., Proske, R. and Albrecht, E. (1980), 'The Asse Salt Mine: The World's Only Test Facility for the Disposal of Radioactive Waste', *Interdisciplinary Science Reviews*, Vol. 5, No. 4, pp. 292–303.

Scheinmann, L. (1965), *Atomic Energy Policy in France under the Fourth Republic*, Princeton, Princeton University Press.

Scheinman, L. (1987), *The International Atomic Energy Agency and World Nuclear Order*, Washington, DC, Resources for the Future, Inc.

Schmid, G. (1988a), *Report Drawn up on Behalf of the Committee of Enquiry on the Handling and Transport of Nuclear Material*, on the results of the Inquiry Parts A and B, European Parliament Doc. A 2–2120/88, PE 123.491/fin, 24 June, 27 June respectively.

SCRAM (1980), *Poison in our Hills: The First Inquiry on Atomic Waste Burial*, September, Edinburgh, SCRAM.

Shrader-Frechette, K. S. (1980), *Nuclear Power and Public Policy*, Dordrecht, D. Reidel.

Shrader-Frechette, K. S. (1988), 'Values and Hydrogeological Method: How Not to Site the World's Largest Nuclear Dump', pp. 101–37 in Byrne, J. and Rich, D. (eds), *Planning under Changing Energy Conditions*, New Brunswick, NJ, Transaction Books.

SKB (1985), *Final Repository for Reactor Waste*, Stockholm, SKB.

SKB (1987), *Swedish Nuclear Fuel and Waste Management Company Activities*, Stockholm, SKB.

SKB (1987), *Briefing on the Swedish Programme on Nuclear Waste Management.* Prepared for Senate Committee on Energy and Natural Resources, January, 100th US Congress, Washington DC, US Government Printing Office.

SKI (1984), *Review of Final Storage of Spent Nuclear Fuel – KBS – 3*, SKI–84:5, Technical Report Series, Stockholm, SKI.

SKI (1984), *Licensing of Final Repository for Reactor Waste – SFR–1*, SKI–84:2, July, Technical Report Series, Stockholm, SKI.

SKI (1987), *Swedish Nuclear Power Inspectorate: A Presentation of our Activities*, Stockholm, Statens Kärnkraftinspektion.

SKI (1988), *Review of Final Repository for Reactor Waste*, SKI–TR–88:5, March, Stockholm, Statens Kärnkraftinspektion.

Slater, H. G. (1971), 'Public Opposition and the Nuclear Power Industry'. pp. 847–59, in *Conference Proceedings of Environmental Aspects of Nuclear Power Stations*, Vienna, IAEA.

Slovic, P., Fischoff, B. and Lichtenstein, S. (1979a), 'Rating the Risks', *Environment*, Vol. 21, April, pp. 14–20, 36–9.

Slovic, P., Fischoff, B. and Lichtenstein, S. (1979b), 'Images of Disaster: Perception and Acceptance of Risks from Nuclear Power', pp. 223–45, in Goodman, G. T. and Row, W. D. (eds), *Energy Risk Management*, London, Academic Press.

Solomon, B. D. (1988), 'The Politics of Nuclear Power and Radioactive Waste Disposal: From State Coercion to Procedural Justice?', *Political Geography Quarterly*, Vol. 7, No. 3, July, pp, 291–8.

Solomon, B. D. and Cameron, D. M. (1984), 'The Impact of Nuclear Power Plant Dismantlement on Radioactive Waste Disposal', *Man, Environment, Space and Time*, Vol. 4, Spring, pp. 39–60.

Solomon, B. D. and Cameron, D. N. (1985), 'Nuclear Waste Repository Siting: An Alternative Approach', *Energy Policy*, Vol. 13, No. 6, December, pp. 564–80.

Solomon, B. D., Shelley, F. M., Pasqualetti, M. J. and Murauskas, G. T. (1987), 'Radioactive Waste Management Policies in Seven Industrialised Societies', *Geoforum*, Vol. 18, No. 4, pp. 415–31.

Solomon, B. D. and Shelley, F. M. (1988), 'Siting Patterns of Nuclear Waste Repositories', *Journal of Geography*, Vol. 87, March/April, pp. 59–71.

Soran, D. M. and Stillman, D. B. (1982), *Analysis of the Alleged Kyshtym Disaster*, LA–9217–MS, Los Alamos National Laboratory, New Mexico, January.

Spence, R. (1957), 'The Role of Chemistry in a Nuclear Energy Project', *Journal of the British Nuclear Energy Conference*, January, p. 590.

Sweet, C. (1983), *Some Lessons from the French Nuclear Power Experience that the UK Might Learn*. Proof of evidence to Sizewell B Public Inquiry, October, London, Centre for Energy Studies.

SWUCO (1986), *Radwaste Report*, No. 2, 15 August.

Texas Department of Agriculture (1985), *Panhandle Residents' Views of High-level Nuclear Waste Storage*, Austin, Texas, May.

Timm, M. (1987), 'Significance of Nuclear for Germany's Energy Supply', *Nuclear Europe*, January, pp. 13–15.

Touraine, A. (1979), 'Political Ecology: A Demand to Live Differently – Now', *New Society*, 8 November, pp. 307–9.

Trabalka, J. R., Auerbach, S. I. and Eyman, L. D. (1979), *Analysis of the 1957–58 Soviet Nuclear Accident*, ORNL–5613, Publication No. 1445. Environmental Sciences Division, Oak Ridge National Laboratory, December.

Treaty Establishing the European Atomic Energy Community (EURATOM) (1957), Brussels.

Trades Union Congress (1986), *TUC Response to the Draft Report on Best Practicable Environmental Options for the Management of Radioactive Waste*, 6 January, London, TUC.

TUSIU (1980), *The Nuclear Triangle: The Miners' Report on Waste Dumping in the Cheviot*, Newcastle, TUSIU.

US DOE (1988), *Environment, Safety and Health Report for the Department of Energy Defense Complex*, Washington, DC, Government Printing Office, 1 July.

US DOE (1978), *Report of the Task Force for Reviews of Nuclear Waste Management*, February, Washington, DC, US Government Printing Office.

US DOE (1985), *Mission Plan for the Civilian Radioactive Waste Management Program*, Vols. 1 to 3, June. Office of Civilian Radioactive Waste Management, Washington, DC, Department of Energy.

US DOE (1987), *Integrated Data Base for 1987: Spent Fuel and Radioactive Waste Inventories, Projections and Characteristics*, DOE/RW–006 Rev. 3, October, Springfield, VA, National Technical Information Service.

US DOE (1988), *Initial Version Dry Cask Storage Study*, DOE/RW–0196, Oak Ridge, Tennessee, US Dept of Energy, August.

USGAO (1981), *Hazards of Past Low-level Radioactive Waste Ocean Dumping Have Been Over-emphasised*. Report for Senator William J. Roth Jr. EMD–82–9, 21 October, Washington, DC, US General Accounting Office.

US House of Representatives (1986), *Staff Memorandum on the Second High-level Radioactive Waste Repository*. Prepared for the Oversight Subcommittee of the House Interior and Insular Affairs Committee, Washington, DC, 30 October.

Van Der Pligt, J., Eiser, J. R. and Spears, R. (1987), 'Nuclear Waste: Facts, Fears, and Attitudes', *Journal of Applied Social Psychology*, Vol. 17, No. 5, pp. 453–70.

Vogel, D. (1986), *National Styles of Regulation: Environmental Policy in Great Britain and the United States*, Ithaca, New York, Cornell University Press.

Weinlander, W. (1987), '*Reprocessing and Waste Management in the Federal Republic of Germany*'. Paper presented at the International Conference on Nuclear Fuel Reprocessing and Waste Management, RECOD, Paris, 23–7 August.

White, I. K. and Spath, J. P. (1984), 'How are States Setting their Sites?' *Environment*, Vol. 26, October, pp. 16–20, 36–42.

Wikdahl, C-E. (1984), 'The Future for Nuclear Power in Sweden', *Nuclear Europe*, October, pp. 15–16.

Woodbury, D. (1955), *Atoms for Peace*, New York, Dodd, Mead & Co.

Zinberg, D. (1979), 'The Public and Nuclear Waste Management', *Bulletin of the Atomic Scientists*, Vol. 35, No. 1, January.

SUPPORT

Albright, D. and Feiveson, H. (1987), 'Why Recycle Plutonium?', *Science*, Vol. 235, 27 March, pp. 1555–6.

Bertell, R. (1985), *No Immediate Danger: Prognosis for a Radioactive Earth*, Sommertown, Tennessee, The Book Publishing Company.

Bojcun, M. and Haynes, V. (1988), *The Chernobyl Disaster*, London, Hogarth Press.

Breach, I. (1978), *Windscale Fallout*, Harmondsworth, Penguin.

Cervenka, Z. and Rogers, B. (1978), *The Nuclear Axis: Secret Collaboration Between West Germany and South Africa*, London, Julia Friedman Books.

Conroy, C. (1978), *What Choice Windscale?*, London, Friends of the Earth/Conservation Society.

Crick, B. and Robson, W. A. (eds) (1970), *Protest and Discontent*, Harmondsworth, Pelican.

DoE (1988a), *The Radionuclide Content of Radioactive Wastes*. Prepared by Electrowatt Engineering Services for NIREX/DoE, October, NIREX Report No. 60, DoE/RW/88.077, London, Department of the Environment.

DoE (1988b), *The 1987 United Kingdom Radioactive Waste Inventory*. Prepared by Electrowatt Engineering Services for NIREX/DoE, October, NIREX Report No. 54, DoE/RW/88.061, London, Department of the Environment.

Dunkelman, M. M., Kearney, M. S. and MacDougall, R. D. (1986), *Plans and Schedules for Implementation of US Nuclear Regulatory Commission Responsibilities under Low-level Radioactive Waste Policy Amendments Act of 1985 (PL 99–240)*. Office of Nuclear Material Safety and Safeguards, US Nuclear Regulatory Commission, Washington, DC

Horizon (1990) 'The 10,000 Year Test', A Report on the Yucca Mountain Project, London, BBC Television, March.

Kuroda, P. K. (1982), *The Origin of the Chemical Elements and the Oklo Phenomenon*, Berlin, Springer Verlag.

Makhijani, A. (1989), *Reducing the Risks: Policies for the Management of Highly Radioactive Nuclear Waste*. Tacoma Park, Institute for Energy and Environmental Research.

Nuclear Energy Agency, *Annual Activity Report*, Paris, NEA/OECD.

Nuclear Waste Technical Review Board (1990), *First Report to the US Congress and the US Secretary of Energy*, Washington, DC, NWTRB. March.

OCRWM (1989), *Annual Report to Congress of the Office of Civilian Radioactive Waste Management*. Washington, DC, US Department of Energy, December.

Palmer, J. (1988), *Europe Without America? The Crisis in Atlantic Relations*, Oxford, Oxford University Press.

Patterson, W. C. (1985), *Going Critical – An Unofficial History of British Nuclear Power*, London, Paladin Grafton Books.

Pearce, D., Edwards, L. and Beuret, G. (1979), *Decision Making for Energy Futures: A Case Study of the Windscale Inquiry*, London, Macmillan/SSRC.

Rhodes, R. (1988), *The Making of the Atomic Bomb*, Harmondsworth, Penguin.

Robbie, D. (1986), *Eyes of Fire*, London, Ravette.

Roberts, L. (1987), 'Atom Bomb Doses Reassessed', *Science*, Vol. 238, 18 December, pp. 1649–51.

Stott, M. and Taylor, P. (1980), *The Nuclear Controversy: A Guide to the Issues of the Windscale Inquiry*, London/Oxford, Town and Country Planning Association/Political Ecology Research Group.

Surrey, J. (ed.) (1984), *The Urban Transport of Irradiated Nuclear Fuel*, London, Macmillan.

Svenson, O. and Karlsson, G. (1988), 'Decision Making, Time Horizons and Risks in the Very Long Time Perspective'. Paper in *Eik Och Kärnavfall*, Seminar held in Stockholm, 8–9 September 1987. SKN Rapport 28, Kasam/SKN.

Town and Country Planning Act (1971) London, HMSO.

US DOE (1990), *Progress Report on the Scientific Investigation Program for the Nevada Yucca Mountain Site, No. 1*, February, Washington, DC, US Department of Energy, Office of Civilian Radioactive Waste Management.

Williams, R. (1980), *The Nuclear Power Decisions*, London, Croom Helm.

Zint, G. (1979), *Gegen den Atomstaat*, Frankfurt, Zweitauseudeius.

FURTHER READING

BNES (1981), *The Environmental Impact of Nuclear Power*. Proceedings of a Conference organised by BNES and UKAEA, 1–2 April 1981, London, British Nuclear Energy Society.

Brown, G. E. (1987), 'US Nuclear Waste Policy: Flawed But Feasible', *Environment*, Vol. 29, No. 8, pp. 6–7, 25.

Commission of the European Communities PAGIS Programme (1988), *Performance Assessment of Geological Isolation Systems for Radioactive Waste*. Summary, EUR 11775 EN; Disposal in Salt Formations, EUR 11778 EN; Disposal in Clay Formations, EUR 11776 EN; Disposal into Sub-seabed, EUR 11779 EN. Nuclear Science and Technology Series. Luxembourg, Directorate-General for Science, Research and Development, Joint Research Centre, Ispra.

Commission of the European Committees, *Annual Progress Reports of the Community's Research and Development Programme on Decommissioning of Nuclear Installations*. Nuclear Science and Technology Series, European Commission, Luxembourg, Directorate-General for Science, Research and Development.

Conway, J. T., Eggenberger, A. J., Case, E. G., Crawford, J. W. Jr. and Kouts, H. (1990), *Testimony of Defense Nuclear Facilities Safety Board, before Sub-committee on Strategic Forces and Nuclear Defense, Committee on Armed Services, United States Senate*, 28 March, Washington, DC.

Environment, Energy and Natural Resources Sub-committee (1989), *The Decommissioning and Decontamination Requirements for Closing Nuclear Facilities*, 3 August, Hearing of the Committee on Government Operations, House of Representatives, 101st Congress, Washington, DC, US Government Printing Office.

Environment, Energy and Natural Resources Sub-committee (1990), *Review of the Status of the Waste Isolation Pilot Plant Project, 12 June 1989*, Hearing of the Committee on Government Operations, 101st Congress, Washington, DC, US Government Printing Office.

Gardner, M. J., Snee, M. P., Hall, A. J., Powell, C. A., Downes, S. and Terrell, J. D. (1990), 'Results of Case-controlled Study of Leukaemia and Lymphoma among Young People near Sellafield Nuclear Plant in West Cumbria', *British Medical Journal*, Vol. 300, 17 February, pp. 423–34.

IAEA (1987), *Packaging and Transportation of Radioactive Materials*, 'Patram 86', Proceedings of a Symposium, Davos, 16–20 June 1986, 2 Vols, Vienna, International Atomic Energy Agency.

IAEA (1986), *Methodology and Technology of Decommissioning Nuclear Facilities*, Technical Report Series No. 267, Vienna, IAEA.

Institution of Mechanical Engineers (1988), *Decommissioning of Major Radioactive Facilities*, Proceedings of an International Conference, 11–12 October, London, Mechanical Engineering Publications, Ltd.

Jungk, R. (1979), *The Nuclear State*, London, John Calder/Platform Books.

McSorley, J. (1990), *Living in the Shadow: The Story of the People of Sellafield*, London, Pan Books.

Murrell, H. (1984), *An Ordinary Citizen's View of Radioactive Waste Management*, Evidence Prepared for the Sizewell 'B' Inquiry, March.

NIREX (1988), *Presentation of the NIREX Disposal Safety Research Programme*, NSS/G108, 1 November, Harwell, UK NIREX Ltd.

NIREX (1989b), *Deep Repository Project*, Preliminary Environmental and Radiological Assessment and Preliminary Safety Report, March, Harwell, UK NIREX Ltd.

Nuclear Energy Agency (1990), *In Situ Experiments Associated with the Disposal of Radioactive Waste*, Proceedings of the 3rd NEA/SKB Symposium on the International Stripa Project, Paris, NEA/OECD.

Nuclear Energy Agency (1988), *Geological Disposal of Radioactive Waste – In Situ Research and Investigations in OECD Countries*, A Status Report Prepared by the NEA Advisory Group (ISAG), Paris, NEA/OECD.

Nuclear Energy Agency (1988), *Feasibility of Disposal of High Level Radioactive Waste into the Seabed*, Vos 1–5, Paris, NEA/OECD.

Nuclear Energy Agency, *Nuclear Waste Bulletin – Update on Waste Management Policies and Programmes*, Paris, NEA/OECD.

Nuclear Regulatory Commission (1984), *Proceedings of the Workshop on Shallow Land Burial and Alternative Disposal Concepts held at Bethesda, Maryland*, NUREG/CP 0055, October, Washington, DC, US Nuclear Regulatory Commission.

Nuclear Regulatory Commission (1989), *Proceedings of the Workshop on Rules for Exemption from Regulatory Control*, held at Pan American Health Organisation, 17–19 October 1988. USNRC/NEA–OECD – NUREG/CP 0101, Washington, DC, US Nuclear Regulatory Commission.

Openshaw, S., Carver, S. and Fernie, J. (1989), *Britain's Nuclear Waste: Safety and Siting*, London, Bellhaven Press.

Rahn, F. J., Adamantiades, A. G., Kenton, J. E. and Braun, C. (1984), *A Guide to Nuclear Power Technology – A Resource for Decision Making*, New York, John Wiley & Sons.

The Royal Society (1986), *The Disposal of Long-lived and Highly Radioactive Wastes*. Proceedings of a Royal Society Discussion Meeting, 30–31 May 1985, London, Royal Society.

Saleska, S. (1989), *Nuclear Legacy: An Overview of the Places, Problems and Politics of Radioactive Waste in the United States*, Washington, DC, Public Citizen Critical Mass Energy Project.

UKAEA (1988), *Research and Safety Assessment*, Safety Studies [for] NIREX Radioactive Waste Disposal, NSS/G100, September, Harwell, UKAEA.

US DOE (1986), *Area Recommendations Report for the Crystalline Repository Project*, DOE/CH–15(1), Washington, DC, Department of Energy, Office of Civilian Radioactive Waste Management.

USGAO (1990), *Nuclear Waste*, Quarterly Report, as of September 30, 1989. March GAO/RCED–90–103, US General Accounting Office.

Wynne, B. (1982), *Rationality and Ritual – The Windscale Inquiry and Nuclear Decisions in Britain*, BSHS Monograph, 3, Chalfont St. Giles, British Society for the History of Science.

Index